The Library of Lost Maps

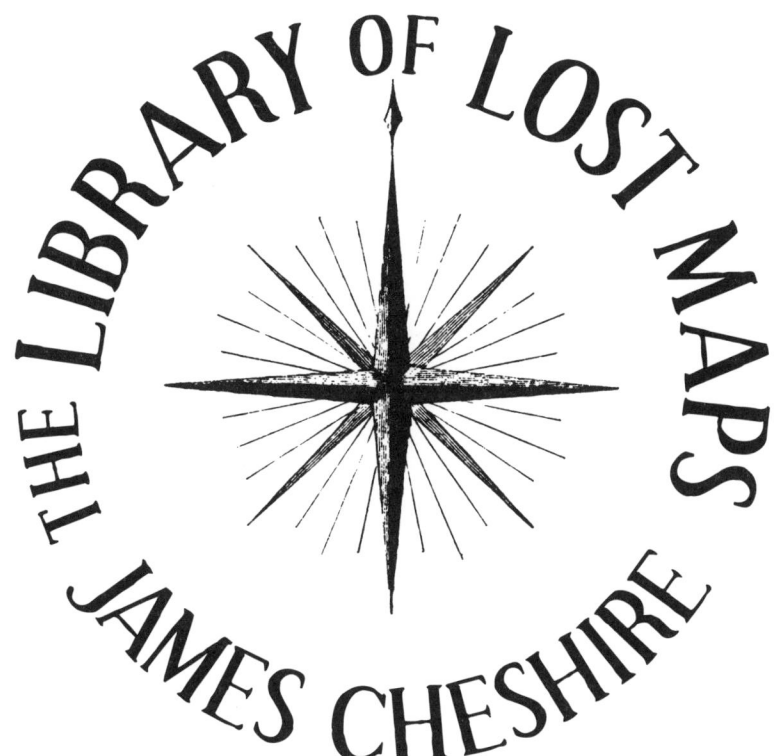

BLOOMSBURY PUBLISHING
NEW YORK · LONDON · OXFORD · NEW DELHI · SYDNEY

CONTENTS

1. WELCOME TO THE MAP LIBRARY 9
2. A ROOM FULL OF STORIES 39
3. GEORGE BELLAS GREENOUGH'S REMARKABLE MAPS 75
4. BLOOMSBURY 105
5. KNOWLEDGE IS POWER 135
6. 'TIDYING' THE MAP 175
7. MANIPULATIVE MAPS 201
8. HIROSHIMA 263
9. A FRESH PERSPECTIVE 269
10. THE OCEAN FLOOR 289
11. MAPS GO DIGITAL 339
12. A SPECIAL PLACE 355

AFTERWORD 361
FURTHER READING 364
PICTURE CREDITS 366
ACKNOWLEDGEMENTS 368
MAP GALLERY 370
INDEX 380

A note on the maps

The largest map I feature is 2 metres across, while the smallest is barely 2cm wide, and there are details I want to share that could be a metre or only a few millimetres in reality.

In each of the captions, I have specified if we are only viewing part of the map by including [Extract]. I have also given the true dimensions of the full version of each map (excluding its marginalia) as width by height in centimetres, so you can get a sense of their size as I lay them out on the central table in the Map Library. At the end of the book, you will find a map gallery that has the complete version of every map shown to scale; they can also be found at libraryoflostmaps.com, where you can also find my comprehensive list of sources.

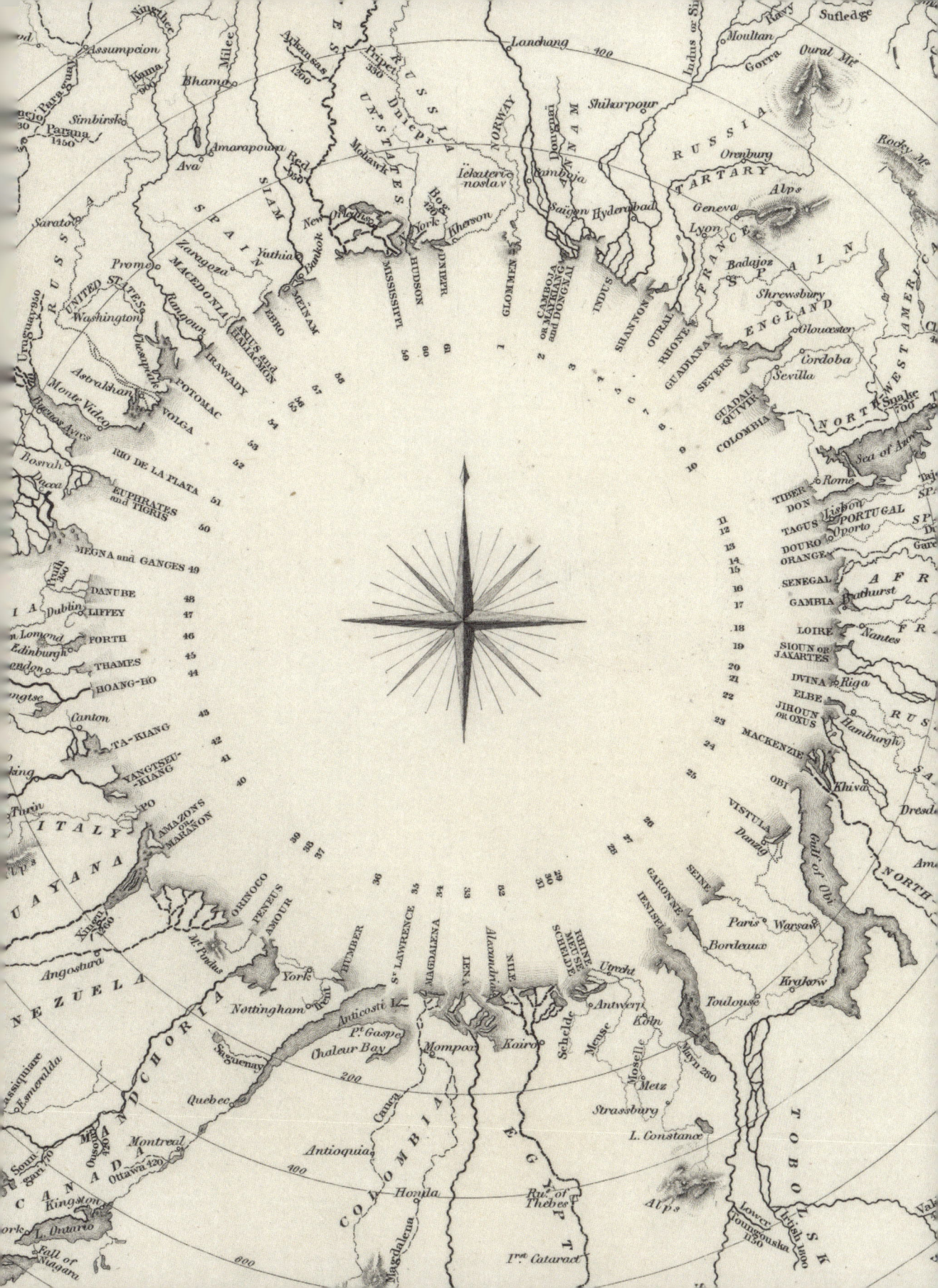

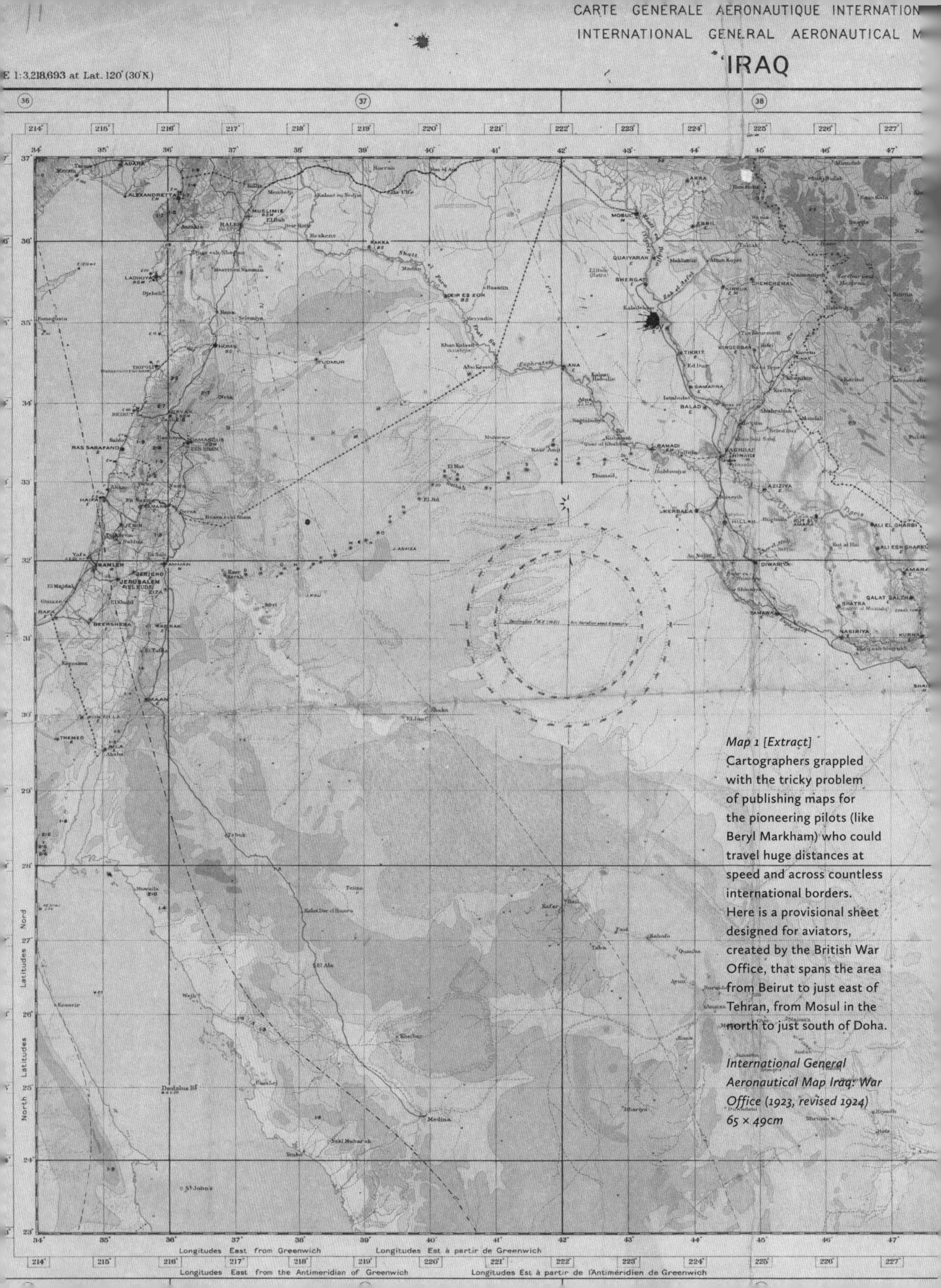

Map 1 [Extract]
Cartographers grappled with the tricky problem of publishing maps for the pioneering pilots (like Beryl Markham) who could travel huge distances at speed and across countless international borders. Here is a provisional sheet designed for aviators, created by the British War Office, that spans the area from Beirut to just east of Tehran, from Mosul in the north to just south of Doha.

International General Aeronautical Map Iraq: War Office (1923, revised 1924) 65 × 49cm

'A map in the hands of a pilot is a testimony of a man's faith in other men; it is a symbol of confidence and trust. It is not like a printed page that bears mere words, ambiguous and artful, and whose most believing reader – even whose author, perhaps – must allow in his mind a recess for doubt.

'A map says to you, "Read me carefully, follow me closely, doubt me not." It says, "I am the earth in the palm of your hand. Without me, you are alone and lost."

'And indeed you are. Were all the maps in this world destroyed and vanished under the direction of some malevolent hand, each man would be blind again, each city be made a stranger to the next, each landmark becomes a meaningless signpost pointing to nothing.

'Yet, looking at it, feeling it, running a finger along its lines, it is a cold thing, a map, humourless and dull, born of callipers and a draughtsman's board. That coastline there, that ragged scrawl of scarlet ink, shows neither sand nor sea nor rock; it speaks of no mariner, blundering full sail in wake-less seas, to bequeath, on sheepskin or a slab of wood, a priceless scribble to posterity. This brown blot that marks a mountain has, for the casual eye, no other significance, though twenty men, or ten, or only one, may have squandered life to climb it. Here is a valley, there a swamp, and there a desert; and here is a river that some curious and courageous soul, like a pencil in the hand of God, first traced with bleeding feet.

'Here is your map. Unfold it, follow it, then throw it away, if you will. It is only paper. It is only paper and ink, but if you think a little, if you pause a moment, you will see that these two things have seldom joined to make a document so modest and yet so full with histories of hope or sagas of conquest.'

Beryl Markham, aviator, *West With the Night* (1942),
New York: Houghton Mifflin, pp. 245–6

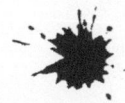

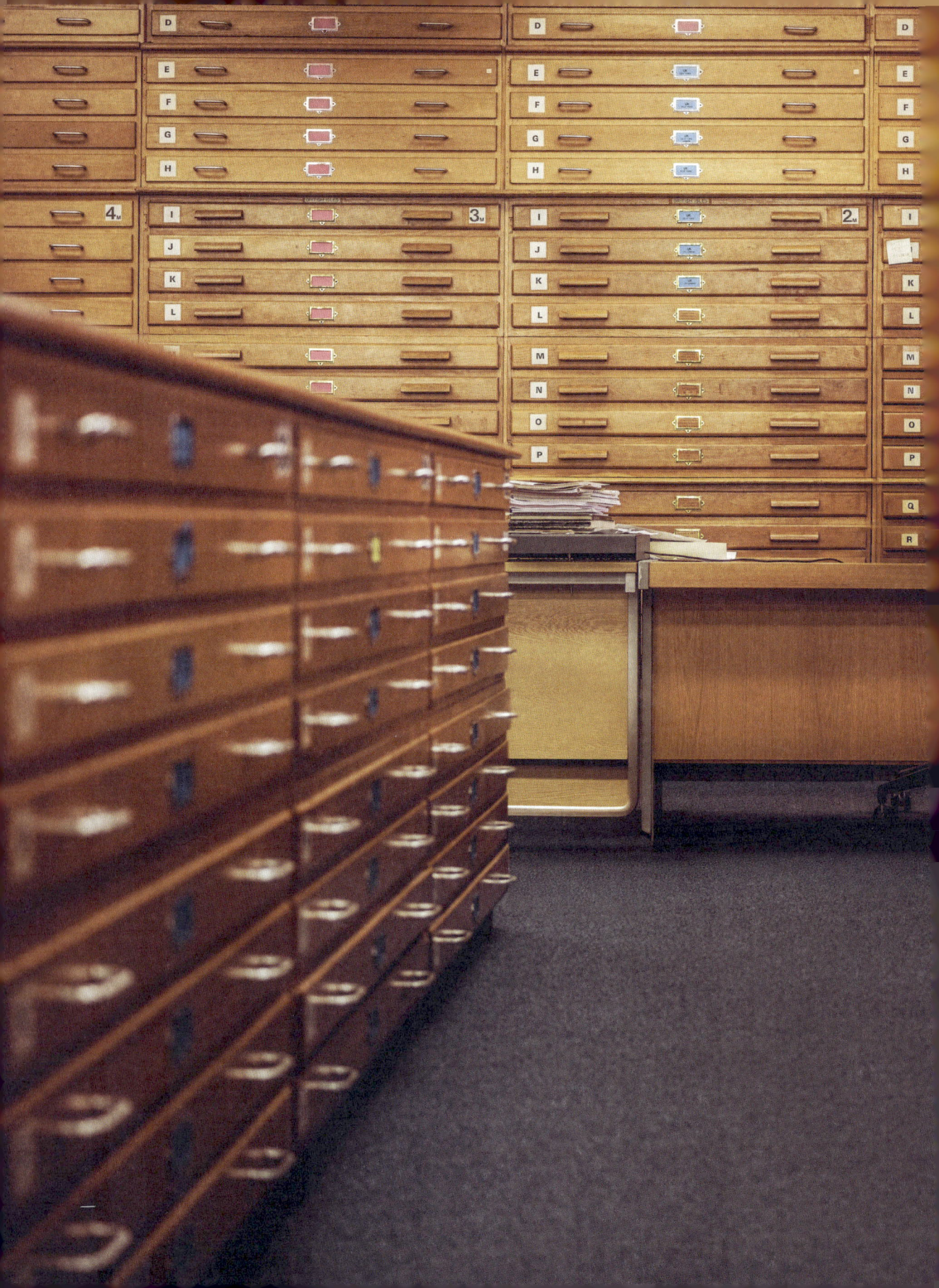

CHAPTER ONE

Welcome to the Map Library

'It is not down in any map; true places never are.'
Herman Melville[1]

In the heart of London's Bloomsbury the scent of ageing paper swirls around a basement floor of University College London (UCL). A few years ago, I followed this distinctive aroma as it took me to the right of the dingy lecture theatre I'd taught in many times, past the damp of a leaky water fountain and on into an unpromising corridor. On I went until I reached a scruffy turquoise door with the words 'Map Room' affixed to it. I pushed at the handle, and, to my amazement, entered an Aladdin's cave of cartographic treasures.

    Bulging out of 440 bespoke wooden drawers and stacked on the shelves of tens of glass-fronted cabinets were thousands of maps and hundreds of atlases. It is a place that had not changed much in the two decades since the retirement of Anne Oxenham MBE, who curated this extraordinary collection in the forty years

The towering drawers – or plan chests as they should properly be called – of the Map Library. There are 440 individual drawers in total, and each can comfortably store tens of maps of almost any size.

Taken in August 2024, Peter Searle

Map 2 The space race of the 1960s inspired many scientific developments, not least the creation of some of the most detailed maps of the moon. Here we see the beautiful Copernicus crater, which is just off centre as we look up at our lunar neighbour.

Lunar Chart 1964: US Air Force and NASA (1964)
74.5 × 55.5cm

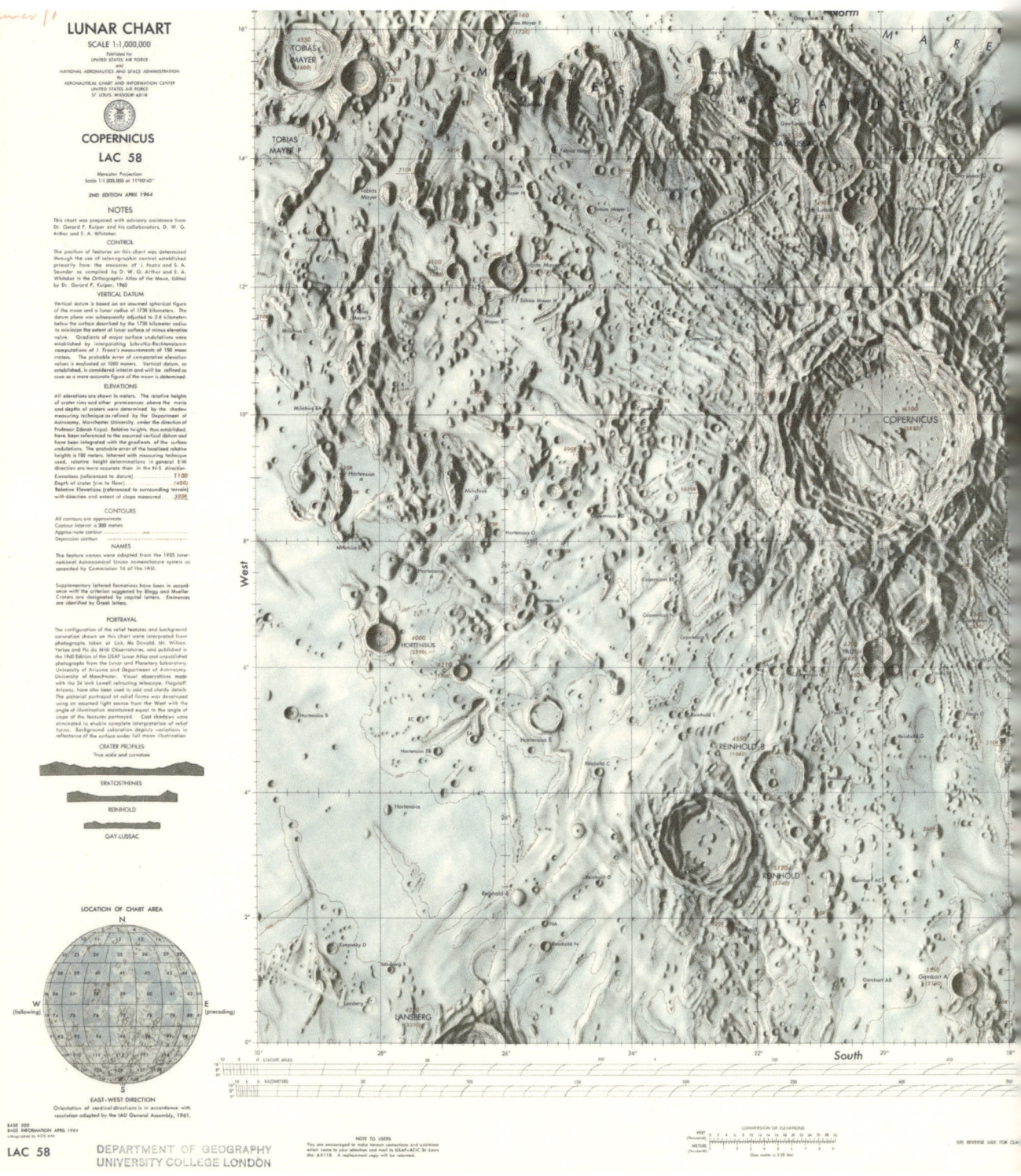

WELCOME TO THE MAP LIBRARY

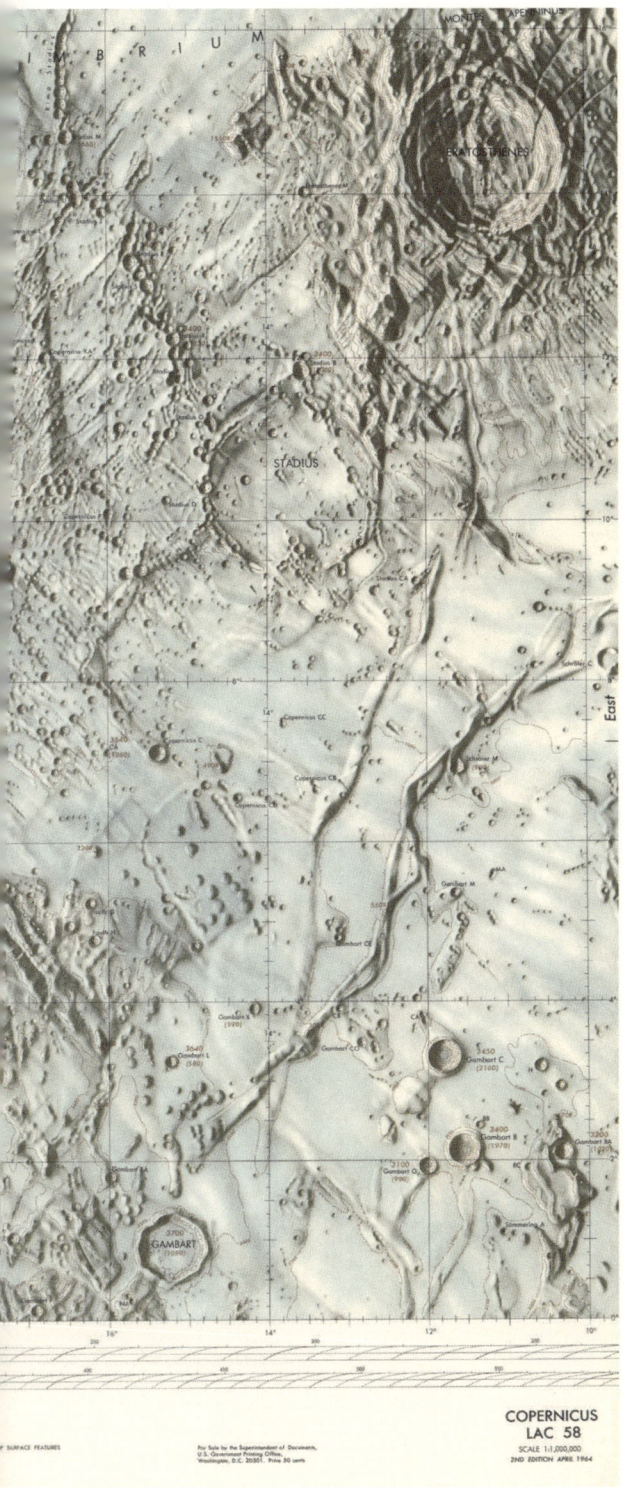

she occupied the role of 'Map Librarian'.

It has been said that the past is a foreign country and perhaps that is the most important destination I was transported to as soon as I stepped inside this library of lost maps. Finding a map of Hiroshima printed just weeks before the atomic bomb was dropped left me devastated by the horrors humans can inflict on one another, while a map of the moon left me elated at the ways maps have been crucial to the pinnacle of human endeavour.

I felt the rush of optimism from the maps of the 'Air Age' that opened new possibilities for travel and global commerce. I was enthralled by the beauty of the maps of the ocean floor, which could not have been created without Soviet scientists sharing data with their American collaborators at the height of the Cold War.[2]

Over and over again I experienced the joy of unfurling a map and knowing that, apart from the occasional fingerprint in the dust and some yellowing of the paper, it was unchanged from the day it rolled off a printing press, and I never tired of witnessing the moment a map was released from the dark confinement of its drawer to come alive and tell its story. Many would share tales beyond

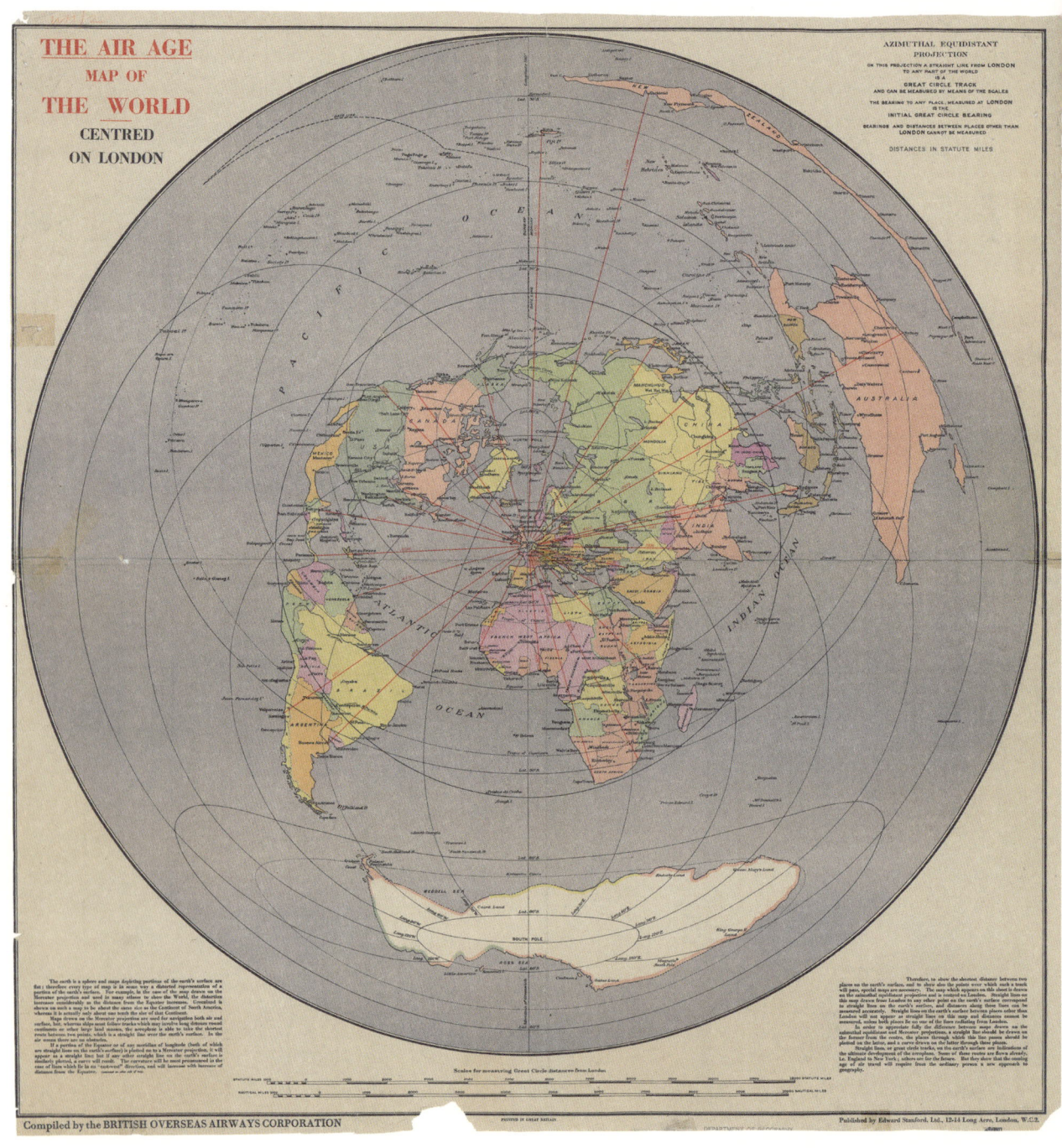

WELCOME TO THE MAP LIBRARY

what was printed on their surface, to reveal the turbulent lives of their makers and of those who held them before settling in the relative calm of the Map Library.

I have never thought so deeply about the moments from our past captured in this room and I have never been more convinced about the importance of maps not just to witness history, but to shape it.

I guessed it had been years since the library welcomed a new map into its collection. The need for paper maps has diminished as distant places are brought closer by cheap travel and, of course, the digital maps in our pockets that allow us to zoom into a street thousands of miles away. In these circumstances, the world sketched by the contours of the Map Library should look antiquated and even redundant but, remarkably, it feels more pertinent than anything we can find scrolling around the maps on our phone.

For example, during the first summer I spent exploring the Map Library's contents, Europe was suffering from extreme heat and drought, and I saw on the news that the melting glaciers in the High Alps were causing national borders to be redrawn between Switzerland and Italy. I reached for a map of the region that was published in 1928 and was in awe of the beautiful rendering of the now shrinking ice formations that surround the unmistakable peak of the Matterhorn and support the international border.[3] Such a map would be no use for navigation now; it is instead an epitaph of a world we once knew.

My first summer was also the year of Putin's full-scale invasion of Ukraine, which began in February 2022, so Europe was experiencing the consequences of war. I saw how Putin's worldview was crystallised by the abundance of maps from the Soviet Union that charted the borders of the empire he wishes to restore. His troops were even seen carrying copies of them as they marched across the border.[4]

Map 3 This large map was a celebration of the spread of air travel at the end of the Second World War. The fantastically named 'Azimuthal Equidistant' projection flattens the Earth around a central point – here it is London – morphing the lines of longitude and latitude into a tangle of different shaped ellipses, but enabling perfectly straight lines to be drawn along the great circle route from the British capital to the far-flung destinations served by the British Overseas Airways Corporation (BOAC).

The Air Age Map of The World Centred on London: Edward Stanford Ltd (1945) 97 × 102cm

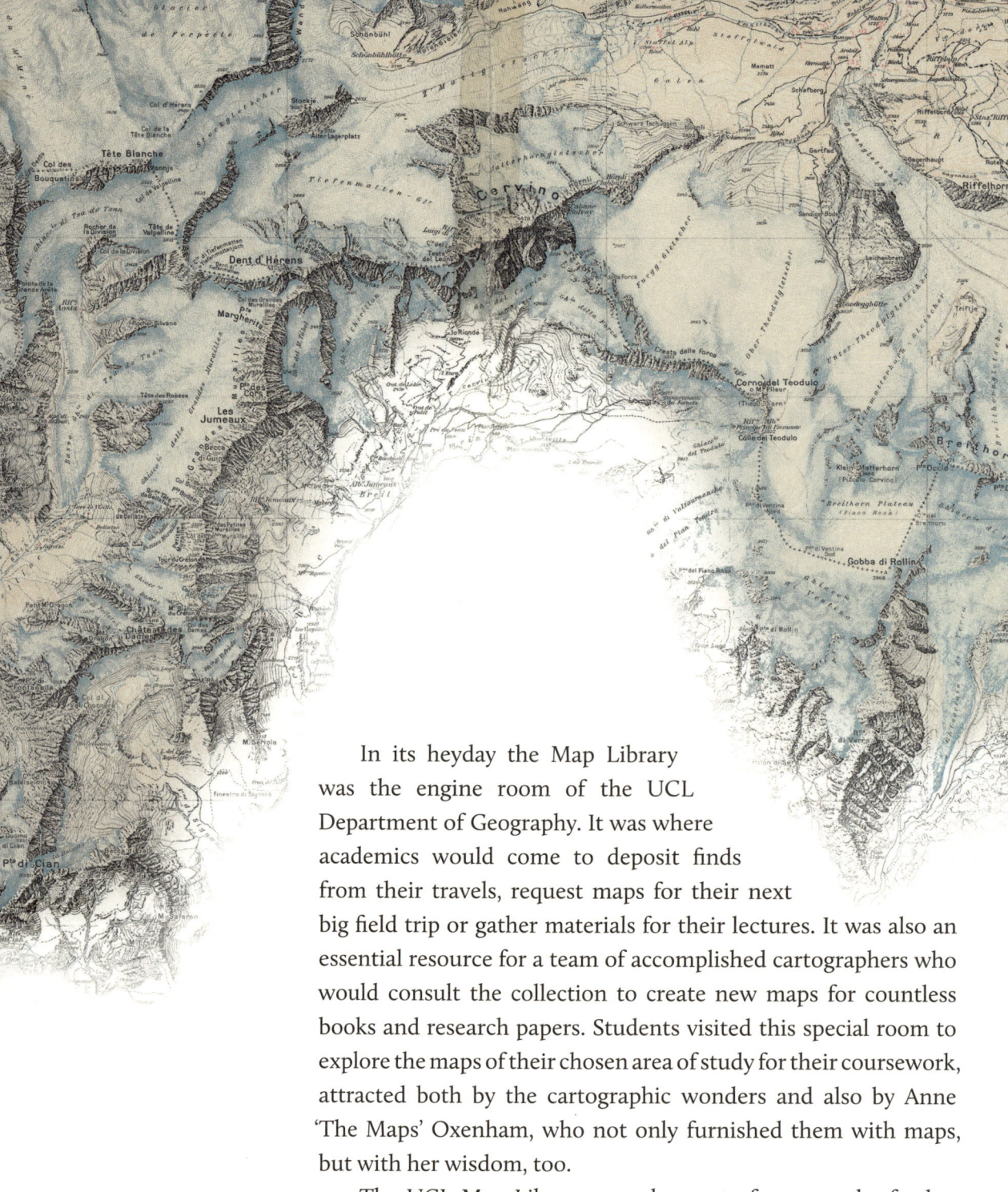

In its heyday the Map Library was the engine room of the UCL Department of Geography. It was where academics would come to deposit finds from their travels, request maps for their next big field trip or gather materials for their lectures. It was also an essential resource for a team of accomplished cartographers who would consult the collection to create new maps for countless books and research papers. Students visited this special room to explore the maps of their chosen area of study for their coursework, attracted both by the cartographic wonders and also by Anne 'The Maps' Oxenham, who not only furnished them with maps, but with her wisdom, too.

The UCL Map Library was also part of a network of other map libraries housed in universities and government departments,

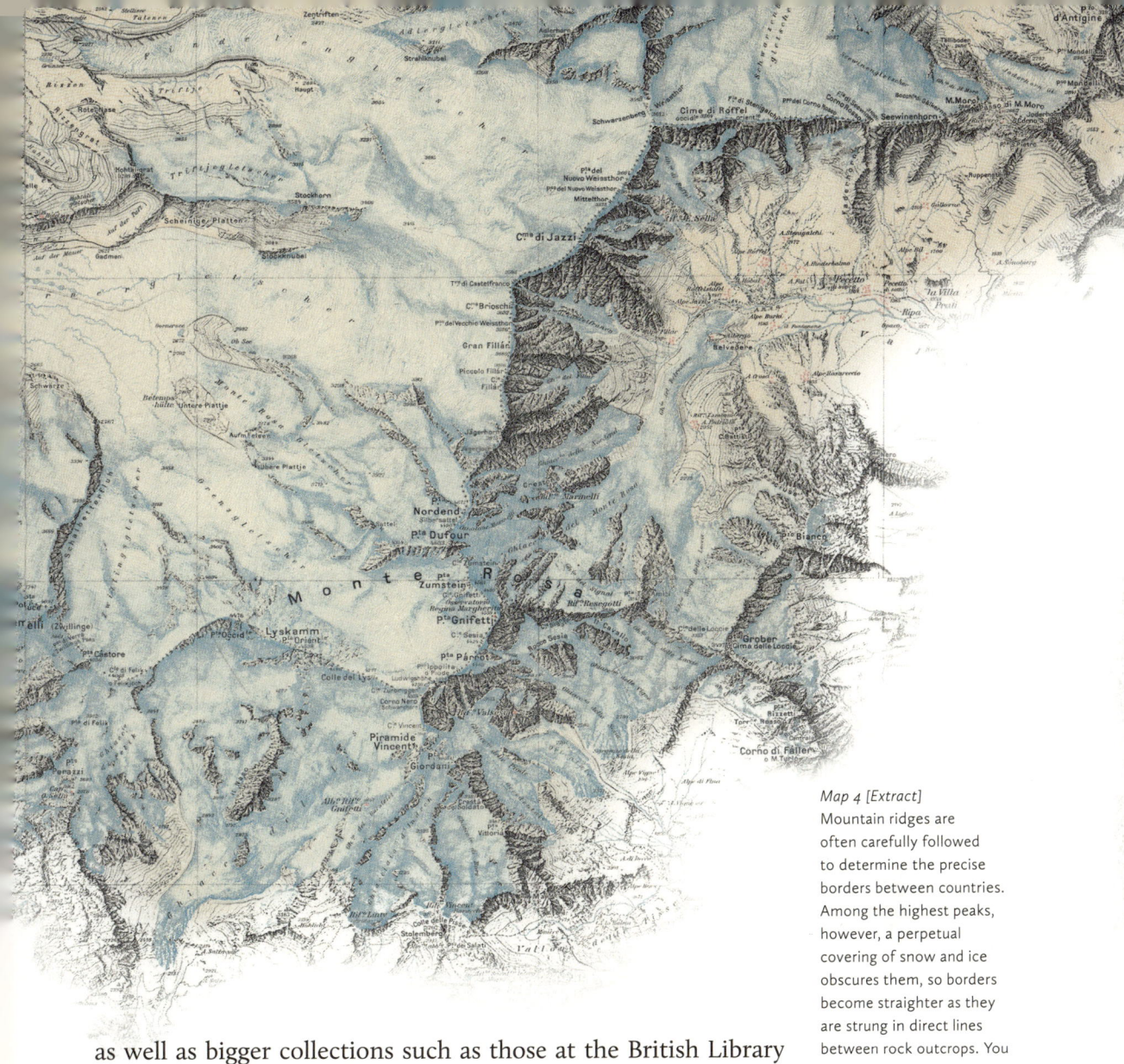

Map 4 [Extract]
Mountain ridges are often carefully followed to determine the precise borders between countries. Among the highest peaks, however, a perpetual covering of snow and ice obscures them, so borders become straighter as they are strung in direct lines between rock outcrops. You can see how the borders in this map – shown as a line of small black crosses – do just that. The snow and ice melting is prompting borders to be reconsidered, as has happened around the Corno del Teodulo peak.

Il Cervino e Il Monte Rosa: Touring Club Italiano, Milano (1928)
76.5 × 49cm

as well as bigger collections such as those at the British Library just down the road, the Royal Geographical Society in South Kensington or the Bodleian Library in Oxford. Maps would be shared and swapped by librarians, so they could add to their ever-growing collections.

The arrival of computer-based mapping meant that there was less and less need to pay a visit to the room and sift through its weighty drawers, as a world that could be seen on a screen no longer needed to be spread out across large sheets of paper.

THE LIBRARY OF LOST MAPS

The shelves of the atlas cabinets were best illuminated by summer sunsets, their glass refracting the light around the reading room. This photo was taken during my explorations, when many of the atlases were brought forward from the backs of the cupboards and into the light for the first time in decades.

The click of keyboards slowly replaced the rustle of paper until the maps fell silent. As Anne and the older academics retired, new lecturers never saw a need to acquaint themselves with the wonders of the collection, so it became dormant.

It is a story that has repeated itself the world over and has led to the closure of dozens of map libraries and, for decades, librarians have been sounding the alarm at the plight of their collections.[5] Fortunately, the UCL Map Library is one of the survivors,[6] but it faces an uncertain future. If a home can't be found for the maps, they will be offered up to map collectors, sellers and enthusiasts, or simply thrown away.

I want to reveal the wonders of this fantastic room while I still can. So, I am delighted to have you for company as we traverse these lost maps, not just admiring their beauty, but meeting the fascinating characters who created them. Together we will use the maps as a portal to the past and contemplate the future they foretold; we will learn how maps are made and the ways that old maps can shed new light on modern events; and we will see proof that cartography can be both essential for scientific discoveries and also a tool of propaganda and empire.

We will pore over maps that have helped transform our knowledge of the ground beneath our feet and clamber to the peaks of the highest mountains to see what maps tell us about our planet today. We will admire the maps that helped educate a newly literate society but also those used by Hitler's henchmen to shape the world to his horrific vision.

I can't promise the smell of the soot deposited by the London smog of the 1950s or the crackle of a map unfurling itself after decades being folded at the back of a drawer. But I can promise that you'll never look at maps in the same way again.

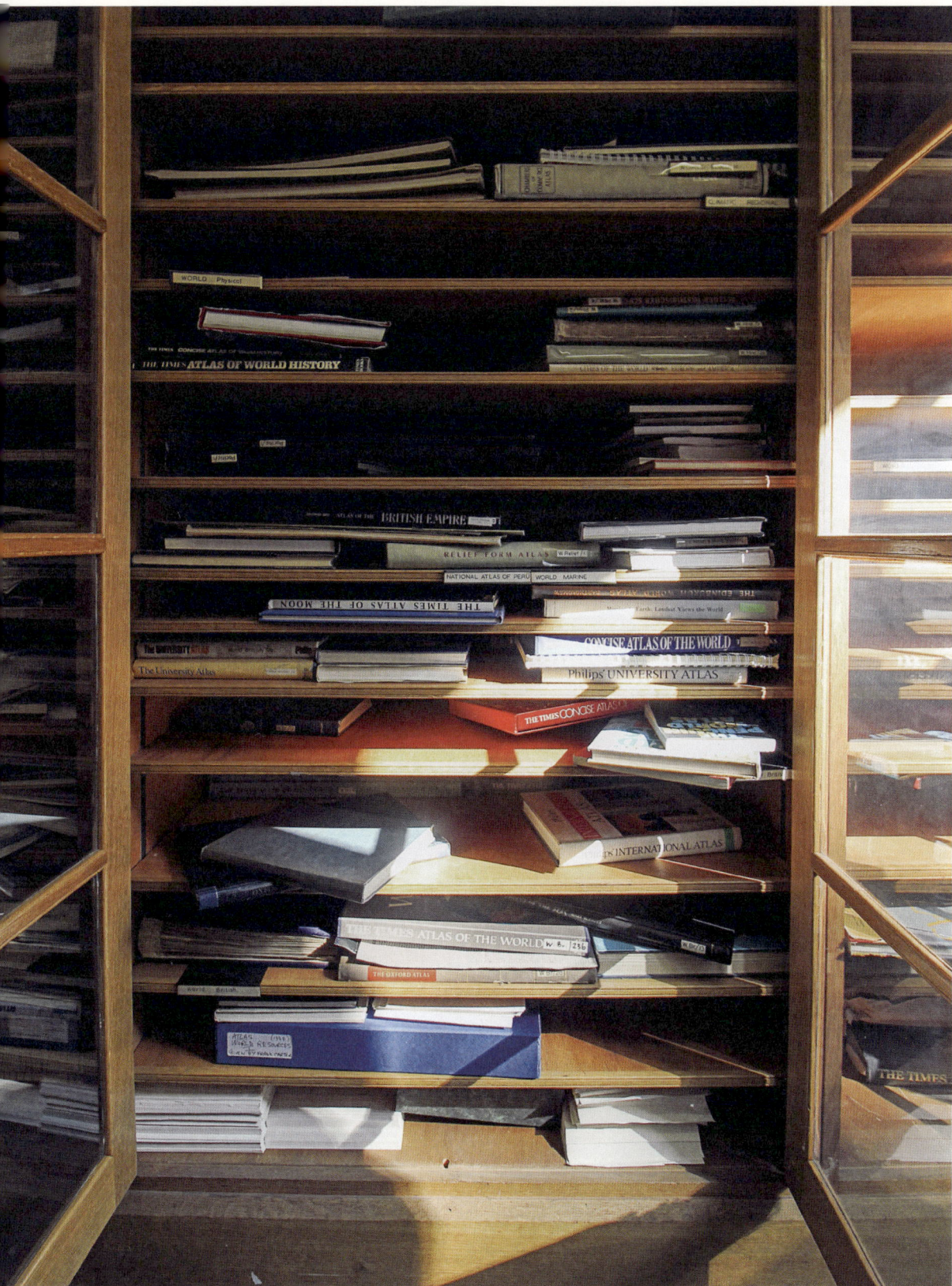

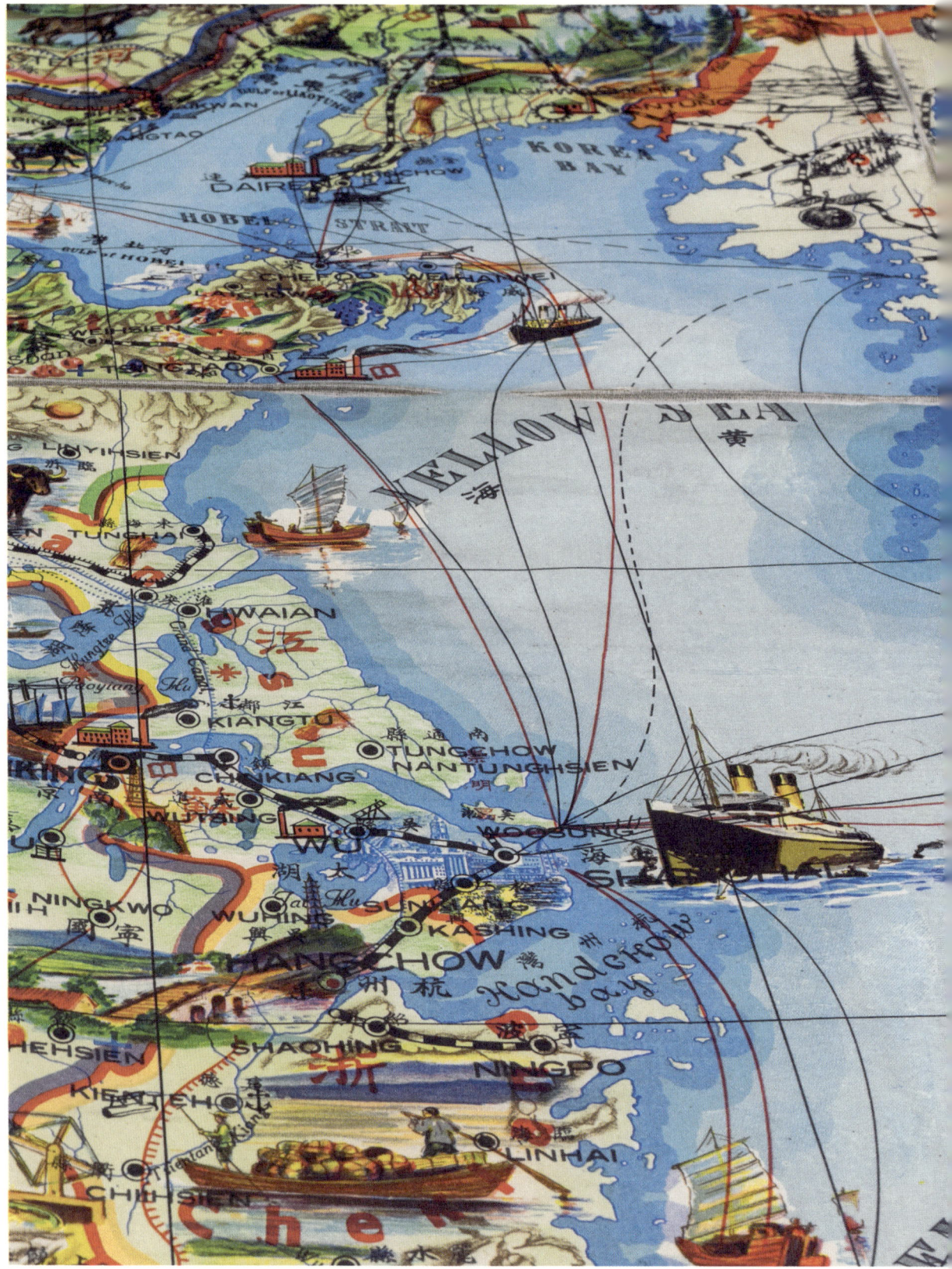

WELCOME TO THE MAP LIBRARY

(Re)Discovery

Most of my research and teaching explores how digital data can be mapped in new and exciting ways. I compile algorithms and databases, not dusty piles of paper, so in my early years as a geography academic I too was indifferent to the plight of map collections. Over ten years passed at UCL before I opened one of the Map Library's drawers to see what was inside.

I was first drawn to the room in search of some historic examples of atlases that I could use in a new class I was teaching about cartography and data visualisation. I had always admired the graphical innovations that the Victorians had developed to share the wonders of their age, but it had never occurred to me that a room full of these would have existed within my department. That is until I began my search for examples in earnest and was directed to the collection by a couple of the department's emeritus professors who'd heard about my teaching plans.

I am privileged to teach students from around the world, so I tried to select maps from as many different places I could. I was excited by what I found but slightly anxious about how well the maps would be received by the group of seventy or so 21-year olds who were taking the class. My anxieties were swept away the moment the first students arrived in the Map Library. I thought – at best – that they would marvel at the cartographic novelties, but instead the room was buzzing with tales of historic border disputes, students comparing what they learned of their country's history at school and how their sense of nationhood was at odds

Map 5 [Extract] One evening, just as I was packing up to leave, I couldn't resist opening the final unlabelled drawer. Inside was a folded linen square that looked more like a grubby dust sheet than a map. I had low expectations as I placed it on the large table in the middle of the room, but I was in for a surprise. It was a vast illustrated map of China depicting every aspect of the country from its people to its wildlife, its trading routes to its climate.

Map of China: The Northern Trading Company and Mr V.F. Yao-Hsuin (1931) 191 × 138cm

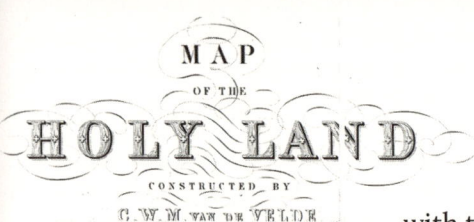

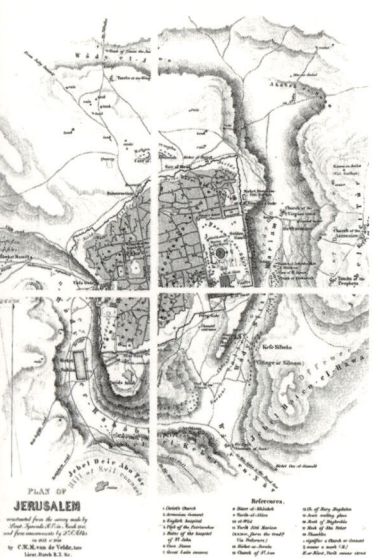

with their friends who grew up in a country nearby.

I'd also underestimated the grip that maps have on the psyche of many countries, as some students would look away as I pulled out maps that would be censored back home because of the borders they portrayed. I was once asked if I had the permission of the government to show such controversial maps.[7] For example, Argentinian students would protest at the maps of the Falkland Islands lacking acknowledgement of their other name, Las Malvinas. There would also be heated debates about peace in the Middle East, the maps of which have long been harbingers of human conflict, and surprise at the extent of the border changes at the end of the First World War.

My students confirmed to me that these old maps were not redundant at all: they still had a lot to say. They also reminded me that the contents of this room only offered a partial view of the world, not least one from a Western perspective that is too narrow compared to how we study geography nowadays. And, beyond the geopolitics, they were in awe of the magic of the room and revelled in the physicality of handling the maps.

Younger generations always see a different world from their parents and grandparents. Today, they do not rely on maps recycled from past wars or approved by governments; they construct their own view of the world online. It's perhaps no surprise, then, that Iceland's tourist industry owes more to Instagram influencers and the writers of *Game of Thrones* than to any tourist map promoting its charms.[8] I should also have expected that the first question I got from that magical first class has been one I have been asked by every group of students I have taken into the Map Library since: 'Can we take photos and post them online?'

Social media etiquette aside, the students wanted answers to where the maps had come from. Why was there an atlas charred by fire (we'll get to that in Chapter 7), and why did so many maps bear the stamps of the great libraries and military map rooms or the names of long-deceased academics? And they kept asking how these maps were made and what were they used for.

My initial answers only scratched the surface of what I knew must be the exciting yet hidden stories of these maps. There were hints and clues left around the place – scraps and fragments for me to piece together to find out about how the maps came to be and the impact they had on the world. I was inspired to begin my own explorations not just to do the maps justice, but to ignite the imaginations of those I was showing them to.

What we'll never know is the unwritten – or even unspoken – histories of some of the maps in the library. I think of the maps that have had the greatest impact on me, or the ones that have had the most eventful journeys to come into my possession that, in times gone by, I would have left without much comment for Anne to file in a drawer, in the hope that they may be of some use to others in the university.

The one that stands out is a tourist map and travel guide for *The Social City of Zaporizhzhia*, which is located on the Dnipro (or Dnieper) River in south-eastern Ukraine. The city and its surrounding region rose to prominence when the nuclear power station to its south – the largest in Europe – was damaged and captured by Russia's invading forces in 2022.[9]

I was given this map in the summer of 2023 by a group of Ukrainian academics from Zaporizhzhia National University whom I met at a workshop hosted by Durham University (UK).[10] Over several days we discussed the ways that maps could be used in preserving the testimonies of those caught up in the horrors of Russia's actions. I could not imagine what they had gone through, and the courage they had not only to travel but also to share their work.

For the UK-based academics, the Ukrainians' presence in Durham was enough, but they also gifted us a bottle of Ukrainian

Map 6 [Extract] This map of Jerusalem is a small inset on a much larger map of the Holy Land that was constructed by C.W.M. van de Velde and published in 1858. The Jerusalem inset was completed in collaboration with Titus Tobler. Van de Velde wrote a memoir to accompany the map, in which he begins 'The study of the Holy Scriptures had made me deeply feel the want of a correct and sufficiently detailed map of the land', a spiritual motivation that few other parts of the world could evoke.[1]

Map of the Holy Land: Justus Perthes (1858) 84 × 131cm

THE LIBRARY OF LOST MAPS

Map 7 A tourist map of the city of Zaporizhzhia, a city on the Dnieper River in south-eastern Ukraine.

Travel Guide Social City of Zaporizhzhia; Consulate General of the Federal Republic of Germany in Donetsk (2010s)
41.5 × 29.5cm

vodka, some chocolate liqueurs filled with Ukrainian spirits and a map of their hometown. This trio of gifts was a small act of defiance, but for me the map was the most powerful among them. We all knew that there was no chance of using the map any time soon, but it was a statement of intent that the day would come when we could pay their 'social city' a visit. As a simple tourist map printed at a time of peace it meant very little, but events had conspired to load it with a weight of symbolism that would have far exceeded the expectations of the original mapmaker.

It's a cliché that history is written by the victors, but we forget that the maps we see most often are made by them, too, since they are the documents that chart the territory gained, or label the places named in honour of the new leaders. The maps made by the 'losers' and those on the wrong side of history may be destroyed or overwritten in official records. But this is not the case in the Map Library because Anne collected the duplicate or slightly out-of-date maps that the British government and major

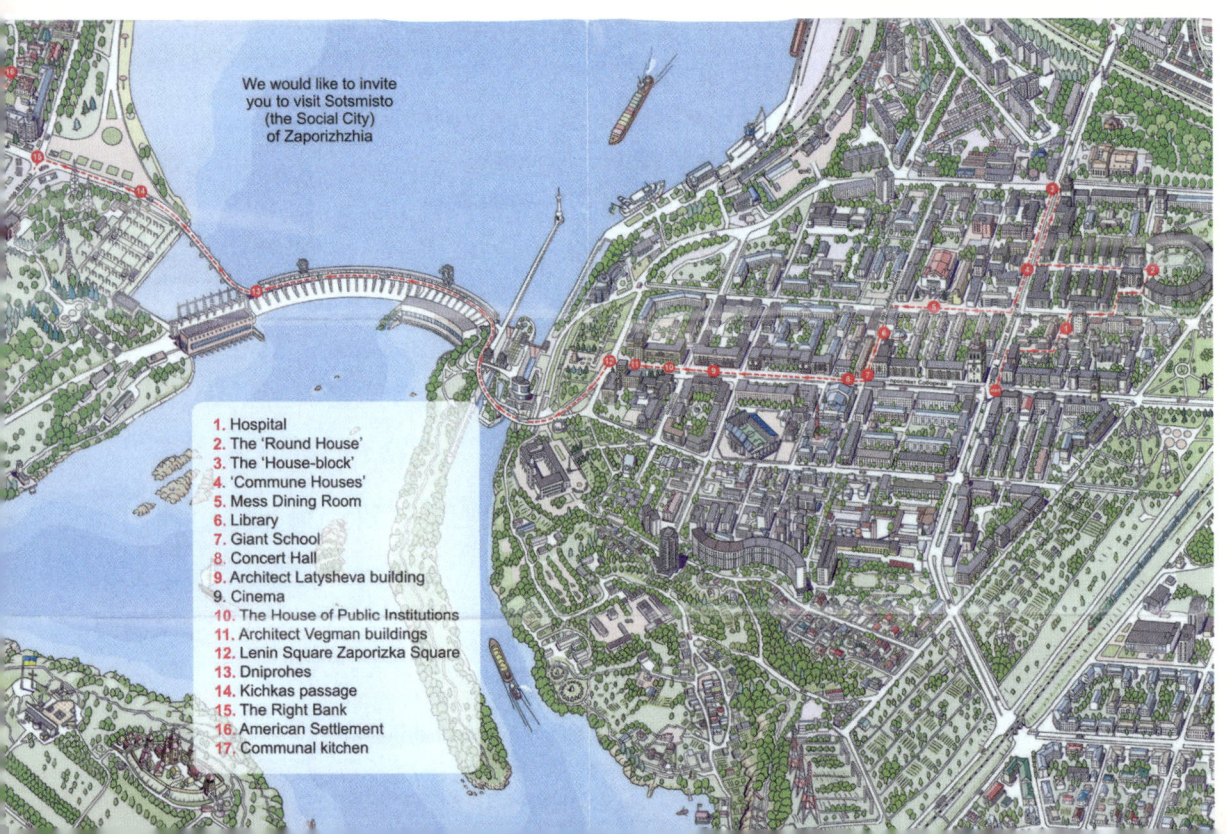

Luftgeographische Orientierungskarte I A–B

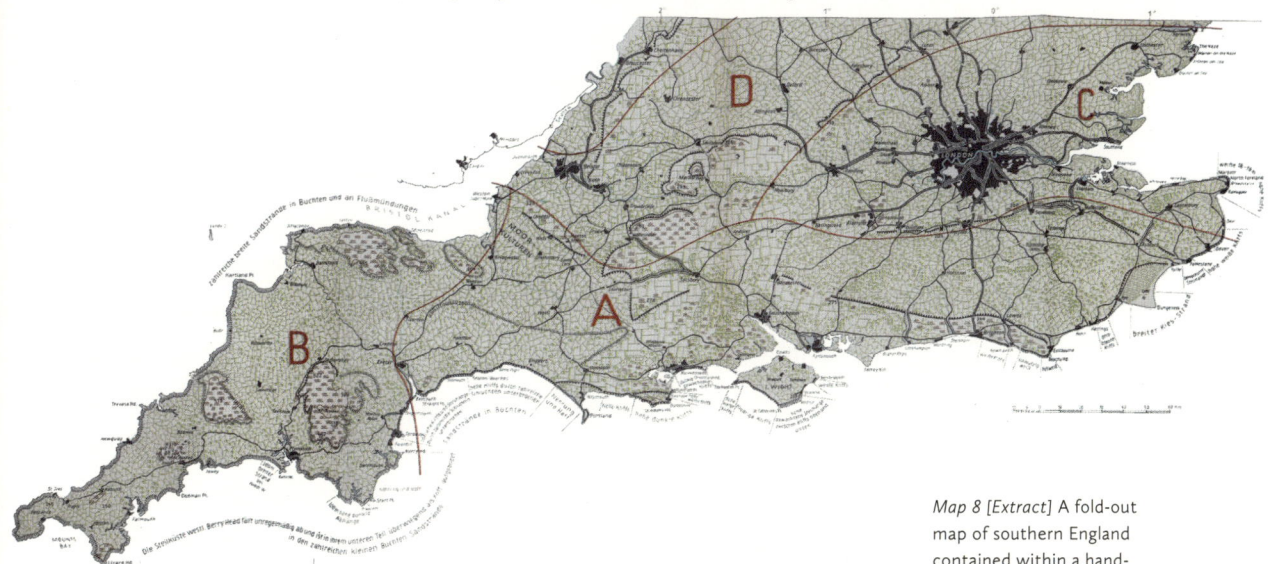

Map 8 [Extract] A fold-out map of southern England contained within a handbook prepared by the German air force (Luftwaffe) during the Second World War that was used to direct pilots to their bombing targets.

Luftgeographisches Einzelheft Großbritannien Band 1: Süd- und Ostengland: Generalstab 7. Abteilung
(c. 1940)
64 × 26.5cm

institutions were disposing of as a frugal way of filling the drawers. When new maps were drawn to navigate a changing world, they were simply placed on top of the old ones. Nothing was thrown away.

This interaction between time and space allows us to step into a world of extraordinary juxtapositions. I have laid out side by side the maps used to bomb Britain, the maps that charted the damage and then the maps to reconstruct it. Compressed into a single pile I would find the makings and outcomes of a major geopolitical issue – such as maps of a new border, the cultures that border dissected and commentary of the unrest that followed. It was moments like these that convinced me that a collection like

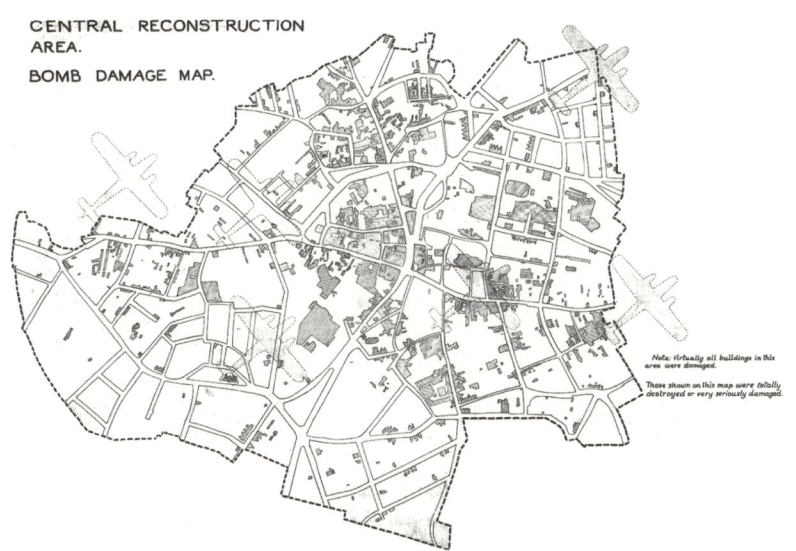

Map 9 This map was produced by the post-war planning officers for Coventry in the UK, who had the daunting task of rebuilding the city after its devastation by the Luftwaffe.

Central Reconstruction Area. Bomb Damage Map: Joint Planning Officers City of Coventry (c.1945)
93.5 × 67.5cm

THE LIBRARY OF LOST MAPS

Map 10 A tiny hand-painted fragment of the 'Land Utilisation Survey of Great Britain'. This scrap would have been part of the bundle of map sheets sent to the engravers who would prepare the printing plates for publication. The handwritten note at the top says 'water not on zinc', which means the zinc printing plate was missing the small lake shown in blue and so it needed to be added for the final printing.

Fragment of the Land Utilisation Survey of Britain: London School of Economics (1936)
11 × 25cm

this one is a vitally important resource for the history it contains.

As we shall see, often it is the maps that at first glance are undeserving of a second look that have the most powerful stories, lurking out of sight from the historians who are beguiled by the grandeur of the great manuscript maps or those that were the 'first' maps of a place. For example, I dismissed some hand-coloured scraps of a map as the workings of a student project, but it transpired that they were watercolours that were part of the creation of the Land Utilisation Survey of Great Britain. This extraordinary project was organised by L. Dudley Stamp of the London School of Economics who enlisted the help of 250,000 volunteers, the bulk of whom were children, to map the landscape of Britain on a field-by-field basis.[11] Survey work began in 1931 and continued for a decade, with the map becoming an important blueprint for the country to manage its agricultural production during the blockades of the Second World War. The project has since been described as a 'modern Domesday Book'[12] and one of the 'great cartographic achievements of the twentieth century'.[13]

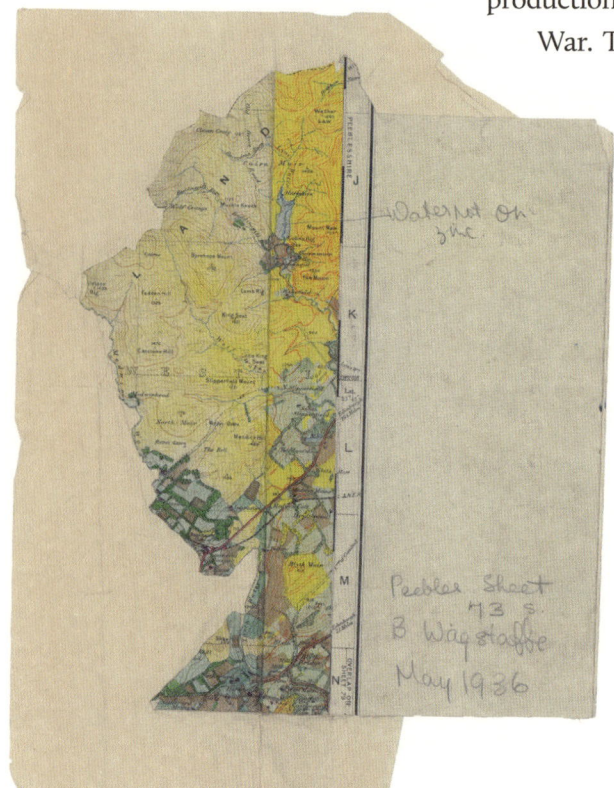

So, from the many thousands of maps sitting idle in their drawers and the hundreds of atlases lying dormant on their shelves in the Map Library, I have given as many maps as I can this second look and have pulled out those that I think you will find the most fascinating. My picks offer us enthralling stories, but they also teach us to see maps in a very different way.

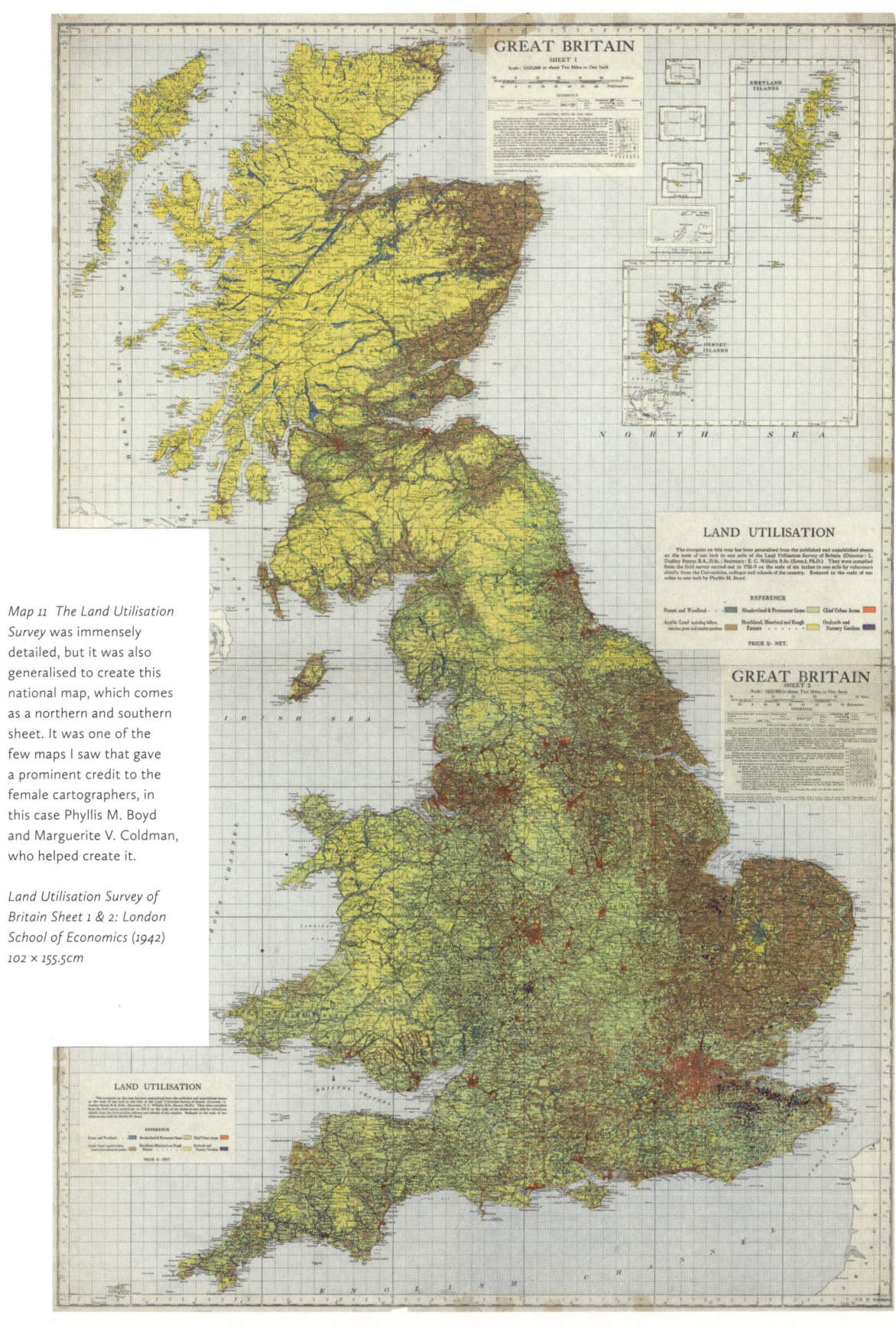

Map 11 *The Land Utilisation Survey* was immensely detailed, but it was also generalised to create this national map, which comes as a northern and southern sheet. It was one of the few maps I saw that gave a prominent credit to the female cartographers, in this case Phyllis M. Boyd and Marguerite V. Coldman, who helped create it.

Land Utilisation Survey of Britain Sheet 1 & 2: London School of Economics (1942) 102 × 155.5cm

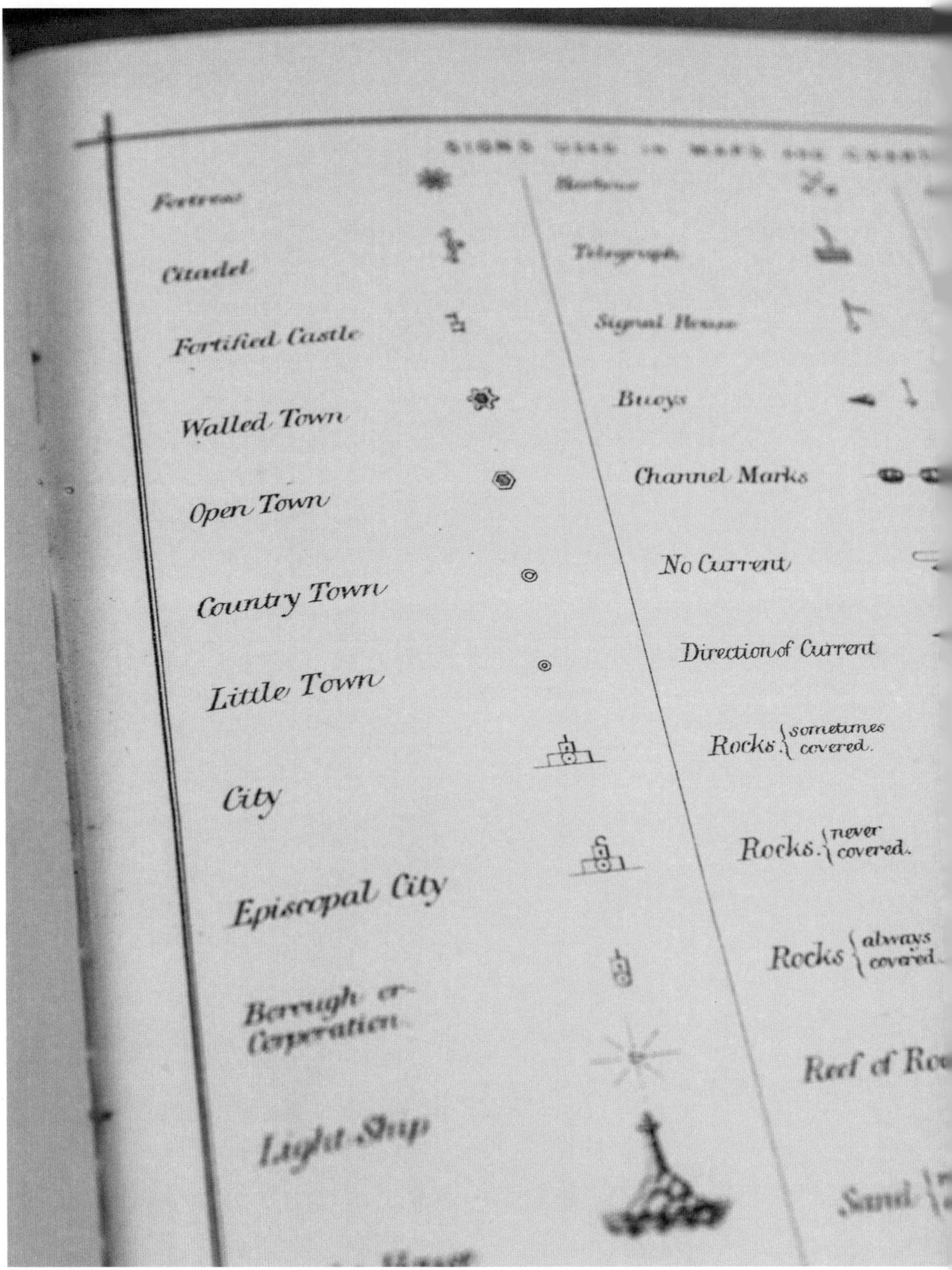

SIGNS USED IN MAPS OR CHARTS

Fortress	Harbour
Citadel	Telegraph
Fortified Castle	Signal House
Walled Town	Buoys
Open Town	Channel Marks
Country Town	No Current
Little Town	Direction of Current
City	Rocks { sometimes covered
Episcopal City	Rocks { never covered
Borough or Corporation	Rocks { always covered
	Reef of Rocks
Light Ship	Sand

WELCOME TO THE MAP LIBRARY

The Secrets of Map-Making

'Are they assigned, or can the countries pick their colours?
What suits the character or the native waters best.
Topography displays no favourites; North's as near as West.
More delicate than the historians' are the map-makers' colours.'
Elizabeth Bishop[14]

Knowing the mapmaker's secrets can help us to understand the maps that we encounter, not just in the Map Library but in our daily lives. Cartographers are keen to standardise the world as much as possible, by using recognisable symbols and conventions that have been established over centuries. An ornate cathedral is stripped back to a simple cross, verdant forests become green blocks of identikit trees and turbulent rivers are becalmed to smooth blue lines. Even a map as beautifully complex and detailed as the map of the High Alps around the Matterhorn is constructed from simple symbols such as the contour lines for elevation and only a few colours – skilfully applied – to distinguish between the ice and rock.

 It's a visual language we've become accustomed to and so, if we put these symbols together, we get something we recognise as a map, even if it is of a place that only exists in our imagination. And it's a language that we take seriously.

 My favourite example of just how seriously map symbols are taken regardless of their context comes from July 2023 when the trailer for *Barbie* the movie was released. It featured the iconic doll,

A page from the *Draftsman's Hand Book*, which was a how-to guide for budding cartographers and map engravers. There were details for all the techniques required for map-making including a series of pages that gave templates for map symbols that could be copied. It is through books like this that cartographic conventions became established, many of which are unchanged today.

The Draftsman's Hand Book of Plan and Map Drawing: E. & F. N. Spon (1874)

THE LIBRARY OF LOST MAPS

Map 12

South China Sea Islands: Cartographic Publishing House (1984)
21.5 × 31.5cm

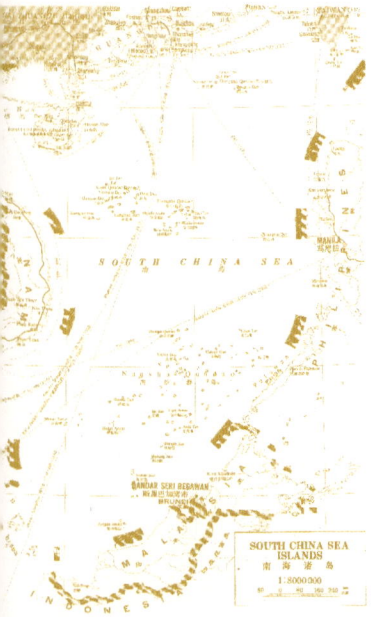

played by Margot Robbie, standing in front of a cartoon map with the words 'Real World Map' visible at the top. Just above a doodle of a pink tortoise was a blue splodge with the word 'Asia' written in it. Coming off the splodge and into the ocean to the right was a line comprising eight dashes forming an inverted S shape.

As a world map it was almost unrecognisable, but the dashes adjacent to 'Asia' were enough to get the film banned in Vietnam, to have the map blurred in the Philippines and to generate a political backlash in the US.

Why? Because the symbology on the map was evocative of China's 'nine-dash line', which is used in some maps to outline China's territorial claims in the South China Sea. These claims are at odds with other countries in the region including Vietnam and the Philippines, hence their visceral reaction to seeing it in the trailer.[15] The accusation was that Barbie was pandering to China and the map was symptomatic of a broader trend in Hollywood to ensure films are well received by the very large Chinese audience.[16] This even triggered the ire of some Republican senators in the US, for instance Ted Cruz, who slammed Barbie's hand-drawn map as 'Chinese communist propaganda'.[17] Had the makers of *Barbie* selected a different symbol – perhaps some meandering footprints or small arrows – instead of those fateful eight dashes for that part of the map, they might have better evoked a journey (which was their stated intention), rather than a disputed border.

Nevertheless, I was pleased to see a debate about map-making get such a high billing in one of the biggest movies of 2023, with *Variety* magazine even commenting: 'Of all the impressive skills Barbie has amassed in her 64 years as a working doll, who knew that cartography would be a focal point of her highly anticipated summer movie debut?'[18]

While the movie will be best remembered for its depiction of a society that had successfully upended the patriarchy, there is a fantastic line, spoken by 'Lawyer Barbie', that really encapsulates why maps can elicit such powerful responses from those who see them and why the lines drawn on them are taken so incredibly seriously: 'I have no difficulty holding both logic and feeling at the same time. And it does not diminish my powers. It expands them.'[19]

Creating this tension between 'logic and feeling' is a superpower that maps have and is what makes them so important in our lives. It is logical not to drive into a deep stream or to try to squeeze a hire car down a road that is obviously too narrow, but we feel we must trust the satnav telling us to do it and so we proceed anyway![20] Maps compel us to act on what they tell us, even if they only offer incomplete information.

I thought of this when I found several early editions of *The Times Atlas of the World* in the Map Library. Each of these weighty tomes, with their spines disintegrating through what must have been decades of heavy use, began with a map showing the parts of the world that had been surveyed thoroughly and those regions that lacked detailed information.

The map itself is particularly interesting because it reveals the priorities of the colonial surveyors who created it. To govern a country, you need to know its limits, so they first turned their attentions to borders, as highlighted in the ribbons of red that track the boundaries that dissect Africa. The continent's interior could be mapped later. There were some exceptions to this, with all of India coloured a confident red thanks to the vast undertaking that was the Great Trigonometrical Survey of India under the leadership of George Everest, who had the mountain named in his honour.[21] For areas coloured in dusky oranges through to faint yellows, *The Times Atlas of the World* had to trust the vague category of 'Fairly Reliable General Maps' and the even more speculative 'Sketch Maps'.

(Overleaf)

Map 13 [Extract] The first map in the *Times Survey Atlas of the World*; an atlas fifteen years in the making and overseen by the 'Prince of Cartographers', John George Bartholomew. It's therefore no surprise that it begins with a map that celebrates cartographic progress. There was still much to survey, but the atlas's index listed 200,000 items, and the whole thing weighed almost ten kilograms, so readers had plenty to explore![2]

Times Survey Atlas of the World: John George Bartholomew/Edinburgh Geographical Institute (1922) 28 × 17cm

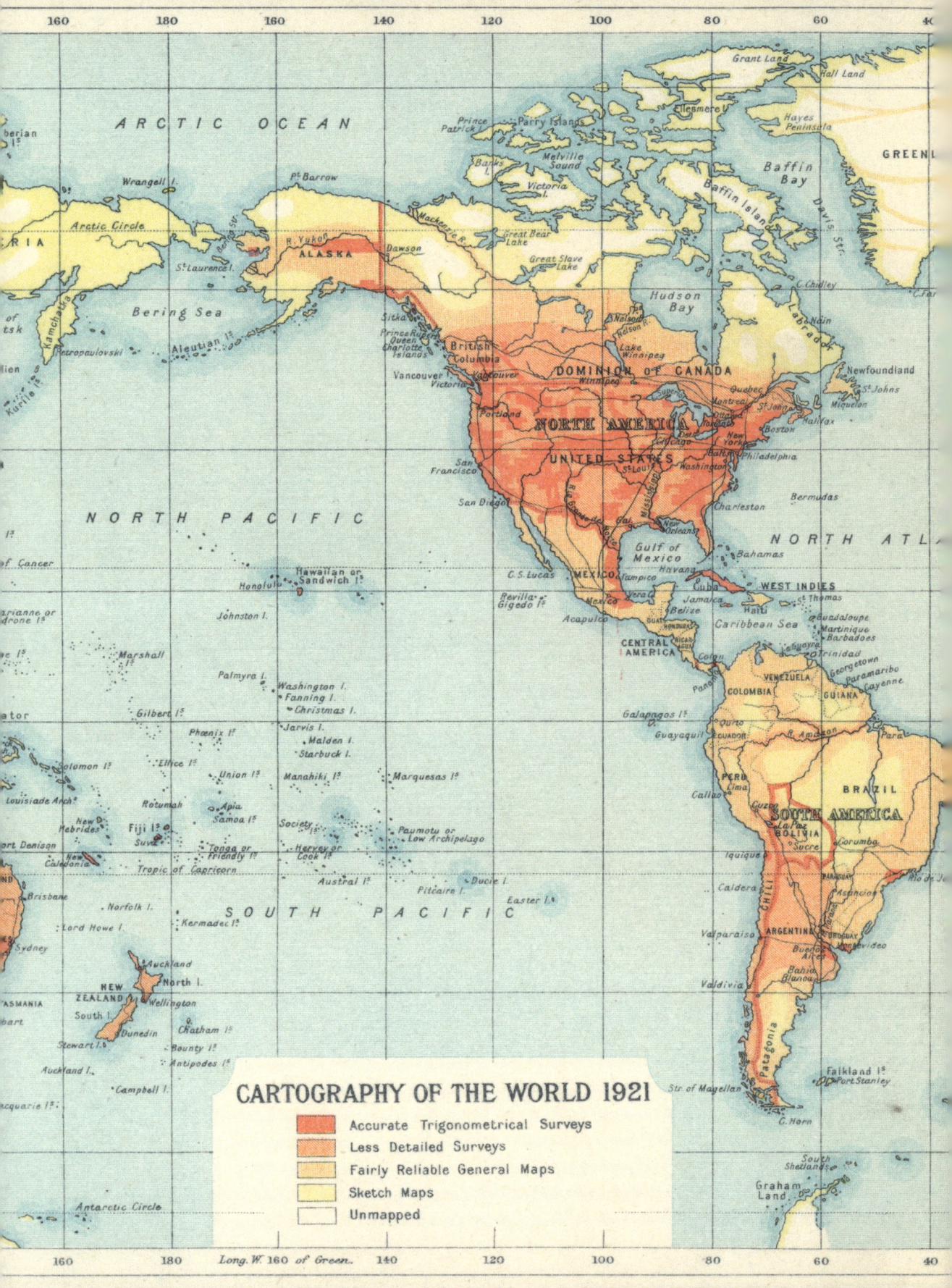

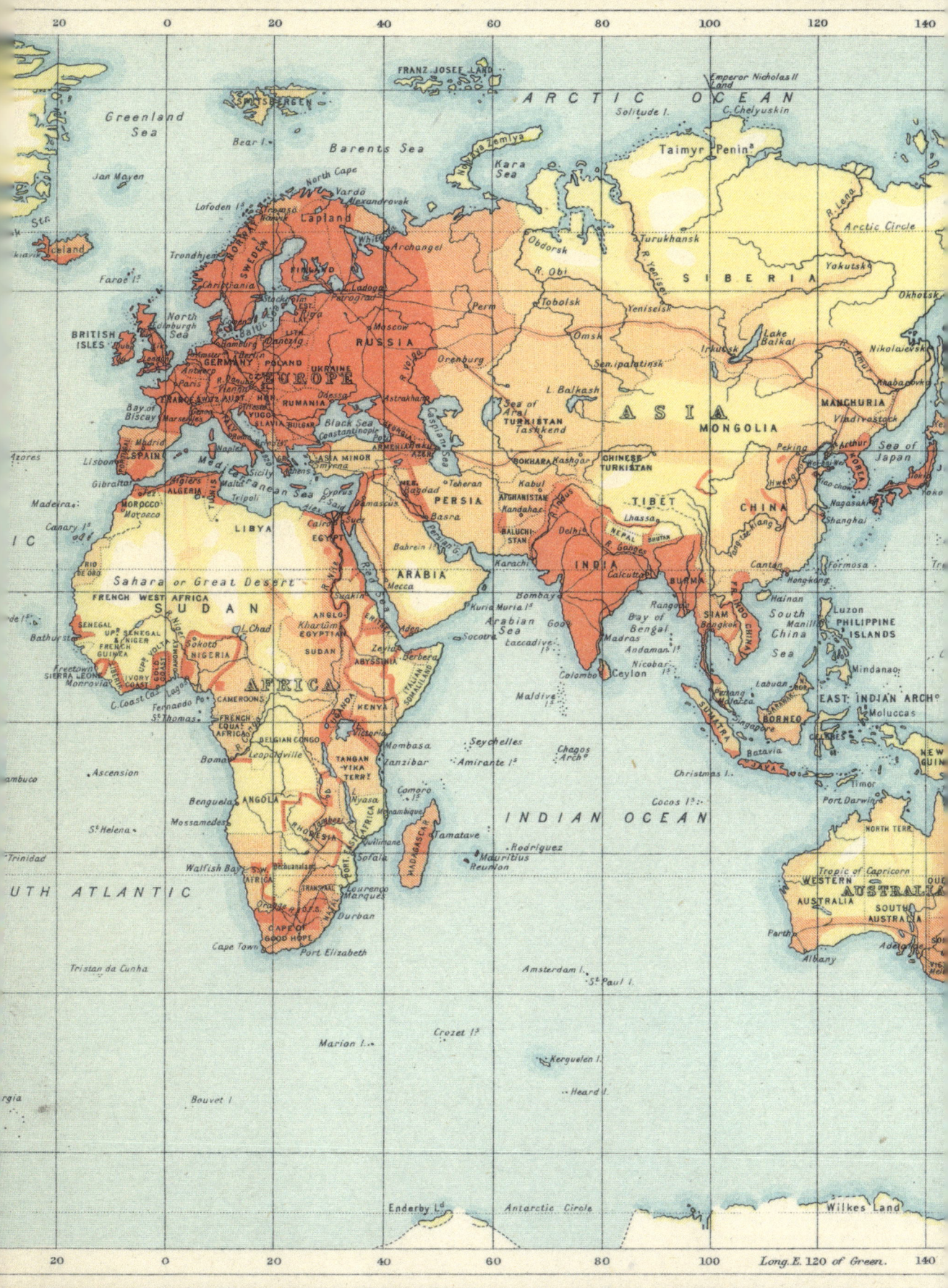

It's a reminder that, well into the twentieth century, in many regions the best source material a cartographer could have hoped to base their maps on was little more than rough drawings, or even handwritten descriptions from an explorer who may have made only a fleeting visit or who had been very creative with the truth to claim a successful expedition. It's therefore fair to say a vivid imagination was needed to flesh out some of the blanks on a map. Acquiring additional information about regions of uncertainty would have been almost impossible, and so they would work with what they had. Lifting entire areas from an existing map would have been hard to resist, even if the facts were indistinguishable from fiction.

Take, for example, the 'Mountains of Kong', which, for over a century, were copied from one map to another.[22] They were first discovered in the late 1790s by a Scottish surgeon and explorer named Mungo Park when he undertook the first of two expeditions along the Niger River. To accompany Park's account of the journey, the cartographer James Rennell produced two maps featuring the Kong Mountains, a vast chain that began in coastal West Africa (Senegal on today's maps) and festooned the interior of the continent by roughly following a line 10 degrees north of the equator. They extended far beyond where Park had travelled.

Over time the mountain chain became smaller as the maps became more refined, but they remained as a great drainage divide between those streams flowing northwards into the mighty Niger River and those flowing south into the Gulf of Guinea.

But the Mountains of Kong don't exist; they were a cartographic mirage.

Park may have realised his mistake on his second Niger River expedition and dispelled the myth of the mountain chain's existence, but it was on this trip that he met a violent end (aged just thirty-five) befitting of the best tales of nineteenth-century exploration gone wrong. Rumours of the cause of Park's demise swirled at the time, but historians have since attributed it to two things.²³ The first was that he set off too late in the year, which meant the river had become shallow, transforming deep, easily navigable water into perilous stretches of rapids that impeded progress. The second was his arrogant approach in making enemies of the local people by not paying chiefs their dues and shooting at anyone who approached him.²⁴

This behaviour, combined with the suspicion that his intentions were to open European trading routes, made him many enemies, some of whom had seen enough. In the words of one historian writing for the *Geographical Journal*: 'This combination of circumstances made escape from interception almost impossible

Map 14 [Extract] The Mountains of Kong as shown on the 'West Africa I' sheet of the Society for the Diffusion of Useful Knowledge's atlas. This was published 1 April 1839.

West Africa I: Society for the Diffusion of Useful Knowledge (1839)
38 × 29cm

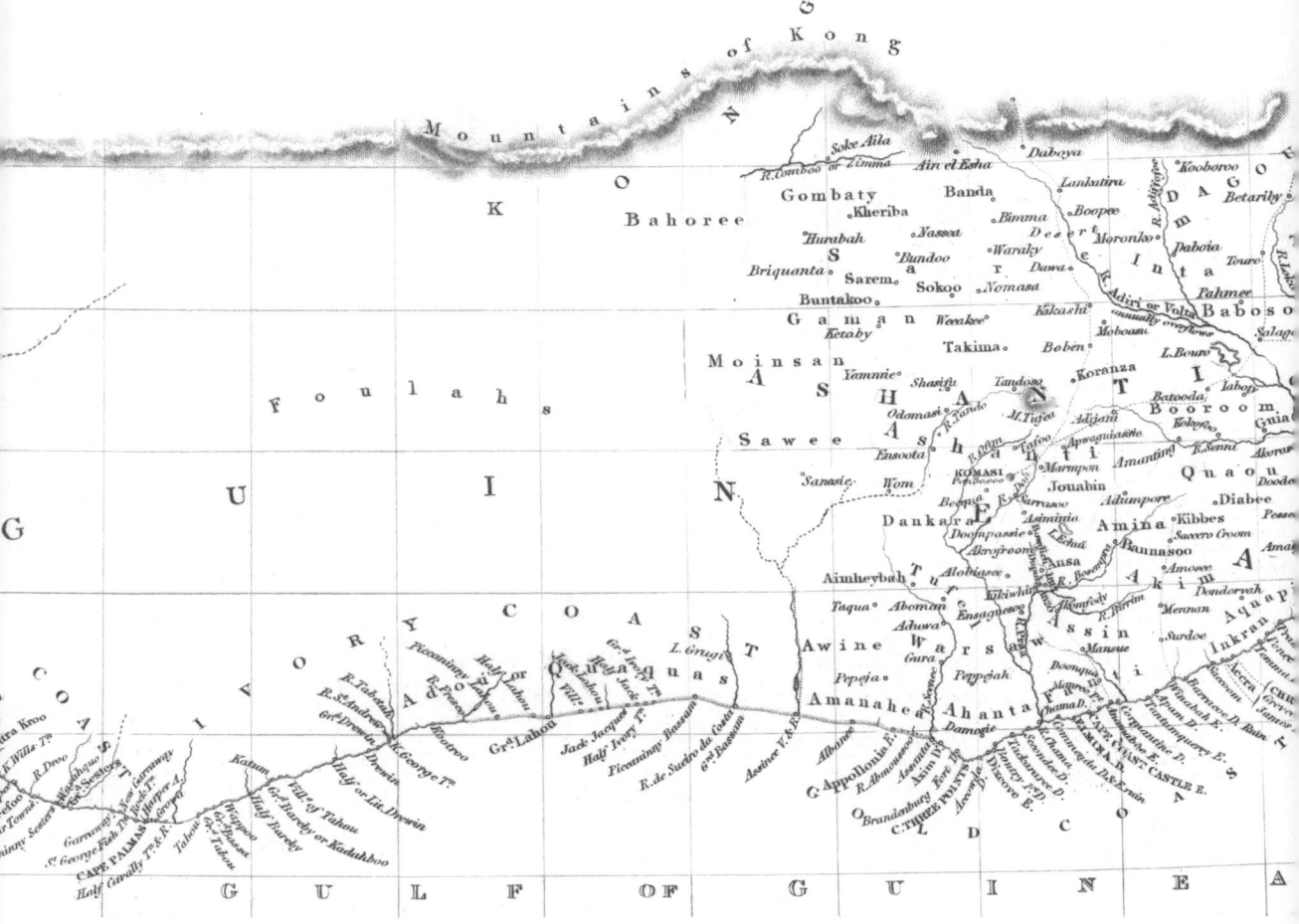

... in [the rapids] of March or April 1806, with a horde of hostile warriors awaiting him, there was no escape.'[25]

Although it was the cartographer Rennell who had extended the Mountains of Kong across Africa, far beyond the Niger,[26] they were seen as a fitting epitaph to Park. Doubts only grew about their existence towards the end of the nineteenth century, and they were finally declared extinct by a French military officer named Louis-Gustave Binger, who undertook an expedition to the area in 1887–8.[27]

As maps became more commercial, cartographers needed to be trusted. They sought out the latest surveys to support claims that they had the best information and therefore the most up-to-date maps. The blurred lines around copyright were sharpened in an Edinburgh court room in July 1853 when the cartographer Alexander Keith Johnston (of the firm W. & A. K. Johnston) accused his competitor, Archibald Fullarton, of copying maps without acknowledgement. There were several aspects to the case, one of the first to be brought under the Literary Copyright Act of 1842,[28] but part of Fullarton's defence rested on the idea that some geographical facts could not be copyrighted: 'If different maps of the same country are differently made, they must agree in their outlines ... It is the filling in of the interior of a country which really constitutes the map.'[29]

However, the weight of evidence – which included witness statements confirming the copying had taken place – was against Fullarton and it took only twenty minutes for the jury to find in Johnston's favour, awarding £200 in compensation (approximately £21,000 today but only a fifth of what he had requested).[30]

Twenty years later, in 1874, W. & A. K. Johnston went on the offensive again and sued a literary magazine, *Athenaeum*, for publishing a negative review of one of their atlases. They perceived the review to be defamatory as it called into question

WELCOME TO THE MAP LIBRARY

the credibility of the maps; after presenting their case, the jury agreed by eleven to one and *Athenaeum* had to apologise and pay substantial damages.[31]

Considering these cases, and others that followed,[32] I'm fascinated that there was so little backlash when the Mountains of Kong myth was dispelled – perhaps because it took decades to wipe them off the map entirely. Today, outed publishers would need to pull their maps within hours and issue rapid corrections. Yet, I spotted the Mountains of Kong in the index of *The Oxford Advanced Atlas*, published in 1936 – fifty years after they were un-discovered!

It is therefore important to remember that, while the first maps of an area might have been based on a visit, they may not have had much more than a cursory glance from the surveyors who sketched their basic outlines. These sketches would have formed part of a collage of maps that were combined into seemingly new editions of maps that benefited from only a few additions and tweaks. This remains true today, with almost all electronic maps being direct descendants of digitised paper sheets. Maps therefore evolve and mature, they aren't born from nothing, and as we shall see they can come close to perfection but will never achieve it.

The index page of the 1936 edition of *The Oxford Advanced Atlas*.

The Oxford Advanced Atlas: Bartholomew/ Oxford University Press (1936)

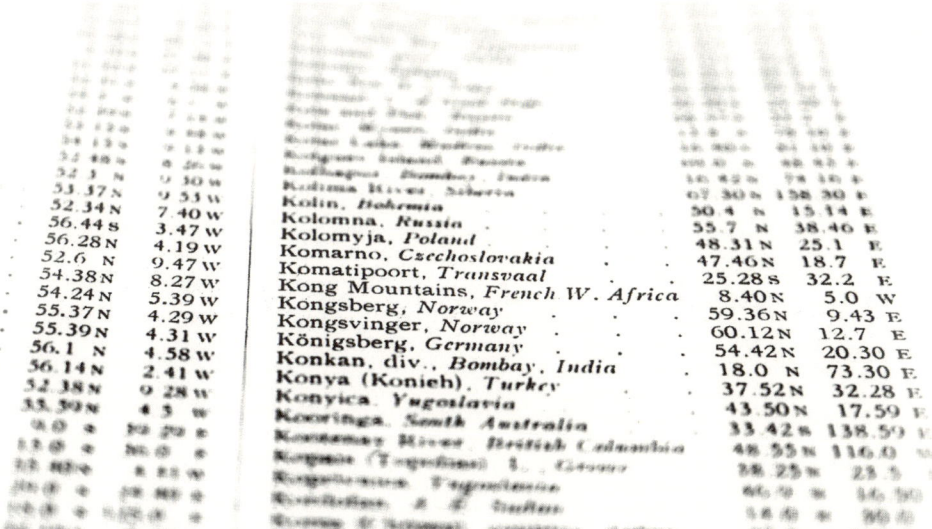

THE LIBRARY OF LOST MAPS

This may sound odd, but it is impossible to create a *complete* map of the world. Maps can be comprehensive, they can even do their job impeccably, but they will never be a perfect mirror of reality. The world is far too complicated for that to happen.

Let's imagine for a moment a cartographer is commissioned to draw a map of your street on a sheet of paper. They would probably begin with two parallel lines to outline the width of the road; they might add some perpendicular lines for any side streets and then probably some rectangles for the buildings. Then what?

More rectangles for parked cars; dots for streetlamps; circles for trees? How will the mapmaker feature your neighbours? They could add coloured shading to the buildings to show how many people are living there, though they'd still be missing important aspects of people's lives such as their age or income. The street on your map would be devoid of litter, the smaller shrubs and the parked bicycles.

They haven't even begun to do the gardens in detail or to add other features like the underground utilities serving the houses, the soil type, the flood risk. On and on they could go, but they'd have to stop when they ran out of time, the page became too full or the details too esoteric. They might take a step back to enjoy their creation only to hear your neighbour rev up their engine and drive off, leaving a gap on the street where their parked car was that had already been drawn on the map. So now the map needs updating. A cartographer's work is never done.

Knowing when to add or remove detail are therefore two of the most important decisions a mapmaker must come to, and they are decisions that are often guided by the map's intended purpose.

If you knew the map of your street was going to be used for a campaign for better street lighting, you might be inclined to ask the cartographer to draw attention to the dark areas, or places where you and your neighbours feel unsafe when walking home

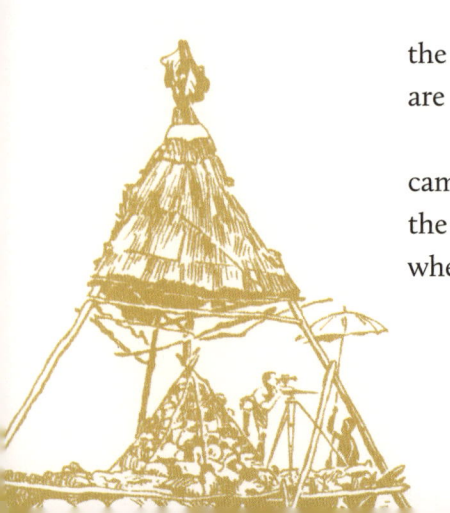

The colonial scene that adorns the frontispiece of the 1932 *Atlas du Katanga*. Katanga was a mineral rich province in the then Belgian Congo and here we see a surveyor peering through his theodolite, shaded from the sun by a person holding a parasol.

Atlas du Katanga: Comité Spécial du Katanga (1932)

at night. However, if you were a member of the local government campaigning for re-election, you might direct the map of the same street to show where the investments have been made to improve road safety. But neither map would suit an insurance company, which would like a map revealing properties that are at risk of being flooded.

Trying to make a map to satisfy street-lighting campaigners, election candidates and insurance companies seems a daunting task and will probably end in failure when it becomes too messy and confusing. Emphasising some aspects of the real world at the expense of others is therefore an essential part of the cartographer's craft.

Like all great works of art, the joy of maps comes as much from the stories of their creation and creators as it does their visual appeal. But maps have the added dimension in that they can be produced with one very specific purpose in mind and then put to something entirely different – perhaps in opposition – to what was originally intended. For example, a map of poverty is of as much use to a loan shark targeting their activities as it is to a charity seeking to open a food bank, while a map of virgin rainforest can aid illegal logging as much as support conservation. So, the life a map leads is also entirely dependent on whose hands it finds itself being held by.

There's a sense of history that comes with use that makes a map all the more fascinating. It can transform it from an unexceptional sheet of paper torn from the pages of a book or magazine into something much more profound. And much of this can be missed as soon as a map is pushed through a scanner and stored digitally, rather than in a physical space like the Map Library.

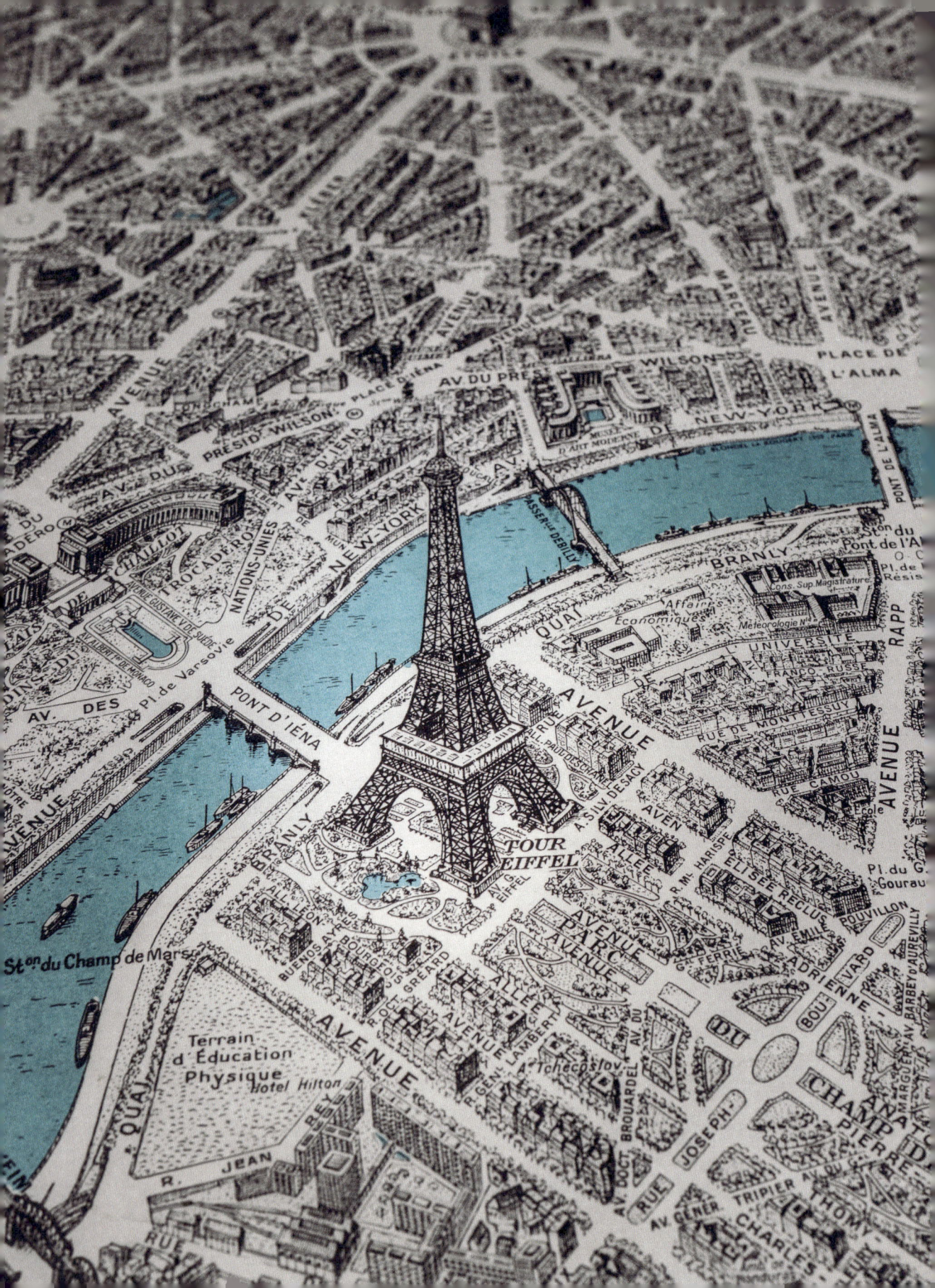

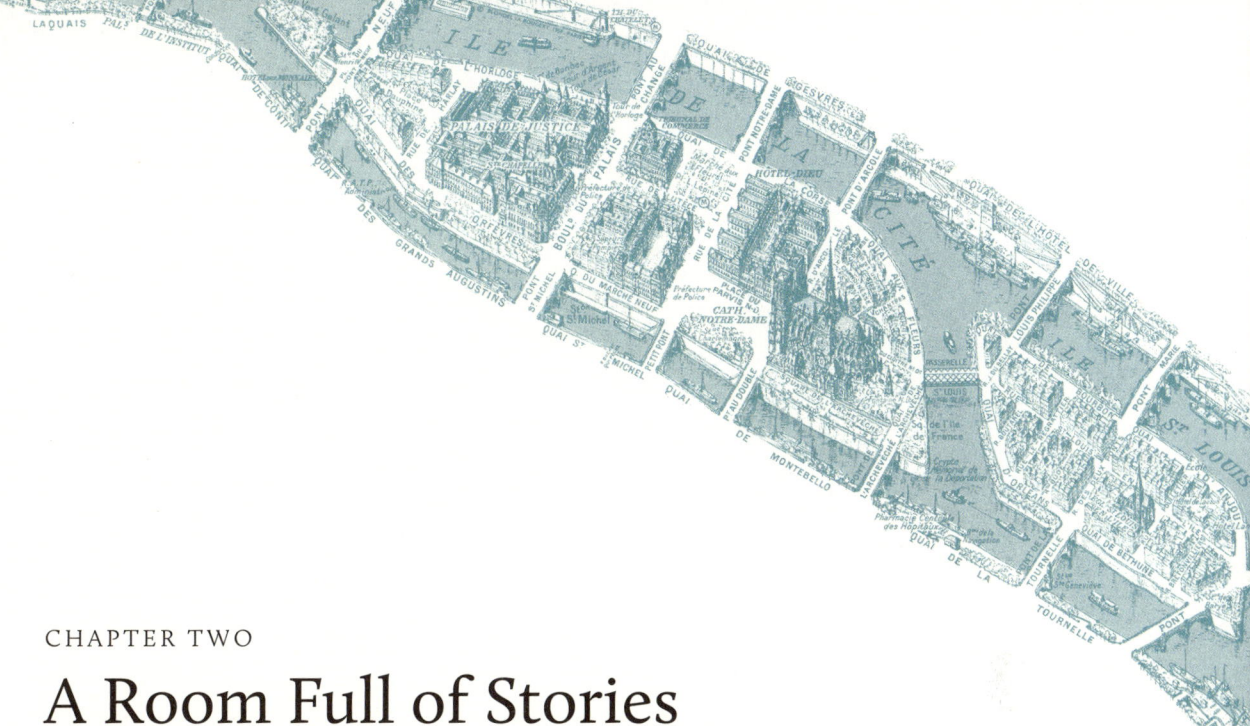

CHAPTER TWO

A Room Full of Stories

'Maps are funny things because they appear to be the reality, and yet they give you a tremendous opportunity to dream.'
Peter Barber[1]

Not long into my searches of the Map Library I grew impatient to make discoveries and weary of clambering up and down the tall yellow stepladder I needed to reach the drawers that towered above my head. To speed things up, I developed the technique of pulling out wodges of maps and then jumping from the ladder, holding the weight of maps as best I could. A few would flutter out of my hands as they caught the breeze of the freefall, while the rest landed on an adjacent table just as my feet hit the ground. I would breathe a sigh of relief at another successful descent without giving myself an ankle injury and then set about collating the maps that had floated off about the place.

Those printed on the thinnest paper stock generally glided the furthest from the table I'd placed in the landing zone. This

Map 15 [Extract] A glorious tourist map of Paris with its iconic boulevards and buildings intricately illustrated by Georges Peltier, who spent 30,000 hours over 20 years drawing it. There have been many editions of the map published, and today it is even celebrated by the luxury French fashion house Dior in its 'Plan de Paris' range.

Plan du Centre de Paris à Vol d'Oiseau : Blondel La Rogery (1959)
95.5 × 65.5cm

THE LIBRARY OF LOST MAPS

became something of a sorting system: the oldest maps turned out to be heaviest as they were often mounted on dense linen so, if they were to fall, they had the straightest descent, landing with a dull thud and triggering a pang of guilt that in my haste I might have damaged something precious. The longest flights were achieved by maps printed at the later stages of the Second World War, when the mass printing of maps was well underway, and paper shortages led to some paper maps being printed on little more than tissue paper.

After one particularly chaotic extrication of maps from high up in a drawer marked 'Portugal/Spain', I was intrigued to see that a pre-war tourist map of Madrid had landed further from the stepladder than most of the others. Tourist maps would usually be on slightly heavier paper stock and often folded to fit inside sightseers' pockets; consequently they tended to land at my feet if I dropped any.

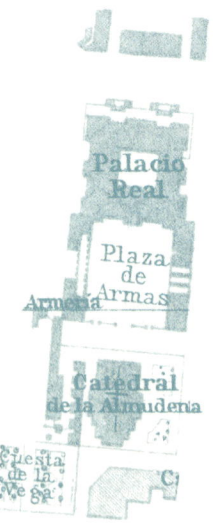

On closer inspection, I saw it had at one stage been folded to an eighth of its full size, and years in the drawer under the weight of other maps had flattened out the creases, as well as yellowed its colour, adding to the grubbiness it must have acquired through use. When I picked it off the floor the backside had been facing up, and I could see that, while it was a map of the Spanish capital, the map's title was in German.

Turning it over, I could see a more detailed view of Madrid at 1:10,000 scale with further annotations and titles in German around the edge (it told me the map on the back was at 1:25,000 scale) and a credit to the famous guidebook publisher Baedeker. The date 1929 was present, but there was another date printed on the bottom right: November 1940.

My gaze was so fixed on the streets of Madrid that it was a while before, out of the corner of my eye, I spotted the stamp to the top left of the sheet. The black silhouette of a swastika was unmistakable. It took a further moment to compute what I was looking at and then to get over the shock. The map I was holding was a witness to one of the darkest periods of history. The dates, the German, the thin paper all made sense and all thoughts of tourists enjoying a visit to Spain left me.

Around the map were key places in the city and the grid references to find them. So, this in fact was a military map, and the faint stamp at the very bottom confirmed it had been captured from the defeated German army, arriving in the map room of the 'Geographical Section General Staff' on 4 August 1945. In this case, the Nazis had obtained a guidebook of Madrid that was published in 1929, and then copied it for use in the war (Spain's neutrality did not make it immune from the possibility of Nazi invasion), a change of purpose that gives the map a whole new meaning.

After tracking down a couple of long-retired military cartographers I learned that tourist maps were often collected to copy for operational charts. Such maps were ideal for this purpose as they were easily accessible, often detailed and regularly updated. I also discovered that the Nazi association with the Baedeker guides, whence the map of Madrid had been copied, had grim consequences for those in Britain who suffered the 'Baedeker Blitz' in the spring of 1942. According to the Imperial War Museum,[2] Nazi propagandist Baron Gustav Braun von Stumm said: 'We shall go out and bomb every building in Britain marked with three stars in the Baedeker Guide.' His motivation for this was revenge for the bombing of Germany's historic towns by the RAF. The subsequent attacks on the cities of Bath, Canterbury, Exeter, Norwich and York became known as the 'Baedeker raids'.

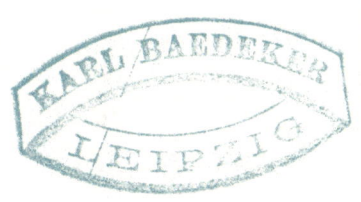

The more I looked at the Madrid map, the more history it revealed. I spotted evidence of another stamp that must have rubbed off from an adjacent map in the same batch. I pondered if this had been a map printed for spies,[3] with its careful folding aiding

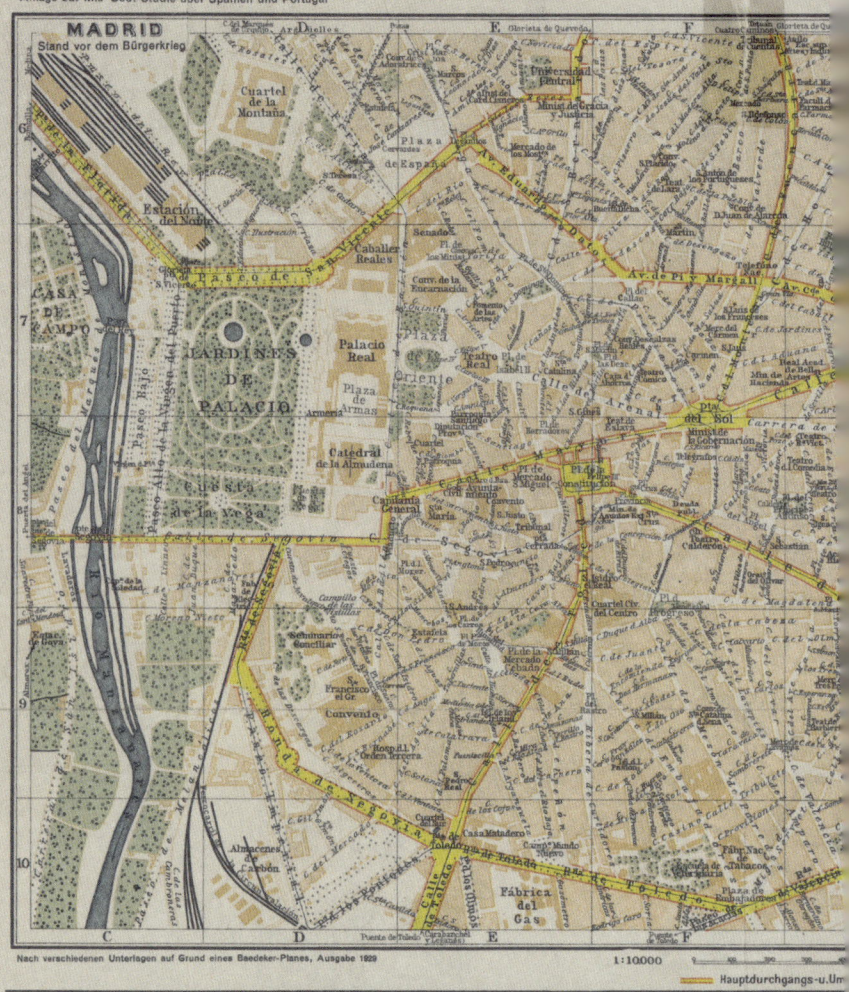

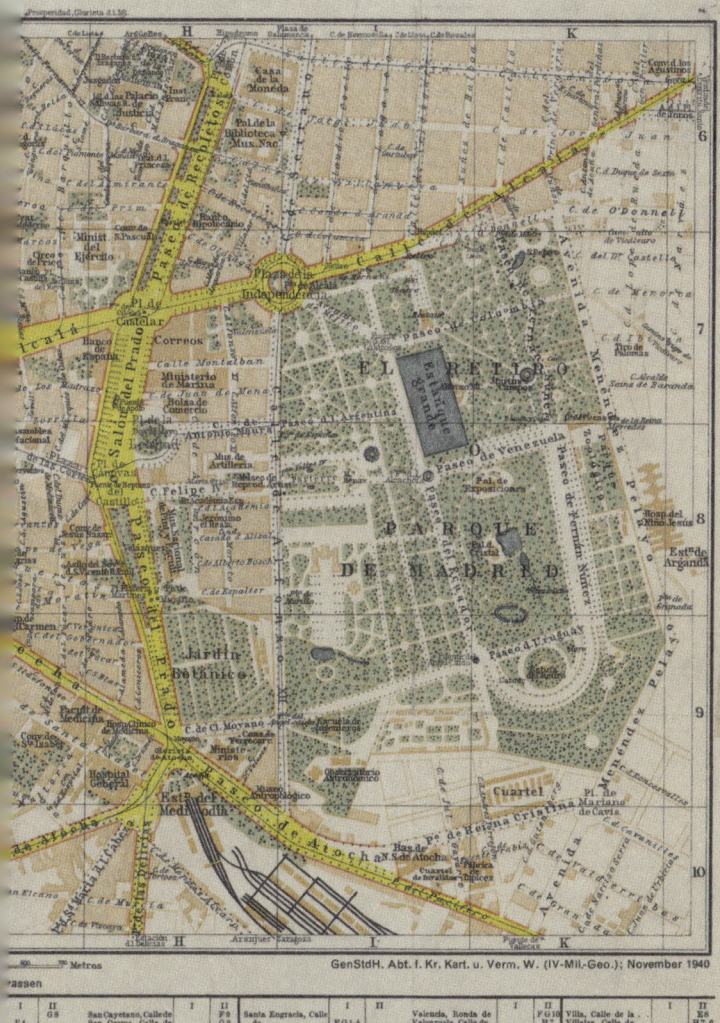

Map 16 The 1929 Baedeker map of Madrid, repurposed by Nazi Germany for use during the Second World War. An index has been included around the edge and in the top right it says 'For official use only'. A 'Directorate of Military Survey' stamp along the bottom states an acquisition date of 4 August 1945, when the map fell into the hands of the Allies.

Zu den Plänen von Madrid: Abteilung für Kriegskarten- und Vermessungswesen (1940)
42.5 × 25cm

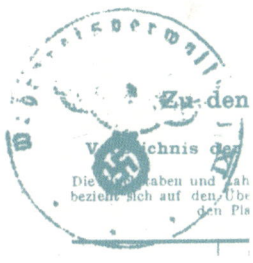

THE LIBRARY OF LOST MAPS

in its concealment as it moved across war-torn borders of Europe.

Looking even closer I saw the faint line of a fold in the top left corner which appears to push the stamp back on itself so as to no longer face outwards. Quite understandably, a later map reader could not stand the sight of it as they were viewing the map. More than once in the Map Library I came across marks and stamps that had been redacted in this way through folds, cut out or coloured over in black ink.

So, a relatively mundane tourist map – first published in 1929 – went on to lead an extraordinary life that we can look back on thanks to the marks and stamps it acquired on each stage of its journey: an instrument of fascism in the early 1940s, a trophy of war from 1945 and then, presumably, an educational resource for the remainder of the twentieth century. After this discovery it took me several months of research to piece together how it might have arrived in the collection, but we'll come to that soon.

We often value more highly a pristine map, but pristine maps

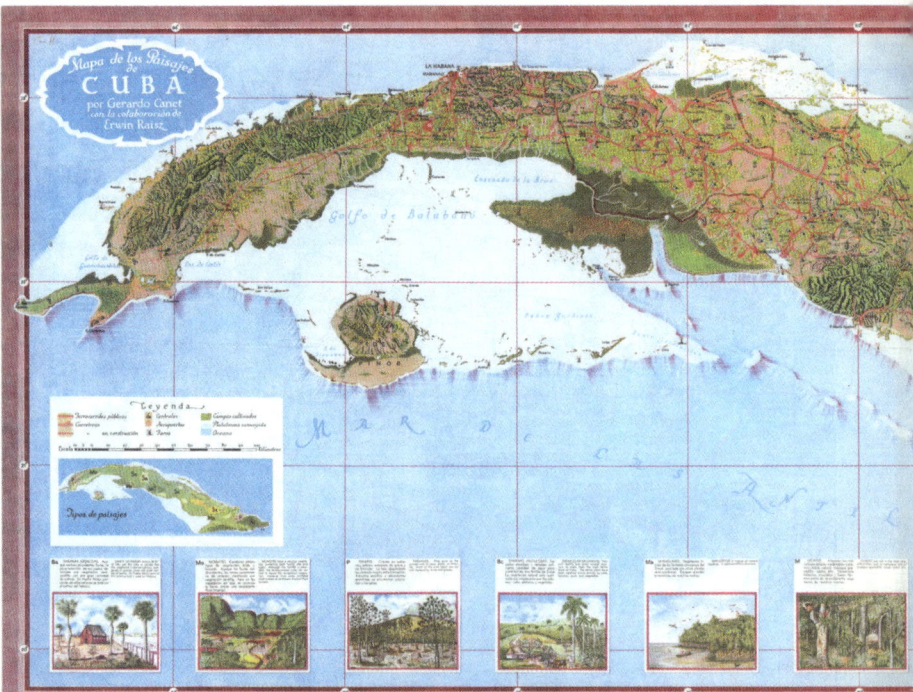

A ROOM FULL OF STORIES

are unused maps and so, while they might be nicer to look at, they are sure to have had much less intriguing histories. Who held them? What plans were made, or courses charted?

As I discovered almost by accident on the Nazi map of Madrid, the answers to those questions can often be found on the back of the map, the scribbles in the margins or a fading stamp. With that in mind, I took to paying close attention to these aspects and encountered some lovely surprises.

My favourite involved a sizeable map of Cuba, created by a Hungarian-born American cartographer Erwin Raisz in collaboration with a Cuban geographer and cartographer named Gerardo Canet. Raisz was at the top of his game at the end of the 1940s and Canet was keen to showcase some of the latest statistics and mapping for his country, so together they created Cuba's first national atlas.[4] As part of that project, in 1945 they completed a large map of the island (1.5 m across and 0.5 m tall) entitled *Mapa de los Paisajes de Cuba* (*Map of the Landscapes of Cuba*) and I was

Map 17

Mapa de los Paisajes de Cuba: Harvard University Press (1945)
146 × 53cm

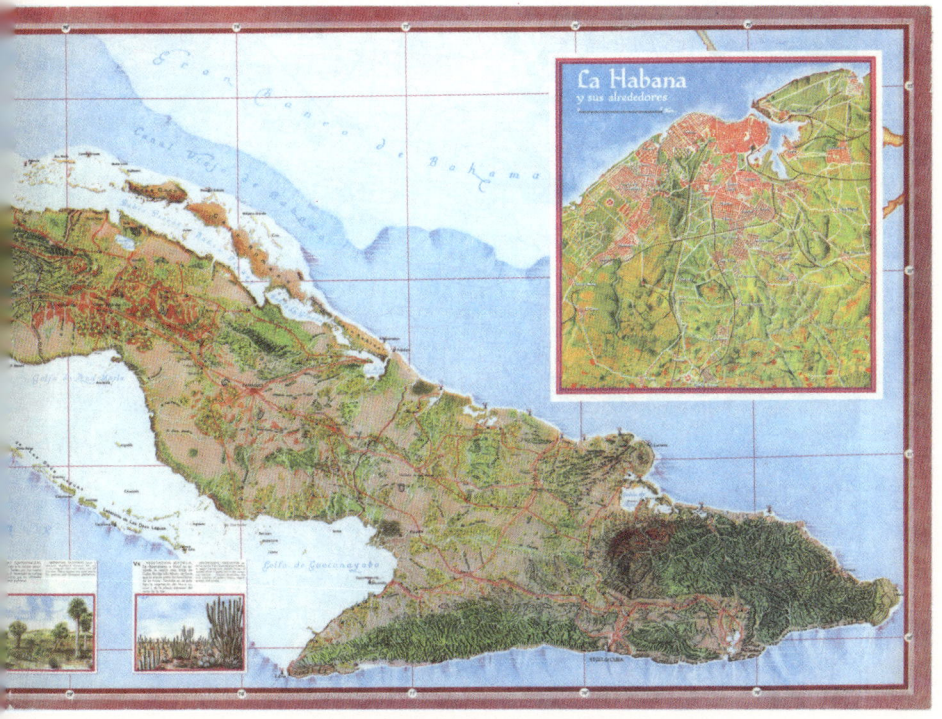

45

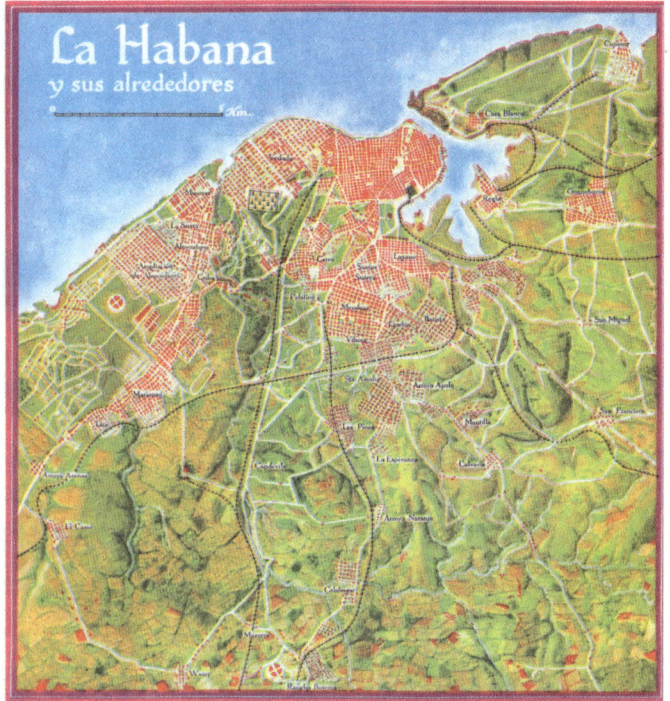

thrilled to discover the Map Library had several copies of it.[5]

Raisz and Canet spread the island across the full width of the sheet and adorned it with a superb inset of the capital Havana. Along the bottom is a series of intricate drawings showing all the different landscape types of the island such as sandy savannahs, pinelands and mangroves. They took inspiration for the colours and symbols from numerous photographs taken on a series of flights over the country. It's gloriously vibrant and unlike any other map published at the time.

The original map was printed on very thin paper and so a couple of the copies I found were in poor condition. I turned my attention to the one that had emerged from its half-century crammed in a drawer in better shape than the others, because it had been mounted on some kind of linen. As I turned it over and held it up to the light, I thought I was hallucinating when I caught sight of what looked like the black edging of a border to another map.

There also seemed to be faint lines casting a shadow against the linen that weren't on the Cuban map on the other side. Putting it face up on the table and looking more carefully in the light areas of the sea, I was in no doubt that the criss-cross of lines was in fact a road network lurking in the shallows. I reached for my phone and switched its torch on. Sliding it between the map and the table it immediately illuminated the word 'Berlin'.

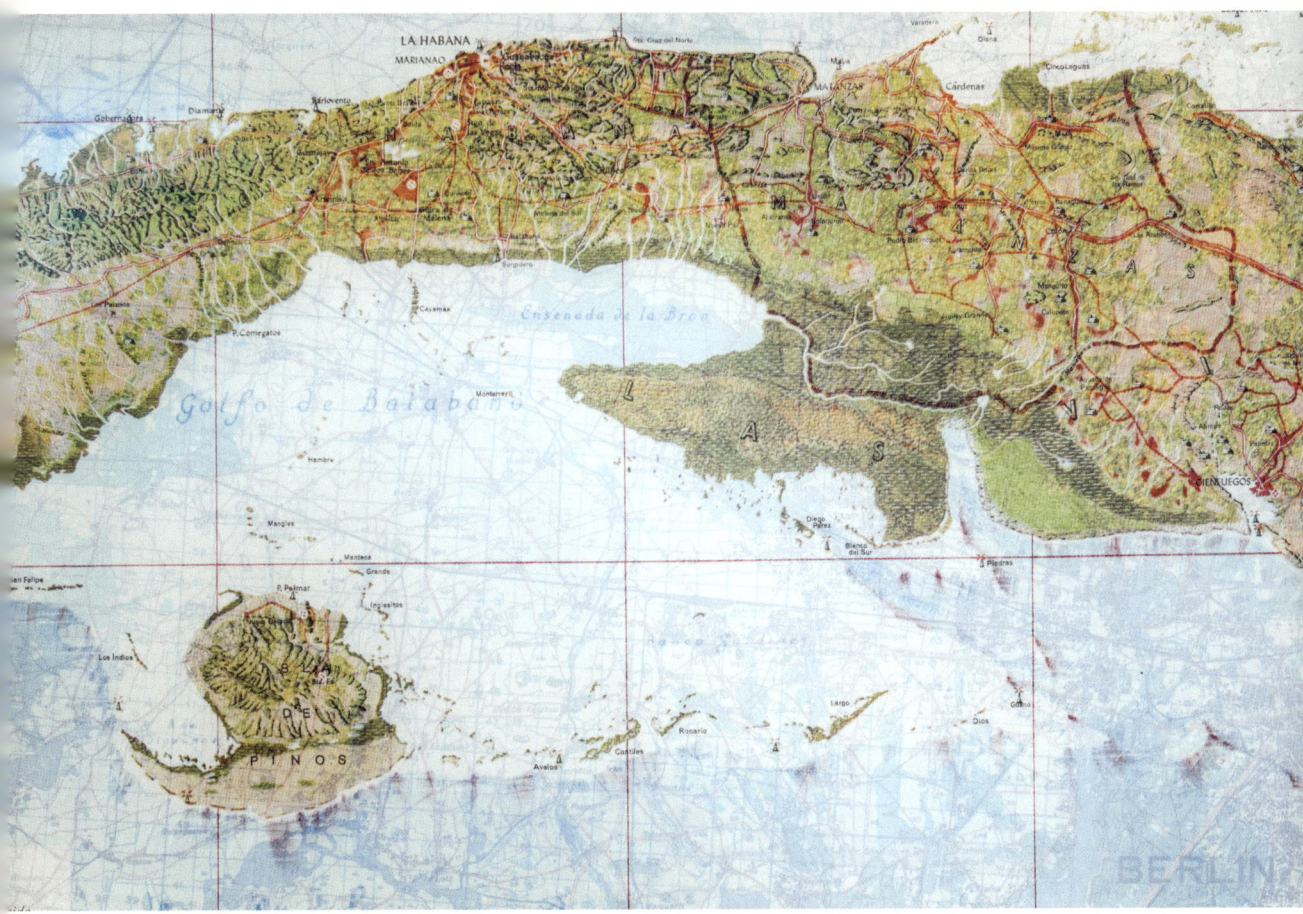

Cuba had been neatly mounted on a Second World War map of Berlin!

Given that it was completed at a time when there were many surplus maps available and a shortage of good paper stock, a map librarian must have decided to strengthen the fragile map of Cuba, simultaneously pasting over the horrors of the war with the warmth and light of the largest island in the Caribbean. When I invite students into the Map Library, this is one of the maps I like to show them. I ask them to see if they can spot what's unusual with the map and, when they can't, I shine my light through with the demeanour of a magician pulling a rabbit from a hat.

Map 17 [Extract] It's hard to do justice to the full span of Raisz and Canet's map of Cuba, but its size isn't the only reason it's special. Look carefully in the shallow water and you might just see the roads of war-torn Berlin shining through.

Once I knew what texture of linen to look out for, I discovered flashes of war maps on the insides of handmade folders that were assembled to hold loose sheet maps on completely unrelated topics. Only in a map library can time and space collide in such fascinating ways that I would have missed had I been viewing digital copies.

Every time I found a gem like this, I had a feeling in the pit of my stomach that I might be the one to dispose of it, should the worst happen to the collection. I cursed the fact that so many other maps with interesting stories to tell have been thrown away in the haste of disposing of map libraries so that they could be repurposed into classrooms or offices in Britain's crowded universities. As one retired map librarian said to me:

> Without the knowledge, people are throwing away maps they have no idea about, and on the one hand, one condemns it as sort of cultural vandalism, but on the other hand, one has a sympathy with the pressures on space and the need to get rid of stuff ... I'm sure the wheel will turn full circle, and, in the future, people will realise what they've lost, but dealing with the present is a huge problem.

Part of the problem with the present is that our attitudes change: what we value now may not be what we value in the future, and the dull and ordinary of today may become highly valued curiosities tomorrow. For example, the contribution that a cartographer named Marie Tharp – I will tell you about her in Chapter 10 – made to the mapping of the ocean floor was overlooked for decades. In recent years she has been given the credit she deserves, and we can understand her much more thanks to the archive she built of her life that now resides in the Library of Congress in Washington DC. If this archive had not been in her house until the final years of

A ROOM FULL OF STORIES

her life, but in a collection that needed to downsize a few decades ago, her maps would have been the first to go to save space for the higher profile – undoubtedly male – mapmakers who will now appear obscure by comparison.

It can also take time to research and appreciate what a map was used for, something I discovered too late when I went to pick over the remains of a map library that was closing for good.

A few of the many different stamps from other collections that I spotted on maps in the library.

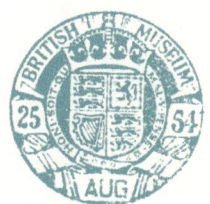

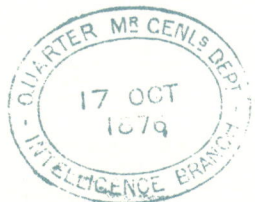

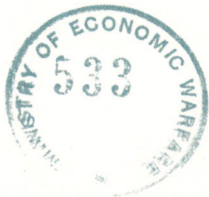

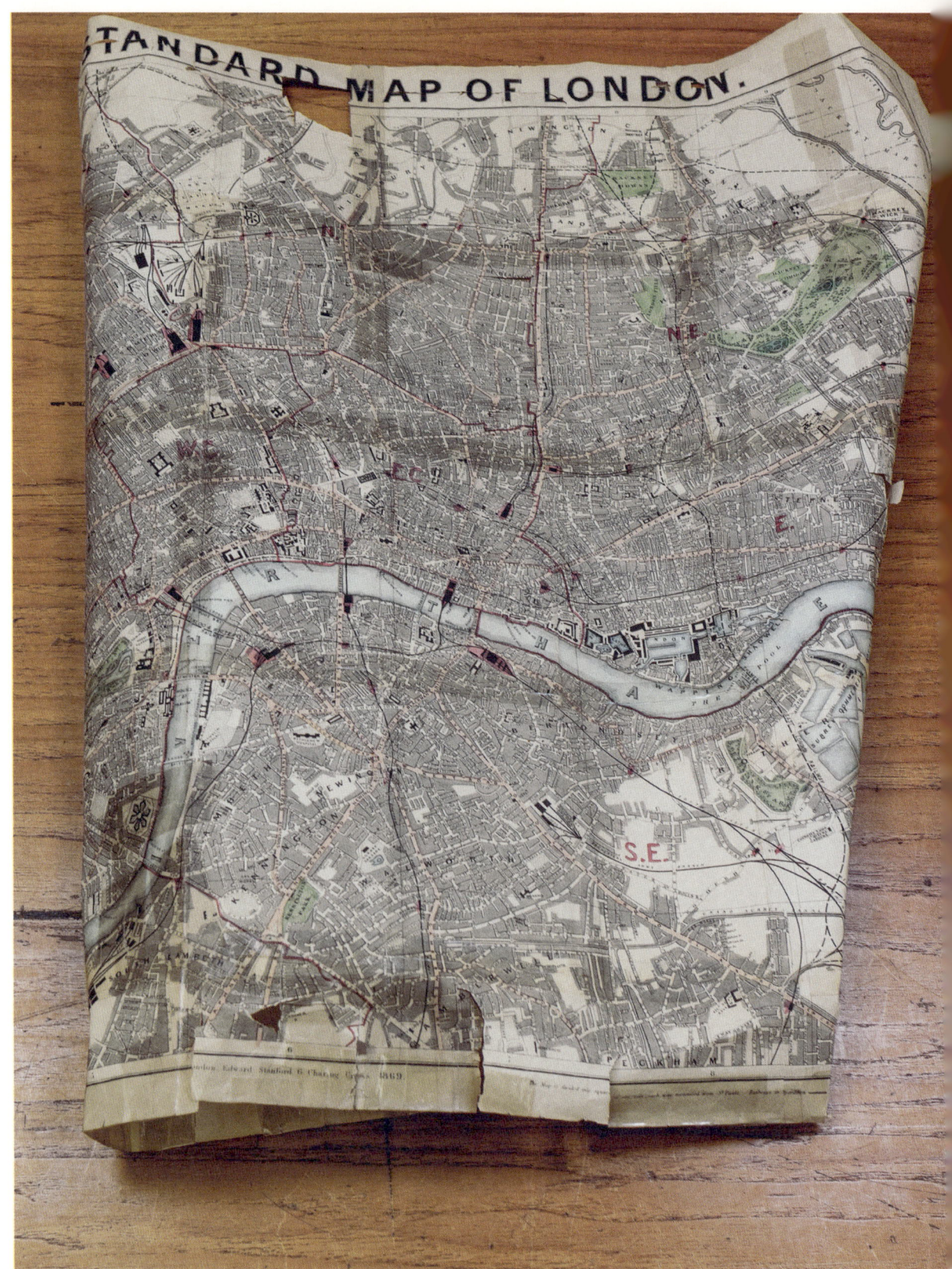

How Map Libraries End

I have heard stories of maps rescued from skips round the backs of the buildings that once held map libraries, from those who have shared the regret they felt at not being able to save more. I have likewise lost count of the stamps of long-gone map libraries I've spotted at map fairs, on eBay or in the few collections that remain. With such a vast quantity of material, I know it is impossible to save it all, but it doesn't prevent the sorrow of the disposals when they happen.

I vividly remember the day, more than a decade ago, that I found out that the London School of Economics (LSE) was disposing of maps from its geography department's map library. I dropped everything and rushed over to see what remained. There were maps coalescing into messy piles, spilling out of half-opened drawers and in the hands of a couple of fellow map enthusiasts who had heard the news. A map dealer was making an offer for a tower of map chests and their contents – a series of beautiful geological maps.

Back then, I had no idea what I was looking for – I was simply on the hunt for maps I could pin to the walls of my flat. So, I just opted to take what I could of places I'd visited before, would like to visit, or maps that particularly caught my eye because I liked the look of them. About an hour after I arrived, I spotted an intriguing pile of linen-backed maps of London that showed the city in extraordinary detail, and appeared to have been coloured in by hand.

Map 18 This *Collins' Standard Map of London* from 1869 is now more sticky tape than it is paper, but that's why I like it. Battle-scarred from 150 years of use, its condition makes it exactly the kind of map that would be thrown in the wastepaper pile during a map library closure. But if given the chance, it can still evoke the streets of Victorian London and, unlike a pristine copy, is steeped in the histories of the countless hands that held it.

Collins' Standard Map of London: Edward Stanford Ltd (1869)
82 × 61cm

THE LIBRARY OF LOST MAPS

I wasn't sure what the colours meant, but they gave the maps the appearance of having been painted by someone using pink, purple and blue highlighter pens. In a few places I noticed mistakes that had been brushed over with a thin layer of whitewash. There was a legend to explain the symbols used but not the colours, and the accompanying text at the bottom read: 'Heliozincographed[6] and Published by the Director General at the Ordnance Survey Office, Southampton, 1916' and 'Scale 1/2500 being 25.344 Inches to a Statute Mile or 208.33 Feet to One Inch'.

I could be sure then that the maps had been printed in 1916 but there were handwritten annotations and extra buildings drawn in, so I reasoned the hand colouring may have happened later, but it was impossible to say. There was also a map seller stamp indicating the price in shillings had increased from '5/- net' to '6/8 net', quite a markup for a map that must have appeared increasingly out of date.

Many were filthy and bore the scars of the pins that had been pressed through their corners either to fix them to a drafting table to be worked on or for display on a wall. They all had holes punched down their left side, as if to be stored in a giant ring binder, and some had frayed edges with bits cut out. Despite their dishevelled appearance, I was drawn to them, not least because I wanted to know who had taken the time to hand-colour every street in London in such detail and for what purpose.

Given their condition, I reasoned they wouldn't have been of any appeal to map dealers or collectors seeking pristine sheets for decorative purposes, so I asked the librarian overseeing the disposal if I could take them. She said I couldn't as the Geography Department wished to keep them. She gave no further explanation so, slightly disappointed, I moved back to admiring a pile of historic maps of London I was working my way through.

As the day wore on, it was just me and a young boy left rifling through the map drawers. Unlike me, he knew exactly what he was looking for and had come with a clear mission to collect every map sheet he could that had been published by Ordnance Survey (Britain's official mapping agency) to complete his map of Great Britain. Purchasing the up-to-date series would be hugely expensive (and no fun), so enthusiasts scour second-hand bookshops and online auctions to pick up what they need for a few pounds a sheet, with those containing errors or rare features being more desirable. To get a complete set at the 1:25,000 scale for a single year he'd have needed to collect 403 individual sheets in all, plus there would have been some updates and duplicates. His mother – who was hovering behind him – was being as supportive as she could while raising concern about how they would get the maps home and where they could be stored. Unperturbed, the lad had assembled a pile of maps that almost reached his eye level from the desk he had placed it on.

This was a once-in-a-lifetime opportunity for a budding cartophile and I was secretly egging him on, not least because I, too, had overdone it and selected more maps than I could physically take back to my flat in a single trip. I carefully rolled what I could carry and set off home with several kilograms of maps under my arm. I left the remainder in their own pile with my name written on a scrap of brown wrapping paper and hoped they wouldn't be snapped up before my return.

The following afternoon I was relieved to see my pile untouched despite a noticeable diminishing of the maps remaining in the room. The librarian, who was tidying the place as best she could, looked up and asked: 'You were after those old London Ordnance Survey maps, weren't you? I can give you two of the sheets if you'd like them?'

I wished I'd been more persistent in asking the day before,

THE LIBRARY OF LOST MAPS

but, grateful for the offer of two, I said, 'I'd love them, but where did the rest go?'

'Oh, the lad who was here yesterday really wanted them, so I let him have them in the end, but these two fell off the trolley as they were loading them into the lift. Maintenance were able to fish them out of the lift shaft first thing this morning!'

She said the kid and his mother didn't leave contact details so there was no way of reuniting the sheets from the lift shaft with the rest of the map. I was free to take the two that covered Greenwich, which sits to the southern edge of the famous final meander of the River Thames as it leaves the city.

I added them to the pile of spoils from the previous day that I neatly rolled into another large cylinder of maps encased in scraps of brown paper, some bearing the stamps of famous London map

Map 19 Three hand-coloured sheets from the *New Survey of London Life and Labour*. They detail Greenwich and part of the Isle of Dogs and are a tiny element of a vast map that covered all of London.

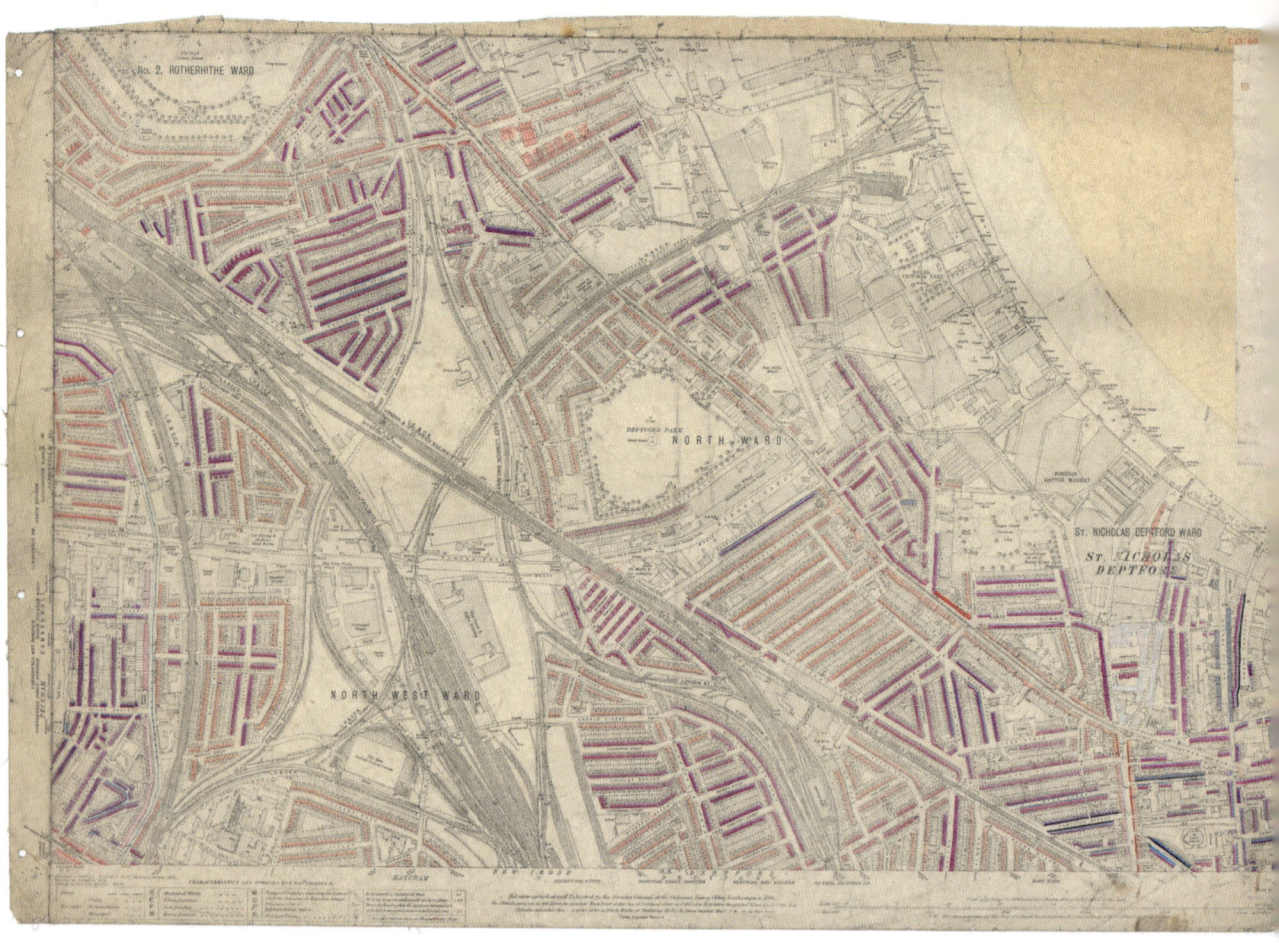

sellers, and wandered back to my office in Bloomsbury.

In the years that followed I didn't give the maps a second thought, but then at the start of 2024 I was scrolling through eBay, on the lookout for interesting maps, and couldn't believe what zipped past on the screen: it was part of the northern bank of the River Thames that was missing from the map of Greenwich rescued from the LSE lift shaft all those years ago. The hand-colouring was identical and even the holes punched on the left side were the same. Better still, the seller had some information about what the maps were showing, and it was clear I was very lucky indeed to have even two from the series as they were an integral part of an important but overlooked study of the social conditions in London in the 1930s.

One of the founders of LSE was Beatrice Webb (née Potter[7]),

On the final printed maps, black denotes 'the lowest class of degraded or semi-criminal population'; blue 'those living below Charles Booth's [Victorian] poverty line'; purple the 'mass of unskilled labourers (and others of similar incomes) who are above the poverty line'; pink is reserved for 'skilled workers and others of similar grades of income'; and red 'the "Middle Class" and the wealthy'. Streets with a mix of different groups were coloured by a line of each colour.

New Survey of London Life and Labour manuscript (1928)
191 × 70.5cm

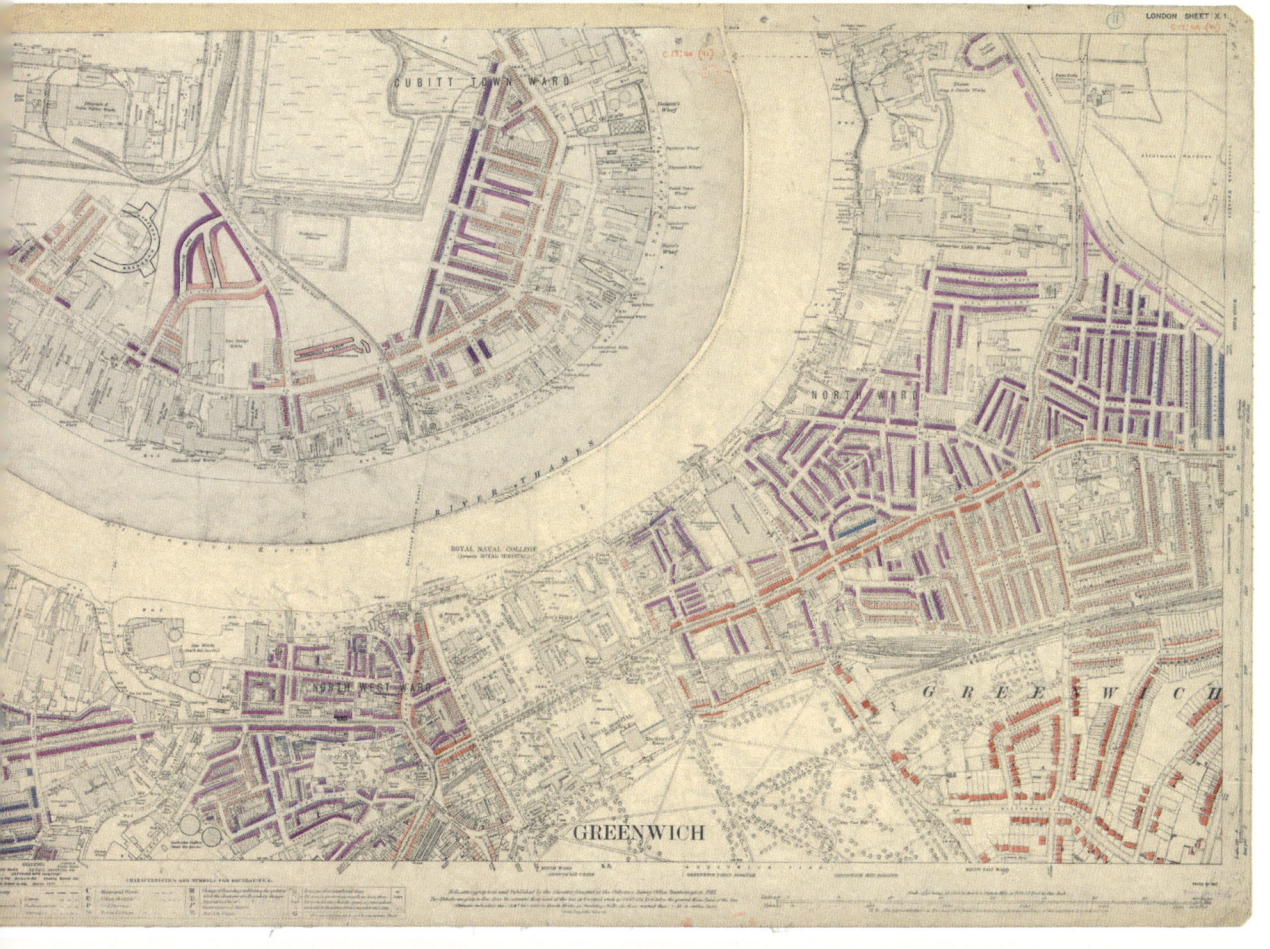

a social reformer of significant renown. Among her formative experiences was the work she did as a researcher for Charles Booth (or 'Charlie' as she referred to him[8]), contributing to his seventeen-volume study *Life and Labour of the People in London*,[9] which was published between 1889 and 1903.

Booth himself did not live in London but had links to the city through his successful business ventures, including creating the Booth Steamship Company that offered some of the first passenger sailings between Europe and Brazil. Despite his wealth he was concerned with the plight of Londoners living in poor conditions and set about doing the comprehensive survey so that it could inform the allocation of funds from the Lord Mayor of London's Relief Fund.

Booth and his study are most famous today for the street-by-street mapping published alongside the statistical and written reports. The maps coloured the streets in London according to a crude social gradient that condemned in black those that were home to the 'Lowest class. Vicious, semi-criminal' and moved through blues ('Very poor, casual. Chronic want'; 'Poor. 18s. to 21s. a week for a moderate family') then reds ('Some comfortable. Others poor'; 'Fairly comfortable. Good ordinary earnings'; 'Middle class. Well-to-do') to the optimistic yellow of 'Upper-middle and Upper classes. Wealthy'. These in large part were judgements made by School Board Visitors,[10] checked and supplemented by reference to other sources, including the police and clergy.[11]

Ensuring these judgements were comparable across the city was tricky so Booth took an approach that would enable him to cover a lot of ground and consult with those who knew their areas best. He employed a team of researchers to gather more detailed testimonies and data, which included Webb, who focused her attentions on the conditions of London's dock workers, those employed in its textile industry and the city's Jewish community.

A ROOM FULL OF STORIES

She provided evocative – although perhaps not terribly impartial – general accounts of the people and places she visited that add colour to the maps:

> This morning I walked along Billingsgate from Fresh Wharf to London Docks. Crowded with loungers – smoking bad tobacco and coarse careless talk, with the clash of a halfpenny on the pavement every now and again. Bestial content or hopeless discontent on their faces – the lowest form of leisure – senseless curiosity of street rows, idle gazing at the street seller low joke – and this is the chance the Dock offers.[12]

Map 20 A section from a slightly later publication in Charles Booth's survey, printed 1902, and covering a section of east London. No streets in this area were considered 'Wealthy' by Victorian standards, and many were 'Very Poor'.

Life and Labour of the People in London: Macmillan and Co. (1902)
44 × 31cm

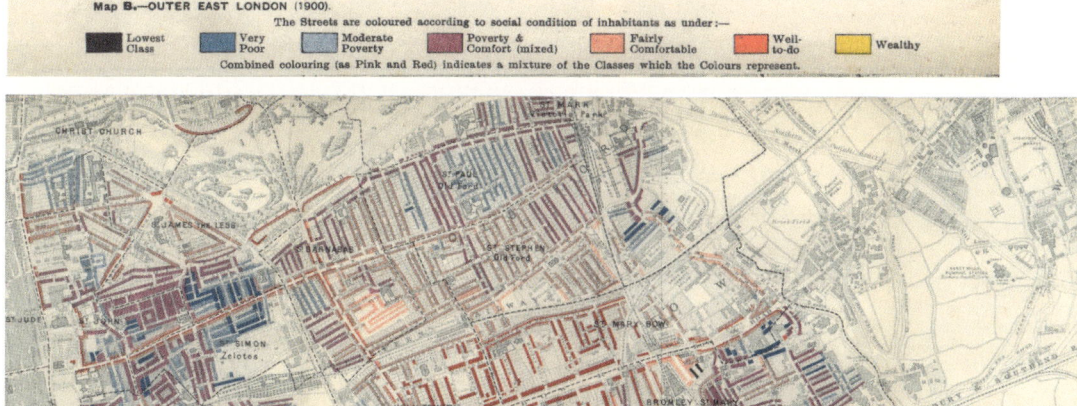

In 1888 Webb reflected on her work with a young researcher named Hubert Llewellyn Smith, jotting in her diary that 'he is formal minded – but has ability – and is generous in his helpfulness to others working on the same line'.[13]

Forty years later, following a successful career in the civil service and as General Secretary of the Ministry of Munitions during the First World War, it was Llewellyn Smith who continued Booth's legacy by leading his own survey to see how the city had changed since he pounded the streets of Victorian London with Webb. Llewellyn Smith was wedded to the idea that his data collection should be directly comparable to Booth's and set out the similarities and differences in his approach before an audience at the Royal Statistical Society in 1929.[14]

The most debated aspect of the new study was the thorny issue of deciding on a definition for 'poverty' and whether it should be relative – that is looking at the range of wealth in 1929 and placing the poorest at one end and the richest at the other – or absolute, offering a minimum living standard applicable to the 1920s, as Booth had done in his original work for Victorian London. Llewellyn Smith adopted the latter, which drew some concern from the esteemed audience. One professor made the important point that 'if the investigation shows that there is an amount of poverty which we may represent by 10 percent, and that there was 10 percent of poverty when Mr. Charles Booth conducted his investigation, then it seems to me that 10 percent in 1929 is a much more serious thing than 10 percent forty years ago'.[15]

The method for creating the maps, which covered some of the suburbs of London omitted by Booth, was almost identical to the original survey. There were 26,000 streets that needed to be coloured[16] and each had their own index card to describe the social conditions of their occupants. These were sent out for validation and checking by those with similar professions to the ones Booth turned to, notably School Attendance Officers (the successors to the School Board Visitors) with additions from Employment Exchanges, Relieving Officers, Police, Probation Officers and others.[17] Once everyone was satisfied, the index card was then used to inform the colours that ultimately made their way on the map.

The most pressing cartographic challenge that exercised Llewellyn Smith was the realisation that the streets themselves might have a blend of different social groups along them; he called these 'composite' streets and developed a two-colour system to represent them. Those few streets that he deemed 'obstinately mixed', where more than two colours might be needed, were depicted with alternating stripes.

Llewellyn Smith used the most detailed maps he could buy at the 1:2,500 (25.3 inches to the mile) scale for the colouring and then transferred these findings in the printing to a 1:10,560 (six inches to a mile). These more detailed maps are the ones I saw in the LSE map library. The colours that looked like they had been created by highlighter pen were in fact those corresponding to the different categories of poverty that Llewellyn Smith had assigned them.

On concluding his work, and because he had used Booth's absolute definition of poverty, he was able to say:

> It is satisfactory to find that the level of poverty in East London is now only about one-third as high as in Charles Booth's time. It is much less satisfactory to learn that ... there are still more than a quarter of a million persons below the poverty line. When we consider how low and bare is the minimum of subsistence of the Booth poverty line, it is impossible to rest content ...[18]

I had been drawn to these scruffy but beautifully detailed maps, hoping they were rough diamonds among the clutter of a map

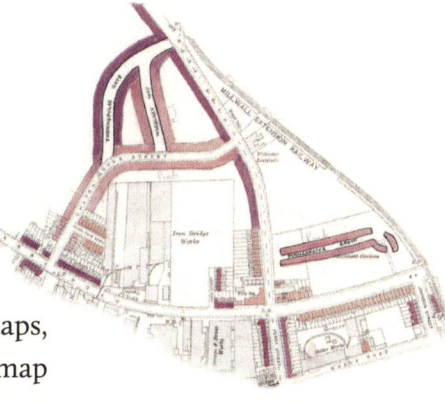

Map 19 [Extract] A section of the Llewellyn Smith map where the surveyor has drawn streets that were built after the base map had been printed by Ordnance Survey.

New Survey of London Life and Labour manuscript (1928) 191 × 70.5cm

library closure. My instincts were right, as we now know these historic maps played an important and beneficial role in helping change the way we care for the poorest in society, by contributing to the case for a more generous welfare state. During his time as a politician, Llewellyn Smith had made substantial changes, such as the introduction of unemployment insurance (a core part of the National Insurance Act in 1911), which were built upon by William H. Beveridge, who is seen as a figure central to the foundation of the modern welfare state in Britain. In his obituary for Llewellyn Smith, Beveridge wrote of the 'immense task' that the updated survey became and remembers him as someone who 'in making new ideas in public administration viable ... can never have been surpassed, and can have had few equals'.[19]

It's also hard to understate the importance of this data for social researchers and historians[20] today because the 1931 census, which would have given detailed societal metrics around the same time as Llewellyn Smith's work, was lost during a fire in 1942, and the 1941 census was not taken due to the Second World War.[21] So, apart from these maps, there is relatively little data available about the social conditions of London between 1921 and the post-war census in 1951.

I contacted the map seller[22] who had the hand-coloured map sheets for sale to ask if he had been the kid that day in the LSE map library (I reasoned he could now be into his twenties) who had fatefully dropped the maps off the trolley. He kindly replied to say the maps he was selling were a much more recent acquisition he'd come by thanks to contact from someone at LSE charged with throwing them out as part of the process of moving their Department of Geography to a new building, but realising their potential value sent them his way instead. I couldn't resist purchasing the piece that so neatly fitted into the two I already had, but, alas, the remainder fell outside my budget.

A ROOM FULL OF STORIES

The original index cards that specified the colours on the maps were digitised in the 1990s,[23] but the geographer in me wishes we could have gained access to the complete set of map sheets to be scanned and stitched together to share in the same way that LSE has done for Booth's maps.[24]

But this opportunity has now been lost as the map has been scattered to at least three places: those that were taken home atop the pile of Ordnance Survey sheets accumulated by the map-loving child; the two copies that were fished out of the lift shaft and I'm relieved I didn't throw away; and the remainder on sale on eBay. Anyone who is lucky enough to find the missing maps will be able to touch the pioneering work of some of those responsible for transforming British society for the better.

To be fair to the LSE, it has done a fabulous job of sharing the original Booth maps and wider archive and saving many other map sheets, like those of L. Dudley Stamp's Land Use Survey, which was based there.[25] But in my explorations of the UCL Map Library, I had the demise of the LSE collection at the back of my mind, not least because so much of what is saved depends on the people charged with saving it knowing what they are looking at. I therefore vowed never to take a map for granted and in so doing uncovered some remarkable stories of not just the maps themselves but how they came to be in the Map Library in the first place. Thinking this way, I soon realised that a map's journey into the archive can be just as eventful and reflective of history as the intended use the map was put to, if not more so. It's an important part of the story, not least because it reveals how the Map Library came into existence.

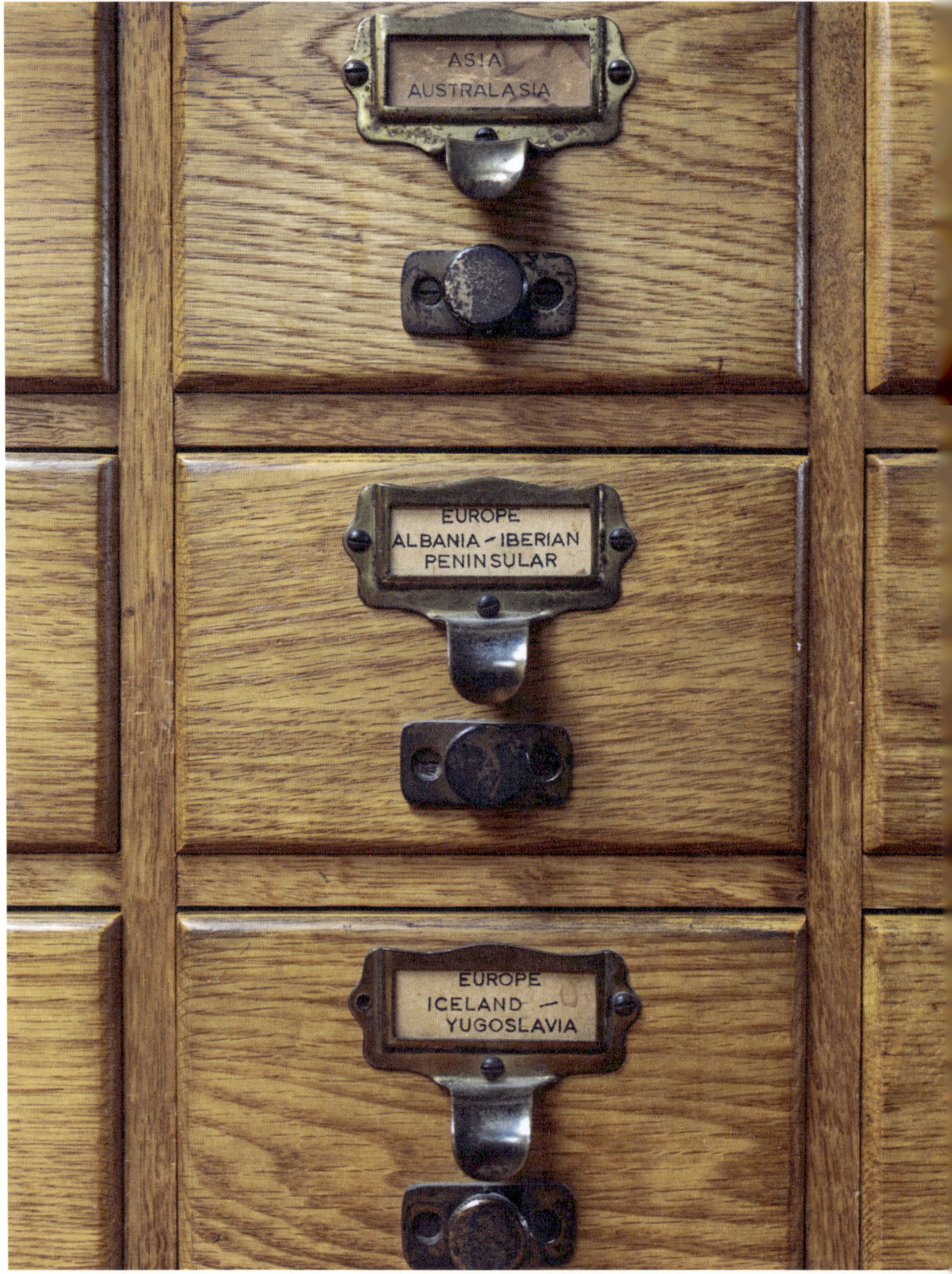

A ROOM FULL OF STORIES

The Birth of the Map Library

With its tens of thousands of maps and hundreds of atlases, the UCL Map Library grew to be one of the largest university collections in the UK. After the Second World War there was a rapid growth in the size of the Department of Geography that housed it, and this coincided with the acquisition of many of the most fascinating maps I found in my explorations. Located in the heart of London and well connected to other universities (not just in the UK, but internationally) as well as major libraries and collections, it was able to thrive, becoming one of the most respected teaching collections in the country.

I soon learned that there is no standard list of maps and atlases that a map library might contain. This is because each is tailored to the intellectual interests of the department it serves, which often had a clear geographical region of focus. Academics at UCL had expertise in Scandinavia, the Balkans and (what was) the Soviet Union. It therefore follows that I encountered relatively few maps of Africa and Asia, for example, because they would doubtless have been covered by the maps that were held by the likes of the School of Oriental and African Studies just a two-minute walk away.

UCL appointed its first Professor of Geography, Alexander Maconochie, in 1833 but his tenure was short-lived and he left in 1836.[26] So it was not until 1903, when Lionel Lyde was hired as Professor of Economic Geography, that the department proper is considered to have been founded. Details of the map holdings prior to the Second World War are fuzzy but we do know that in

Three of the drawers that housed Anne's cataloguing system. Each map, map series or atlas would be detailed on an index card and placed in the drawer that covered the appropriate part of the world. When requests came in, Anne would first consult these cards to direct her to the drawer that housed the map.

1939, under the threat of German bombing, the UCL Department of Geography decamped to the relative safety of the west coast of Wales and Aberystwyth University. The staff took with them the department's maps, atlases and wall charts. As the threat of air raids began to lift, the Head of Department, Charles Bungay Fawcett, returned to London to discover that some of the buildings of the university had been reduced to rubble in the Blitz. He needed to revive teaching, and a note[27] written around 1944–5 reveals his priorities for moving the department back to Bloomsbury.

Alongside a request for the ceiling of a still-standing lecture theatre to be raised by six feet, next to which someone had scrawled 'how?', there is reference to the need for a 'map store and work-room (for a Technical Assistant)' as well as a room in which 'maps and books could be best used together – which is essential for students of geography'. He goes on: 'It has not been possible to keep the collection up to date during the war. And even now it is possible to obtain British maps only by special permit and in very small quantities; while it is practically impossible to get any maps of other countries.'

Fawcett called for 'an almost complete replacement of the maps' and requested £2,000 in funds (equivalent to approximately £73,000 in 2025[28]) to be spent over five years (on the basis that there were shortages of complete map series so time was needed to build them up) to help replace the 3,000 sheet maps and 200 wall maps that were in the department prior to the outbreak of war.

It's not clear whether he got his request for a raised ceiling, but he did get a budget for maps: I discovered the ledger used to purchase them, with multiple requests for maps therein. Most of the purchases listed were for routine maps from the Ordnance Survey as well as copies of the Land Utilisation Survey (shown in Chapter 1), although the sums spent are nothing like the £2,000 requested, so I doubt he got the funds he wanted.

But, as it turned out, this didn't matter, because around the time that Fawcett was asking for cash there were developments in mainland Europe that would mean the department was going to get more maps than he could ever have imagined.

As the Allies swept across the continent, they were raiding the map archives and interviewing the captured cartographers and geographers of the retreating Axis powers. This was essential work to determine if there was any 'live' military intelligence that could be parsed and then communicated to the field of battle. The rest of the contents were then checked and gathered up if they were deemed of potential use in the longer term. This might have included high-quality maps to be copied but also technology that the enemy had developed in survey and printing techniques. Old maps were seized, too, dating back to the First World War and even before, many of which were never fully documented.[29]

A first-hand account of the work undertaken by the US's 'Office of Strategic Services' (OSS, the precursor to the CIA) was published by geographer Leonard S. Wilson in 1948. In it he sets out the sheer quantity of maps being acquired. For example, when the OSS arrived at the Justus Perthes printing plant – a major German map publisher – they found it was still operational with its staff on the premises. Wilson notes that the 'map stocks were intact. Copies of all publications and maps, to the amount of nine tons, were removed, and interviews were conducted with officials.'[30] On the thin wartime paper weighing only a few grams per sheet, nine tons was a lot of maps!

And this was just from one location. Wilson goes on to say 'the same group made detailed examinations of the geographic holdings of the German Foreign Office in Berlin ... The map collection of the Prussian State Library, discovered in a salt mine, was thoroughly utilized.'

The stamp of Justus Perthes publishers, from a map that possibly came from the OSS raid on the printing plant. The same map has a British military stamp which states the map came into their possession on 1 August 1945.

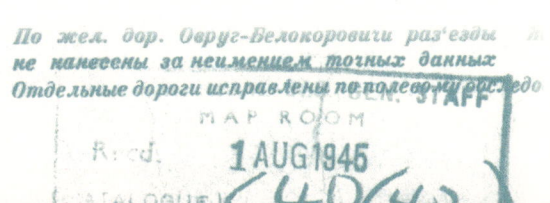

THE LIBRARY OF LOST MAPS

In an article for the *Smithsonian* magazine,[31] journalist Greg Miller researched the extraordinary stories of another group of men, led by a Major Floyd W. Hough, who made some of the most significant acquisitions for the US, and returned with a weight of materials even more startling than Wilson's.

In Bamberg's city hall, Hough established a new headquarters for the team, and commandeered nearly an acre of storage space for sorting the captured material. The team culled this to 90 tons of maps, aerial photographs, high-quality geodetic survey instruments and reams of printed data, which they packed into 1,200 boxes to be shipped to the Army Map Service in Washington.

The haul included complete geodetic coverage of more than a dozen European countries and states, including Russia, and several more in North Africa and the Middle East. Hough later estimated that 95% of this data was new to the U.S. military. It also included approximately 100,000 maps covering all of Europe, Asiatic Russia, parts of North Africa, and scattered coverage of other parts of the world.

Acquisitions were happening across the theatres of war in North Africa and Asia, too. After capture, maps were sent to London before being allocated to the services that might have been interested in them in the UK and US. Simultaneously, the Soviet Union's Red Army was approaching from the east and hoovering up what it found, so there was a concerted effort to get to places ahead of them to avoid the best material falling into their hands. Less is known about the extent of their gains, but we do know the Red Army had its own group of specialists who comprised the 'Trophy Commission' charged with gathering up anything of

CAPTURED MAP FILES

value (including machinery and food) and sending it back to the Soviet Union.

In *The Rape of Europa*, Lyn H. Nichols explains that the removal of the surviving works of art and artefacts began from Berlin's museums five days before the final German surrender. The Red Army started with the museums and galleries which would later be in the sectors of Berlin assigned to the Western Allies, and, for example, the best pictures from the Nationalgalerie were selected to be sent back to the USSR 'and the rest were left to whoever wanted them and soon began to appear in Berlin shops'.[32]

To do their own preservation a group was set up by the Allies: the now famous Monuments Men. The 'Monuments, Fine Arts, and Archives Section Unit' (MFAA), to give it its full title, was tasked with working, often behind enemy lines, to save the cultural treasures looted by the Nazis and under threat of the bombardments of the approaching armies. Among them was Raleigh Ashlin Skelton, who was Assistant Keeper in the Department of Printed Books at the British Museum (which became the British Library in 1973), a few blocks from the Map Library, on the edge of Bloomsbury.

Skelton was an expert on antique maps but spent most of the war with the Royal Artillery in the Middle East and Italy. In early 1945 he was then assigned to the MFAA in Austria. We may never know the extent of his efforts, but we do know that he investigated a monastery that had been used by Hitler as a storage depot for looted treasures destined for his planned Führermuseum in Linz, Austria. It housed eighty items of the Rothschilds' furniture and more than one hundred German paintings and sculptures. The former were handed over to the Austrian government but returned to the Rothschild family only in 1998.[33]

In addition, there was a vast programme of gathering, denazification and – later – restitution of books taking place that

(Overleaf)

Map 21 Here we see a map of 'Greater' Germany's caves and cave regions. If it was printed today we might see it as a leisure map, something that cavers and explorers could be interested in, but this was a map produced by OSS in 1945 for a very different purpose. It was used by the likes of the Monuments Men to target their searches for documents, art or other important artefacts squirrelled away by the Nazis towards the end of the war.

'Greater Germany' Caves and Cave Systems: Office of Strategic Services (1945) 75 × 60cm

major libraries and their librarians were supporting, and which would certainly have swept up maps and atlases with it.[34]

I was to discover that the Map Library has a good number of historically significant atlases from Germany and Austria that feature the stamps of the British Museum. Although we cannot be certain, Skelton is the most obvious source of these as, following his service with the MFAA, he was promoted to Superintendent of the British Museum Map Room in 1950, which became a hub for geographers and central to the network of map libraries across Britain.

It has been estimated that over the course of the war well over 3 *billion* map sheets were printed by the British, Americans, Soviets and Germans. Some 70 to 80 million map sheets were printed by the Americans for D-Day and the Normandy operations alone.[35] And while huge numbers never returned to London, due to being lost on the field of battle or discarded by those wanting nothing more to do with war, there were still millions of maps in British and American military storage that needed to be disposed of.

University map libraries were the perfect solution[36] not least because the years of embargoes and shortages had

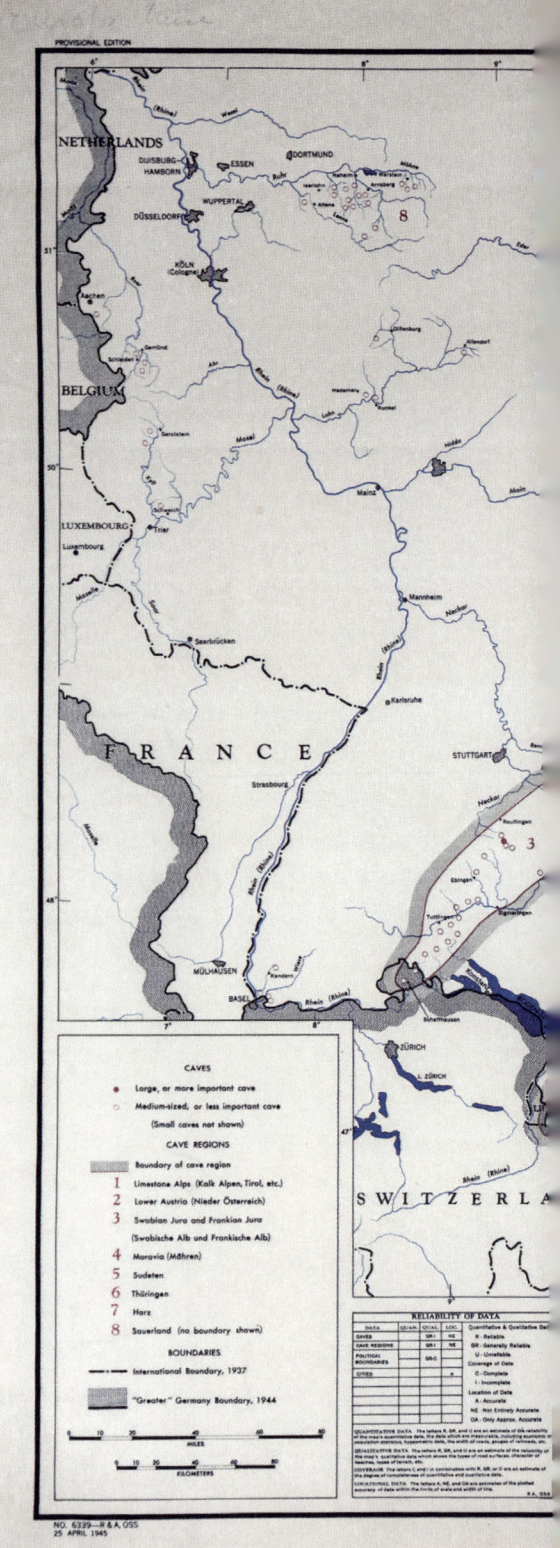

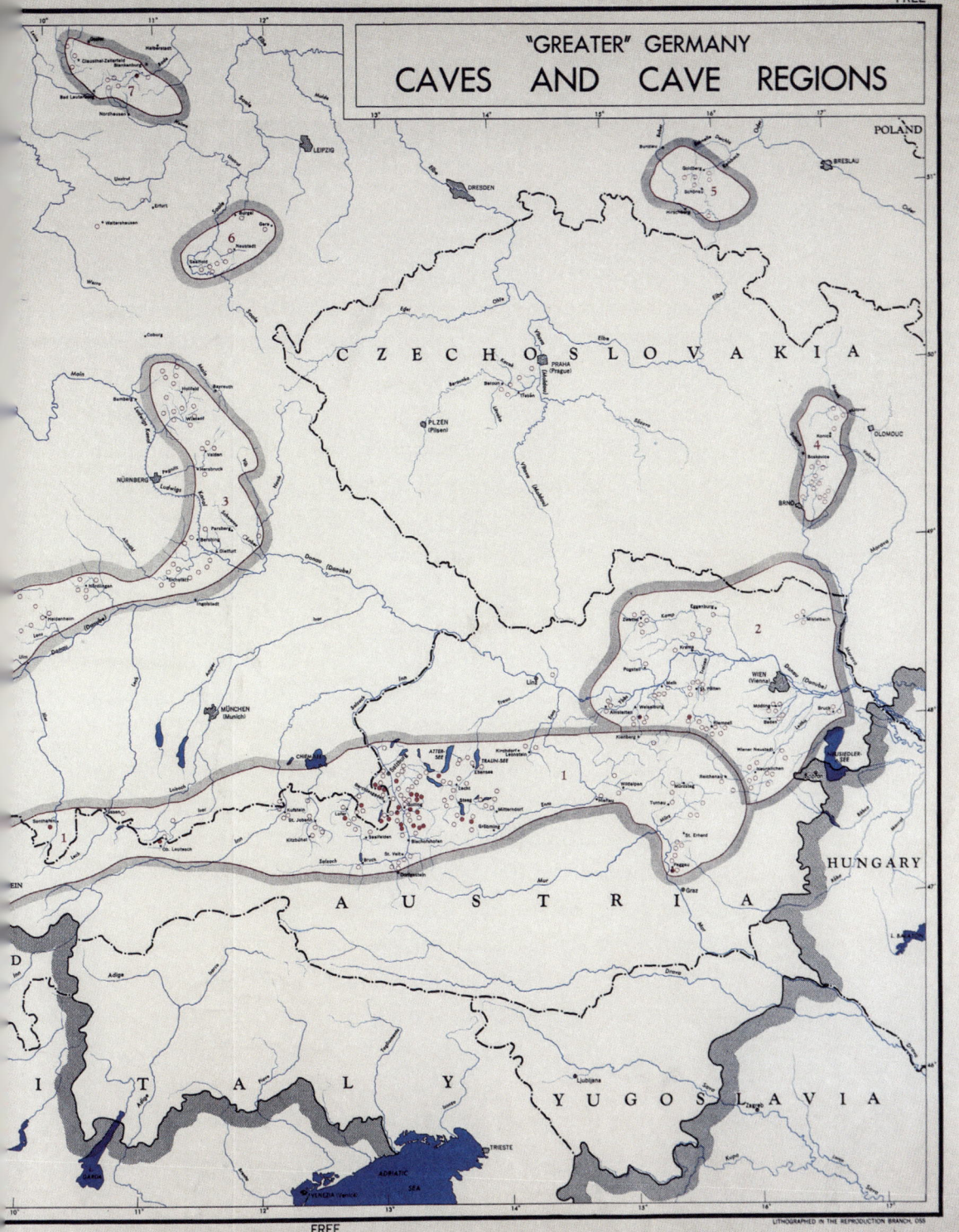

left them desperate for material. Maps from the theatre of war were also gladly accepted by the academics and map librarians who had seen military service and were therefore familiar with what was on offer. For example, at UCL two academics – R. Ogilvie Buchanan[37] and Robert E. Dickinson – had senior intelligence and mapping roles in the Royal Air Force and Fawcett's successor as Head of Department, Henry Clifford Darby, had direct experience with maritime intelligence by preparing handbooks for the British Admiralty.[38]

The library owes much to Darby, who headed the Geography Department between 1949 and 1966 and who had a keen interest in maps – himself editing *The University Atlas* in 1936 and then *The Library Atlas* the following year which, between them, sold 600,000 copies.[39] It was Darby who professionalised the Map Library set-up by employing Anne Oxenham to oversee it, and ensured the department had a well-qualified team of draughtsmen to create new maps for the academic staff. He also continued to procure maps and initiated the library's collection of 800 atlases, among them some of the finest examples of their craft, many of which are extremely rare.[40]

In addition, the Map Library continued to receive maps for decades directly from the military as part of a scheme organised by a curator named Peter Clark, who was appointed as the Chief Map Research Officer of the Ministry of Defence (MoD), between 1964 and 1983, overseeing the principal government map library in the UK.[41] In an era before sophisticated electronic mapping and widespread satellite surveillance, Clark had to ensure the MoD library was kept up to date, and therefore received a constant stream of the latest maps from around the world, with the more recent maps displacing slightly older editions that then needed to be disposed of. Rather than throwing them away, Clark set up a programme to share them among other collections, including university map libraries.

This sorting process played out in a room full of large cardboard boxes, one for each receiving library. Once full, Clark would send the boxes out or ring up map librarians to tell them to come and collect them from the headquarters in Tolworth, in suburban south-west London. Anne would get the call about once a year and she told me: 'We never knew what would be in the boxes until we opened them ... some of the stuff we got from them hadn't even been out of the drawer. It was just perfect, pristine.'

On one memorable collection trip, Anne was dispatched in her tiny Mini Cooper to pick up several boxes for UCL and a couple of the neighbouring map libraries. She loaded the boxes into her car, had a cup of tea and a biscuit with Clark and then headed back to London. Anne recalls: 'It was a very precarious drive back. I hadn't realised how heavy the boxes would be. And I was really floating gently up the A3, making sure to stay on the inside lane. I don't know what it did with the springs. But, eek, I was quite glad they were all delivered!'

I was also able to hear other acquisitions stories from Anne or from the emeritus professors in the department who knew some of the maps' creators and owners personally. There were tales of maps and atlases being retrieved from suitcases after a professor's trip abroad, donated by passing dignitaries and even emerging from behind the Iron Curtain, either as gifts from Soviet academics or by being obtained through less 'formal' means.[42]

Anne likes to tell the story of Frank Carter, who would bring maps back from his travels, tightly folded or rolled into tubes, tasking her to mount them on cloth so they could be hung on the wall when he was teaching. He is remembered by his former office mate Professor Hugh Clout as a 'master geographer' who 'opened up the mysteries of Eastern Europe'[43] by his uncanny ability to assimilate into the countries he studied. Fluent in their languages, Carter was 'Bulgarian in Bulgaria, Polish in Poland',

THE LIBRARY OF LOST MAPS

and would even get his hair cut in the local style of the time to help blend in. In 1968 he witnessed the Prague Spring, a period of political liberalisation that was suppressed by a Soviet invasion in what was the Czechoslovak Socialist Republic: Carter's photos and maps told an important part of the story.[44] Anne's view on his acquisitions was unequivocal: 'He'd have been shot if he was seen taking stuff. I mean, you couldn't even bring a cigarette out!'[45]

When I embarked on my explorations, I was expecting piles of the relatively mundane combined with the occasional miracle discovery of a map or atlas that was particularly unique or

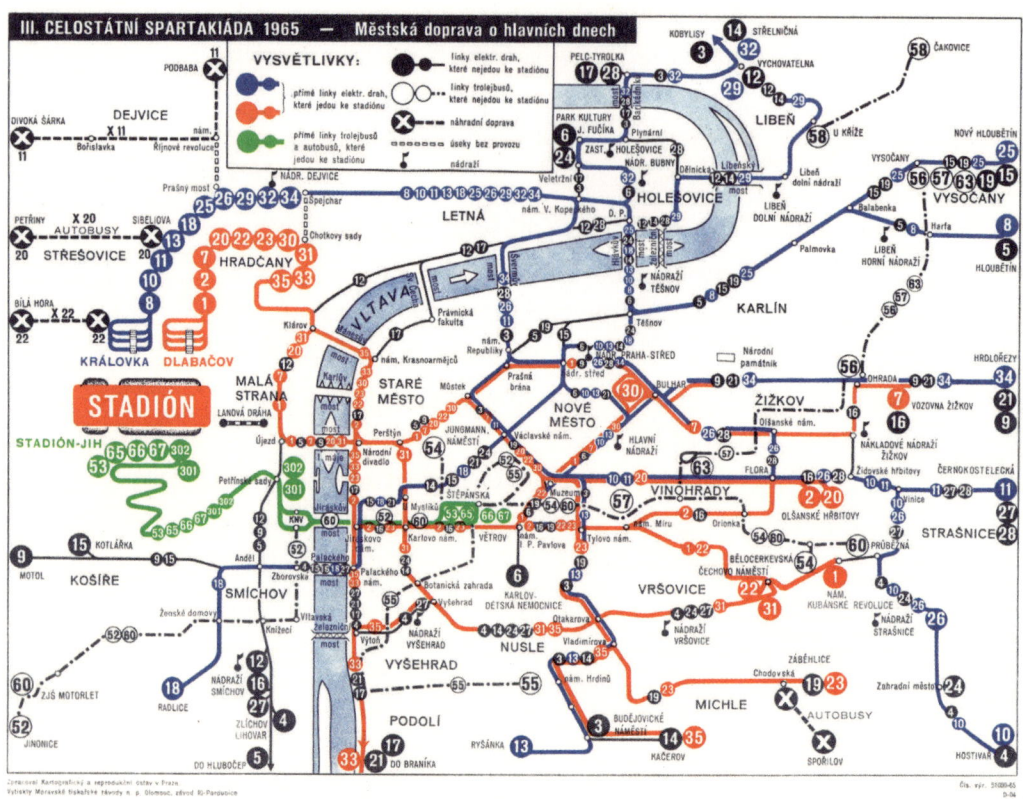

significant. And yet I have found maps that both take us back to the most important moments in history, like Carter's maps of Prague, *and* are worthy of the grandest collections. These maps aren't just exquisitely produced, they are also intimately tied together as the extraordinary people who created them took inspiration from those who came before.

What's wondrous is that the maps I have pulled from the drawers and shelves can be arranged chronologically into a grand cartographic relay race we can replay now. It begins with the trailblazing intellects of nineteenth-century Bloomsbury, who pass their batons on to the geopolitics-obsessed geographers of the turn of the twentieth century, who then – alas – hand over to the fascists and propagandists of 1930s Europe before they, thankfully, push their batons into the palms of some of the greatest scientists and visual communicators of the last century. Today the race is being run by artificial intelligence-fuelled cartographers who power the maps on our phones.

The maps that tell this story have taken me not only to the unexplored personal archives of pioneering cartographers but also to unexpected corners of history, such as the interrogation reports of high-ranking Nazis and a legendary Soviet spy. And while I have sneaked in one or two maps I picked up elsewhere, amazingly all routes lead back to the UCL Map Library. Even I had never realised quite how important maps and atlases have been in the shaping of our world today.

Map 22 A transportation map of Prague from Professor Frank Carter's collection. This shows routes to the city's stadium for the 1965 'Spartakiada'. These were extraordinary mass gymnastics events held in celebration of the Red Army's liberation of Czechoslovakia in 1945. The event in 1965 was a downsized version compared to its predecessors, but would still have attracted hundreds of thousands of performers and millions of spectators, with Carter likely among them.

Celostátni Spartakiáda 1965 – Mětská doprava o hlavnich dnech: Kartografický s reprodukčni ústav v Praze (1965)
24 × 18cm

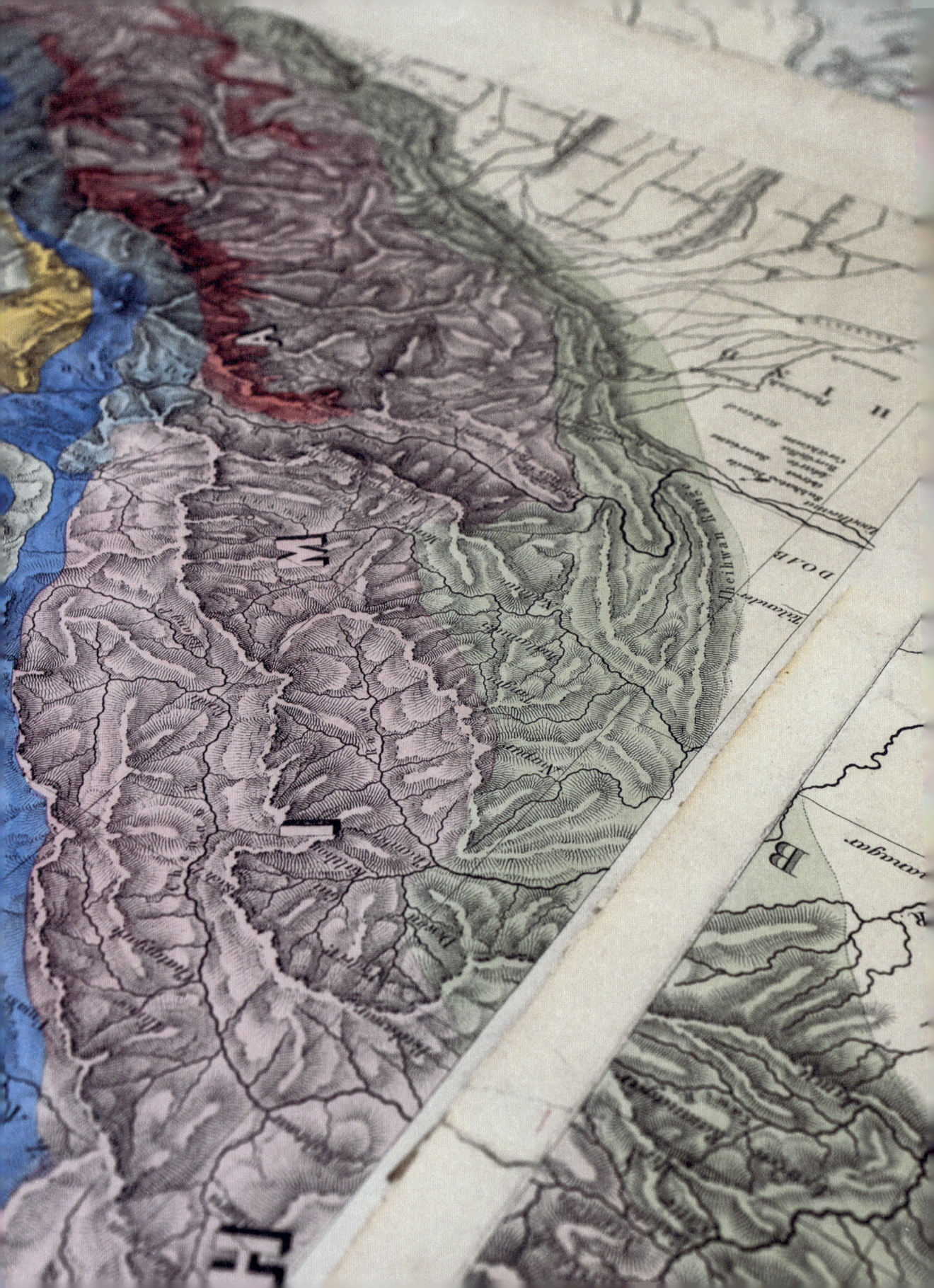

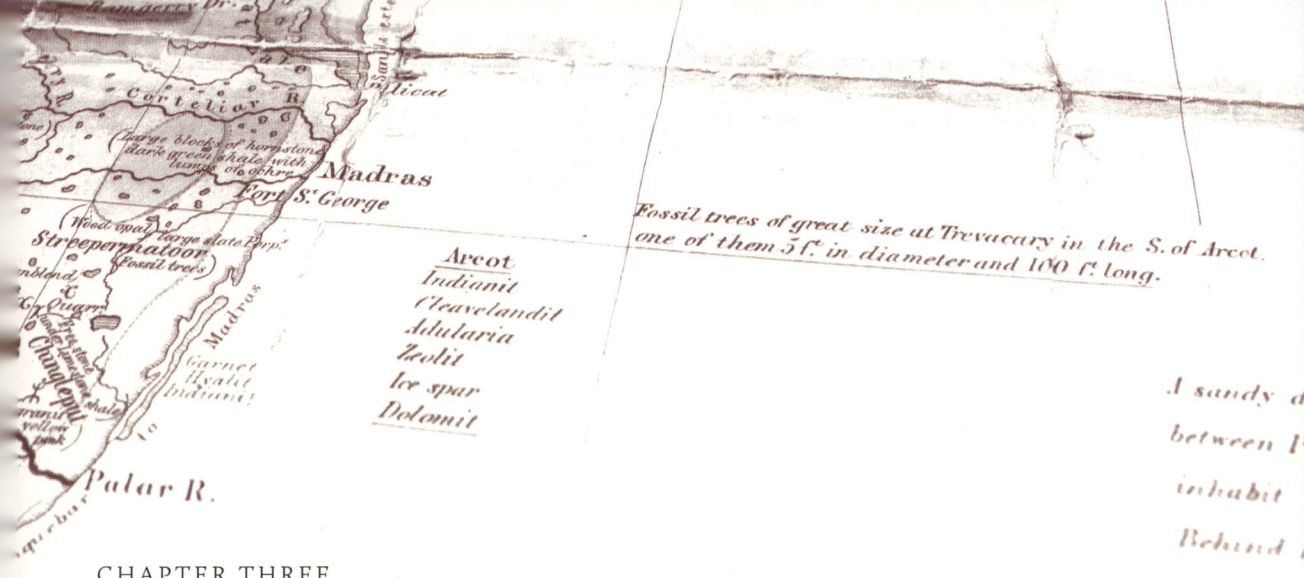

CHAPTER THREE

George Bellas Greenough's Remarkable Maps

'[Mapping is] that beautiful contrivance which I know not whether to class with the fine arts or with the exact sciences, so intimately is it connected with both.'
George Bellas Greenough[1]

My induction into the realm of nineteenth-century map-making pioneers came in the form of an unpromising discovery: a map that had faded and been torn from too many years spent pinned to a wall. But I now know it to be one of the most significant treasures in the Map Library: it was the first map to show the geology of the entirety of the Indian subcontinent and I was amazed to find out that only thirty-four copies were known to have survived the passage of time.[2] This scrappy copy added one more to that modest total.

I was also to discover that the map's creator, George Bellas Greenough, was one of the first to see the potential for maps as something more than a way of showing people where to go. To

Map 23 [Extract]
[Left] A view looking down from the top of the first geological map of India.

[Top] The coastline of the same map, where the sea has been filled with annotations detailing interesting features and discoveries.

General Sketch of the Physical and Geological Features of India:
G.B. Greenough (1854)
164 × 194cm

Greenough, maps were an essential tool for scientific discovery and debate that could reveal new understandings of the world. He is an overlooked character in our history but, as we shall see, he was a transformational figure who elevated the status of maps to be drivers of industrial development, educational reform and essential tools of government.

Published in 1854, the map I had found slowly took shape as I pulled it – in pieces – from the India drawer. The first part to emerge was a torn sheet of paper with the words

GENERAL SKETCH
OF THE
PHYSICAL AND GEOLOGICAL
FEATURES
OF
BRITISH INDIA
BY
G.B. Greenough Esq. F.R.S., F.G.S, &c. &c. &c.

Map 23 Greenough's geological map of India, published in 1854. The copy in the Map Library has seen better days, but its intricate details are still legible, and so too are the brush strokes of watercolourists who added the sweeps of pinks, reds, greens and blues to denote the different rock formations across the vast country.

General Sketch of the Physical and Geological Features of India:
G.B. Greenough (1854)
164 × 194cm

written on it, with each line in its own ornate typeface. The next sheet I pulled out covered the Bay of Bengal and only hinted at the rest of the map to come, with a list of Indian coalfields nestled below the 'Mouths of the Ganges and Brahmaputra'. The sheet after that detailed part of the Himalayas which were coloured in shades of pink[3] that transitioned to greens.

The third section was barely held together by a yellow strip of long-degraded tape. It was double the height of the other sheets, with tears that were expanding before my eyes to evoke newly formed rock fractures. It charted the southern portion of India, including Ceylon (now Sri Lanka). In this part of the map, the pinks become abundant, and the darker reds are confined to the mountainous areas.

GEORGE BELLAS GREENOUGH'S REMARKABLE MAPS

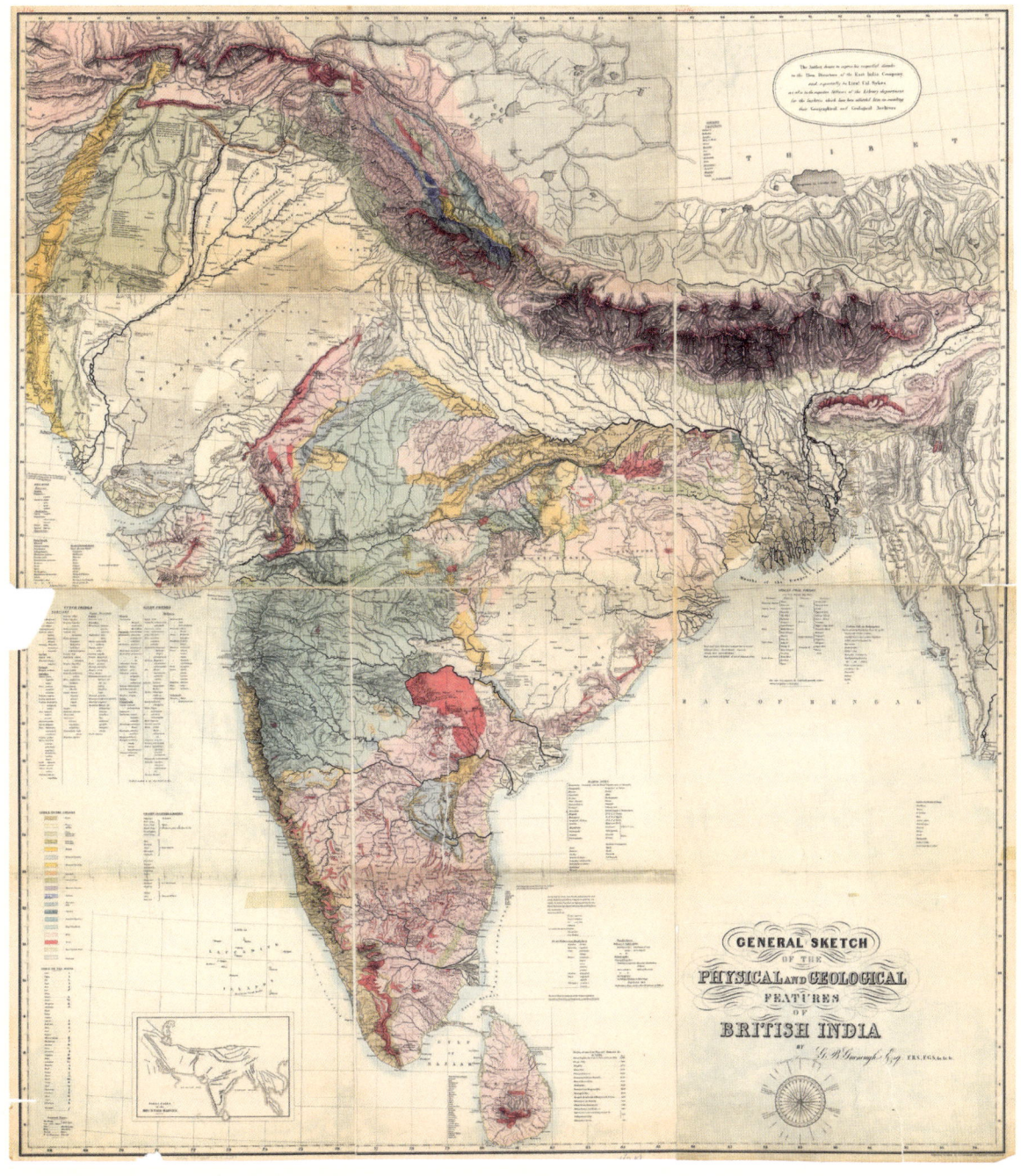

Once I'd carefully extracted the remaining sheets from the drawers, I was able to assemble Greenough's immense map. From a distance it looks like it's made from broad brushstrokes, but then up close you see the level of detail in the annotations. In tiny text there are countless notes and, in particular, references to the fossil record.

Greenough has also included a dazzling list of 'Minerals found in Ceylon', which features diamond, ruby, amethyst, topaz and sapphire. On this sheet, too, is a list of Indian diamond mines, leaving no doubt about the resource value of the continent to the East India Company.

It is a map that both commands a room and also requires reading up close for the enormity of the endeavour to be fully appreciated. I felt blessed to be able to handle one of the few copies in existence.

With the continent's mineral wealth, a map such as this would have been appreciated at the time by India's colonial rulers. However, it's not clear why Greenough himself picked India for this geological mapping project, especially as he was already quite elderly when he embarked upon it, and he had never visited the continent.[4]

It took Greenough over a decade to compile the data he needed and to transfer it onto an initial sketch map,[5] but in 1853 he was finally ready for the map to go into production. He passed the drawing to the 'Physical Geographer to the Queen', Augustus Petermann, who was to oversee the engraving of the printing stones. Petermann sped things along by cutting the master drawing of the map into sections so each could be worked on simultaneously by a different engraver. He also had them working on the project 'till 12 and 1 o'clock every night',[6] etching away by candlelight.

Even working such long hours, the engravers still took the best part of a year, with the first proofs available for Greenough's detailed inspection in April 1854. It was printed just a fifteen-minute walk from the Map Library by William Henry Rauschen of 12 Gower Street North, under the 'care of Mr Stanford',[7] about whom we shall hear more in Chapter 4.

Petermann's final invoice from June 1854 gives a sense of the labour that went into his part of the project alone:

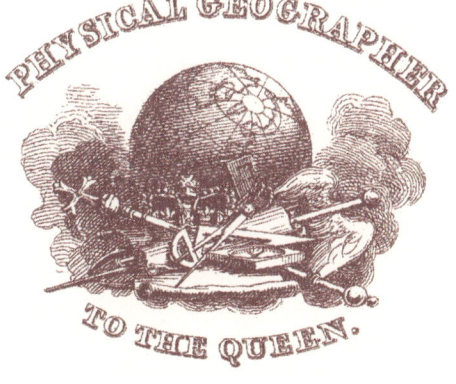

Augustus Petermann's letterhead.

> Engraving of Geological Map of India in 9 sheets ... £410.
> Material of the 9 [printing] stones. £30.
> Printing 100 copies of the Map. £12.
> Paper for 100 copies. £12.12s.
> Total = £464.12s[8]

With no institutional backers, Greenough paid for this expensive[9] labour of love from his own pocket. He was also on his own for the publicity and marketing operation so, despite his poor health and mental exhaustion from the project, Greenough had literally to doorstep those in positions of influence to ensure they had sight of a copy. The East India Company was seen as his principal customer and Greenough had already name-checked its 'Honourable Directors' in the top right corner of this map to thank them for enabling him to access their archives to research the map. It should have been an easy sell, but it transpired that the map and its letter of introduction had been mislaid.

> I delivered the map ... to one of the Porters in Sir James Melvill's anteroom [Melvill was the Secretary to the East India Company] ... and of course I thought it would be submitted to the board on the day following. This did not happen nor did I receive information upon the subject

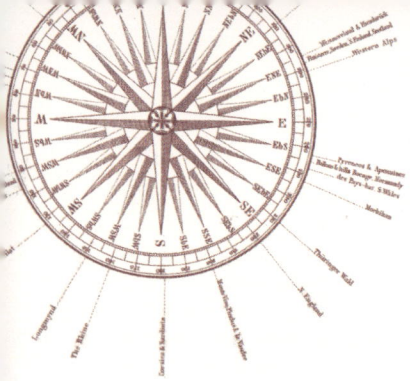

from the Secretary or anyone else ... On Saturday last therefore I walked upon the Chairman who gave me a very gracious reception, stating however that the map had never been shown to him ... The search was somewhat tedious but eventually I did find it lying upon a chair in one of the offices where no one was in attendance.[10]

His persistence paid off and the East India Company purchased sixty copies of the map priced at four guineas each (recouping roughly half of Petermann's bill).[11] It's worth noting that, while Greenough was chasing sales, he was in no way hawking a finished product and was quite transparent about this. The pioneering map had 'Sketch' in its title, a deliberate reference to it still being a work in progress and therefore subject to change. He expected to have made mistakes that needed correcting. For example, I spotted a question mark beside 'Inoceramus', one of the fossils he listed as being found in the Himalayas. Greenough also uses the word 'perhaps' ahead of some labels. Indeed, thirty of the East India Company's copies were forwarded on to the Government of India with an invitation to share with those who might be able to provide constructive feedback.

The map generated a lot of positive reactions and correspondence for suggested corrections, but there was sadly no time for Greenough to enact them, nor to pursue his even bigger vision to map the geology that connects Europe to Asia. He died from 'an attack of dropsy'[12] on 2 April 1855 while travelling in Italy.[13] His death so soon after the publication of the India map also meant that he was not present to defend his masterpiece when critics with their own vested interests rounded upon it.

According to Christopher Toland, who has written the only comprehensive history of Greenough's map,[14] some of the most negative comments came from the 'Sub-Committee of the Asiatic Society of Bengal' headed by a Thomas Oldham. It listed fifty-eight errors, which, given how minor some were and the size of the undertaking (both in terms of geographic extent and volume of data), is a trifling number that could have been easily corrected in a revised edition, as Greenough had intended. Oldham's committee also cited concerns about the quality of the topographical base map, but there was no better alternative at the time. Nevertheless, the committee concluded that the map was so imperfect 'that it should be allowed to remain in its present state, to form a memorial of the condition of our geological knowledge at the time of its publication'.[15]

This indictment was passed on to the Geological Society, which agreed. At that moment, Greenough's extraordinary map was condemned. Oldham promptly stepped forward with his own proposal for a better map to be completed by his employer: the newly formed Geological Survey of India. As Toland notes, Oldham was 'hopelessly optimistic' about how quickly he could match Greenough's effort. His initial estimate was 'two to three years' to produce an improved, but smaller, geological map of India, 'but it was to take more than twenty calendar years and 300 man-years of effort before the Survey published its *Preliminary Sketch of the Geology of India*', statistics that also gloss over the loss of a large proportion of his men to injury and disease.[16]

If Greenough's map had been updated in the way that he had intended, with incremental improvements suggested by those working in India, it might not have achieved perfection, but it would have had a use for the twenty years it took for the Geological Survey of India's map to come to fruition.

The map's impact is therefore hard to assess, with roughly only half of the 202 maps (there were three print runs in all) entering general circulation, since the remainder were either internally used by the East India Company or given as gifts to private

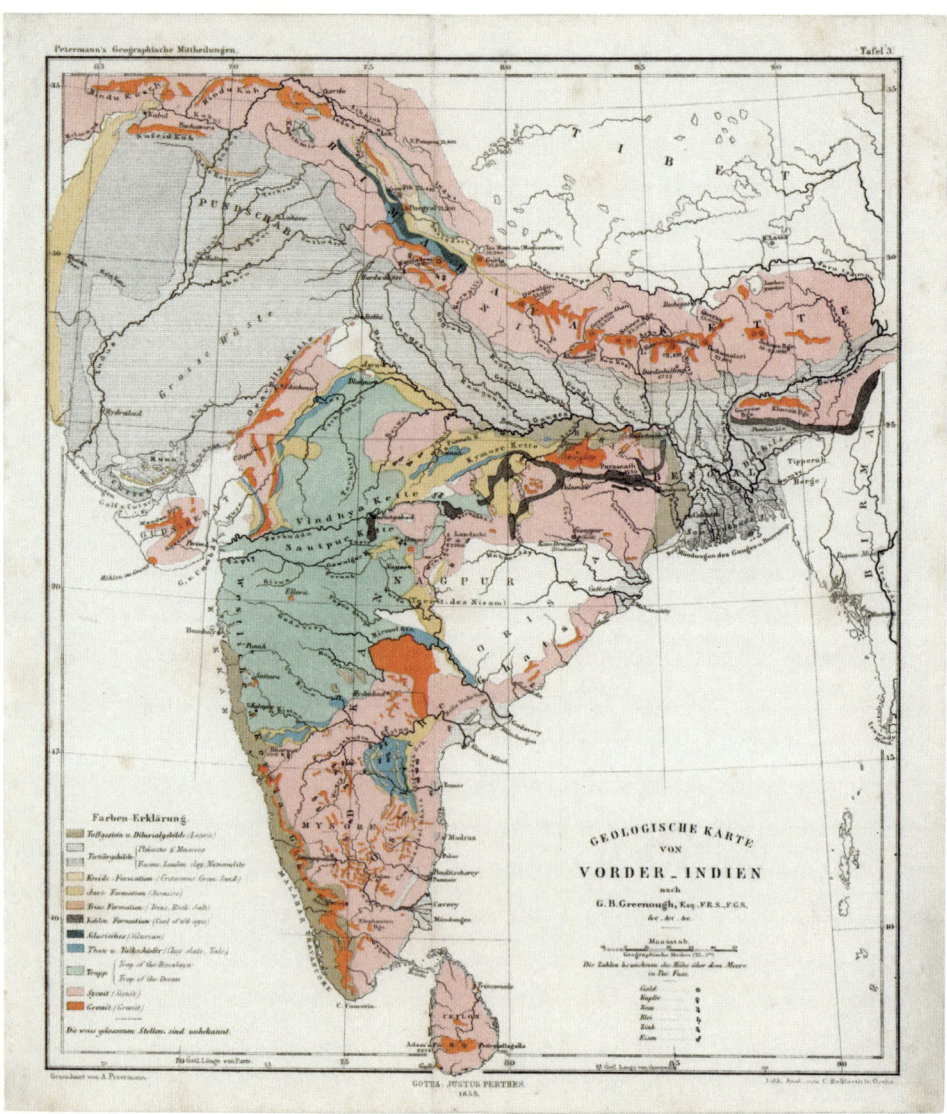

Map 24 Petermann's much-reduced version of the original Greenough masterpiece.

Geologische Karte von Vorder – Indien: Justus Perthes (1855)
20 × 24cm

individuals. That said, the East India Company was a hugely powerful force in the region and so there is no telling how many consequential decisions were informed by the map. For example, an understanding of geology was considered an important part of military education, and it was taught to officer cadets of the East India Company's own army at the time the map was published.[17] It wasn't for explicitly military objectives that such knowledge was deemed necessary, but because soldiers were tasked to be on the lookout for new sources of mineral wealth.[18] Other copies of the map may have been accessed through the map libraries of the likes of the Royal Geographical Society and the Asiatic Society of Bengal, but it seems it came nowhere near fulfilling its potential as the most definitive depiction of India's geology at the time.

However, thanks to the map's engraver, Petermann, Greenough's efforts weren't completely in vain. Petermann had moved back to his native Germany in 1854 and started working on a periodical called *Petermanns Geographische Mitteilungen*. Before his death, Greenough had agreed that Petermann could adapt and drastically reduce the size of the map for the first edition of this publication. With a print run of 4,000 copies (including 1,000 outside Germany) Greenough's labours were appreciated by a larger audience than perhaps even he had dreamed for his preliminary, but pioneering, sketch map.

In addition to his map of India, I was able to assemble a pile of ten other map sheets linked to Greenough. But unlike the India map, these were not maps he had created but, rather, maps he had owned. Greenough's name was written on an important-looking name plate and accompanied by a Royal Geographical Society of London stamp. These were some of the finest maps I had found in the Map Library, mounted on linen and in such good condition that, to my naive eye, I thought they were reproductions. It was only after I ran them past some of the UK's leading map curators, who confirmed their age and provenance, that I realised these ten sheets were very special indeed.

Enthused by this, I took to researching their owner in more detail, and it quickly became apparent that Greenough was a man who didn't just create one of the most priceless maps in the library, but also profoundly changed how we conduct science and, most significantly, pioneered how maps could be used to share knowledge in a rapidly changing world.

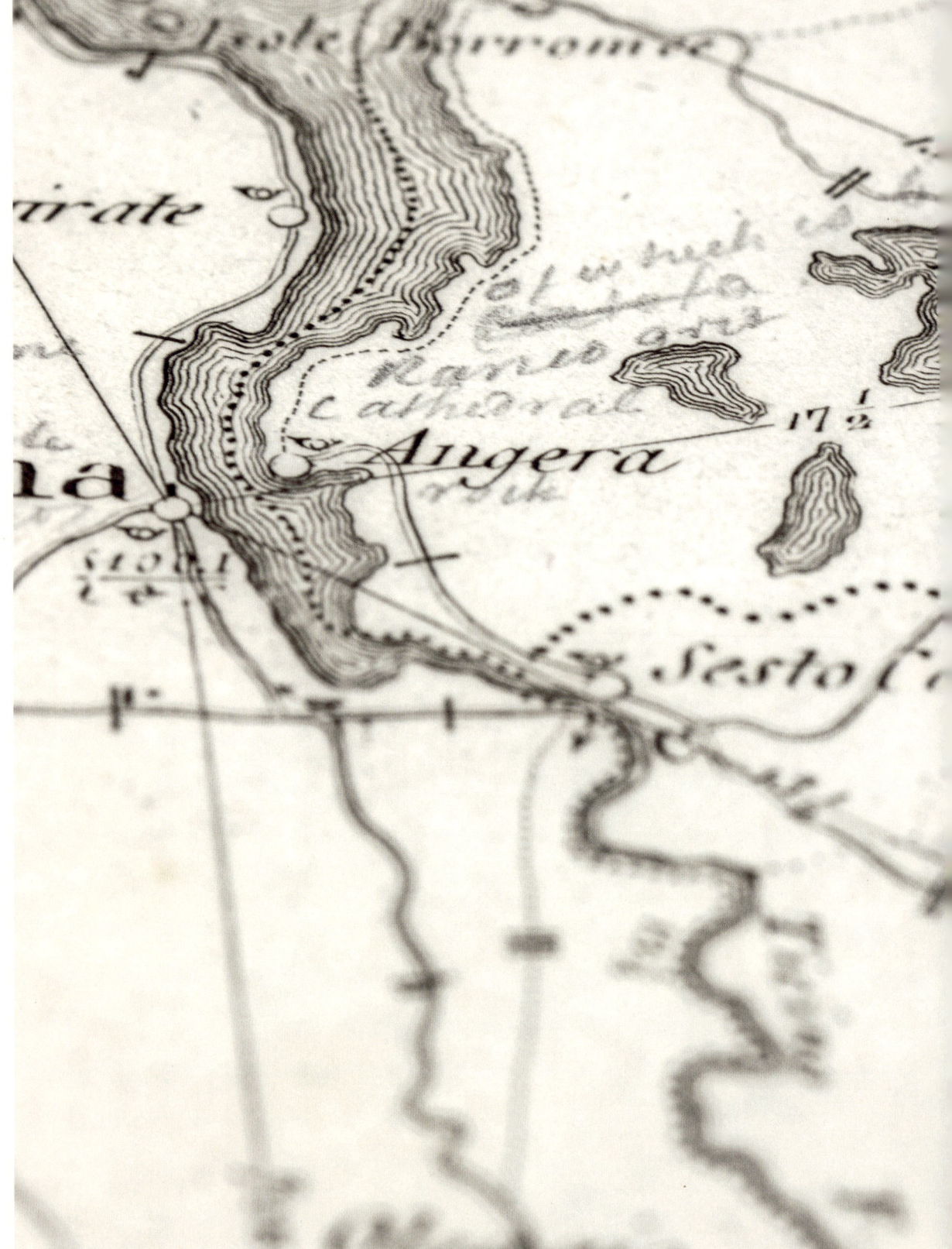

GEORGE BELLAS GREENOUGH'S REMARKABLE MAPS

Maps Reveal the Man

Through my research in the Map Library I discovered that George Bellas Greenough was a member of the highest echelons of Georgian and Victorian Britain and was able to surround himself with an extraordinary group of people. There are archives of his correspondence with some of the most famous names in the arts and sciences, including Samuel Taylor Coleridge, Michael Faraday, Marc Isambard Brunel, Charles Darwin and John Herschel. But it is his maps that have taught me most about him, and the way that men of his status and influence saw the world.

George Bellas was born in 1778, orphaned at the age of six, educated at boarding school and raised by his maternal grandfather, Thomas Greenough. Greenough Snr was a wealthy apothecary who had a global[19] pharmaceutical empire selling all manner of fantastically named products and cures from 'Pectoral Lozenges of Tolu'[20] (to stop tickly coughs: nowadays 'Tolu' is used as a fixative in perfumes) to his patented 'tincture for cleaning teeth and for curing tooth-ache'.[21]

Thomas Greenough died in 1795, passing his vast wealth and the Greenough surname to George, who at the age of seventeen knew he would never have to work a day in his life. But rather than kick back and cut loose in Georgian London, Greenough went first to the University of Cambridge to study law (but was not awarded his degree as he refused to accept the Church doctrines required to graduate[22]), and later to the University of Göttingen, where his passion for geology was ignited after attending lectures

Map 25 [Extract] Greenough would cover his maps with tiny handwritten annotations, which often speculate about the geology of an area. This is a small part of a longer note that refers to the area on the map being the source of materials for Pavia Cathedral (100 km to the south-east). It should also be said that he was sometimes mistaken in what he wrote!

Carta delle Stazioni Militari Navigazione e Poste del Regno d'Italia : Ministro Della Guerra (1810)
124.5 × 94.5cm

on the subject to help improve his German. During his studies, Greenough embarked on geological tours across Europe, collecting minerals and studying geological collections.

He returned to England in 1801, when he began enthusiastically pursuing the natural sciences. Given his wealth and status, Greenough was able to gain recognition very quickly: for example, he became an active member of the Royal Institution in 1804, while continuing to embark on grand tours of Britain and Europe in search of knowledge. If there wasn't already a society catering for his interests, he'd have a hand in creating it. Greenough helped found the Geological Society of London, the Royal Geographical Society and the British Association for the Advancement of Science (now the British Science Association). He was therefore prominent at a time when science was getting organised and developing at a rapid rate, and so many of the most prestigious scientific societies still in operation have been directly or indirectly influenced by Greenough's ideas.

Given this legacy, it is remarkable that Greenough has been largely forgotten. He was unknown to me, despite having sat next to his marble bust many times at meetings of the Royal Geographical Society, and he was close to the founders of UCL, where I have worked for over fifteen years. If his maps hadn't been so beguiling I would have looked past him again in the Map Library, but they drew me to him and implored me to recognise his astonishing legacy.

The first map from Greenough's collection that I was able to piece together comprised five separate sheets that could be seamlessly laid out on the large table in the centre of the Map Library. Placed together, they spanned the full length of the Pyrenees, the mountain range that frames the border between France and Spain, from the Bay of Biscay to the Mediterranean, embracing the tiny state of Andorra along the way. I could see from minuscule lettering along the bottom that it was dated 5 January 1809, making it one of oldest maps I'd encountered.

On both sides of each map sheet is the stamp of the 'Royal Geographical Society of London', indicating that they had been part of this collection before coming to the Map Library. The maps were gifted to the Society upon Greenough's death in 1855. As a founding member in 1838 and past president, Greenough was a major supporter of the Society and bequeathed it his map collection, plus some funds to care for it. This map, though incomplete, was part of that donation.[23] Quite why the Royal Geographical Society let this go remains a mystery,[24] but the most likely explanation is that a curator at one time or another decided they were duplicates that could be of more use to the UCL Department of Geography.[25]

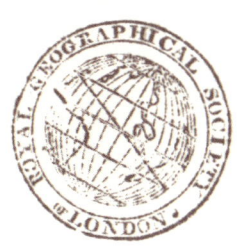

The Royal Geographical Society's loss, though, is our gain, and I was delighted when another map from the same bequest emerged from the drawers. Again in separate sheets, the united map doesn't just create a stunning depiction of northern Italy, but it reveals much about Greenough the man.

Published first in 1808 and then reprinted in 1810 by the Italian military, the map is titled *Carta delle Stazioni Militari Navigazione e Poste del Regno d'Italia* (Map of the Military Stations, Inland Navigation and Postal Stations of the Kingdom of Italy). It shows a grand sweep of Italy northwards from the Apennines to just beyond its land border that threads along the Alps. The mountains are even more beautifully rendered than the Pyrenees map, by masterful shading that pushes the major ridges out of the page and has the valleys delicately cradling the great Alpine lakes of Garda, Maggiore and Como, and buttressing the rivers that sustain them.

Off to the margins the military cartographers have sketched in the main roads to other cities, so in the north-east we can follow the route all the way to Vienna. It is also possible to plan a

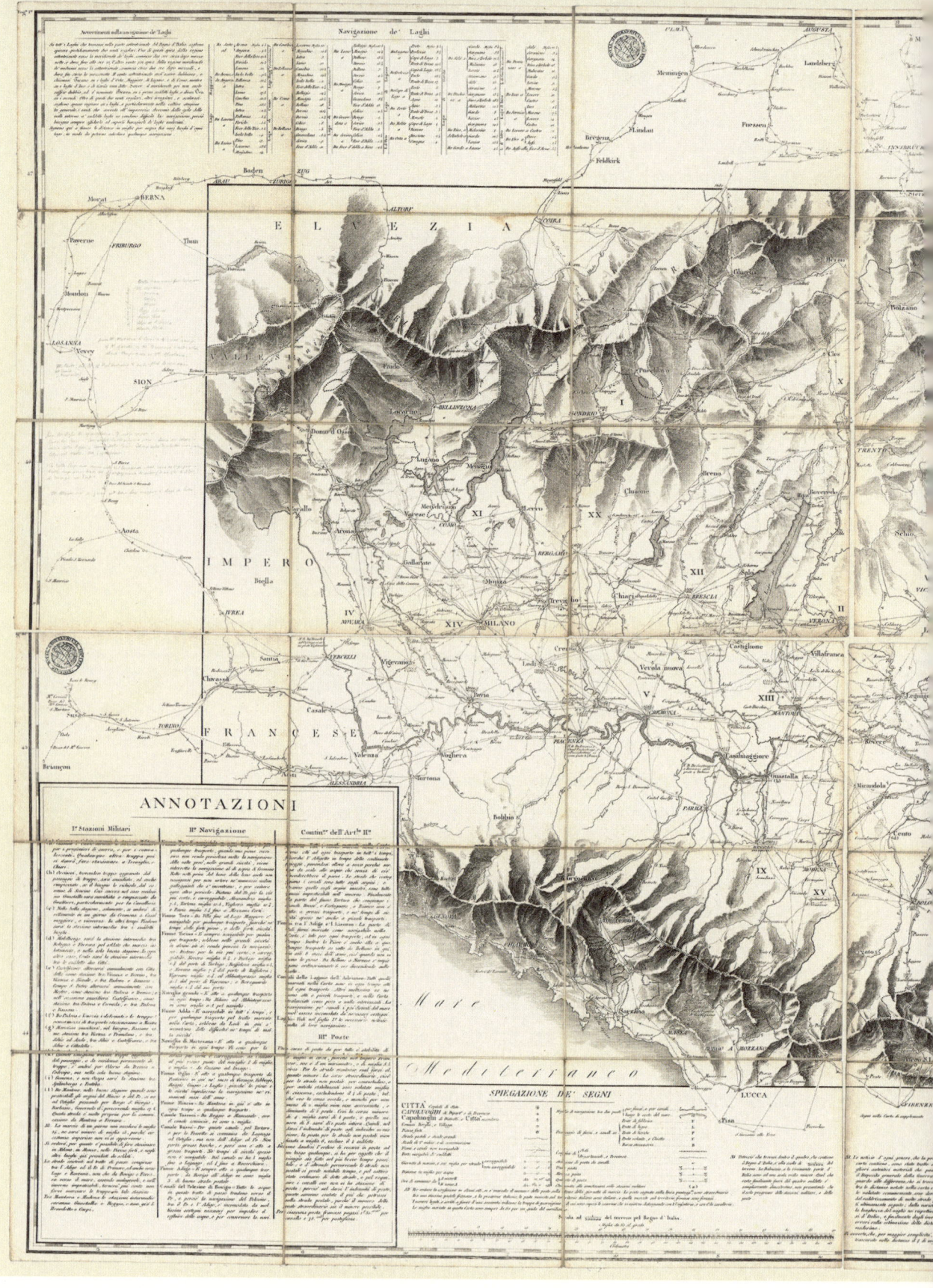

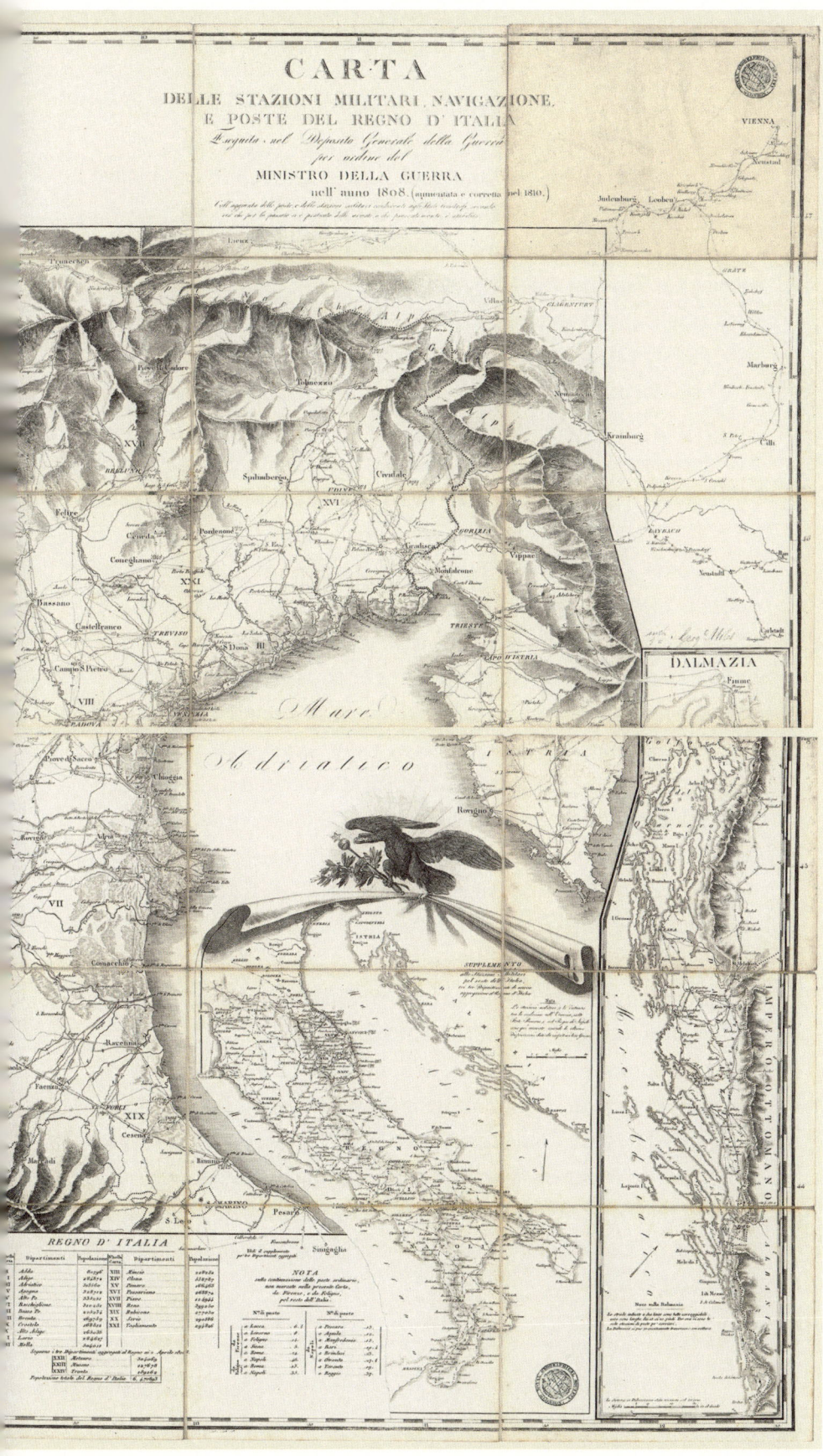

Map 25 The full translated title of this stunning map reads 'Map of the Military Stations, Inland Navigation and Postal Stations made in the General Deposit of the War following the order of the Minister of the War in the year 1808 (extended and corrected in 1810)'. The symbols and annotations explain how long it takes to travel from one place to the next, accounting for time of year – with some rivers only navigable in winter, for example. There are local recommendations too, including a warning to travellers not to cross the Po River in the total darkness of a new moon due to many hazards, such as floating water mills.

Carta delle Stazioni Militari Navigazione e Poste del Regno d'Italia : Ministro Della Guerra (1810)
124.5 × 94.5cm

THE LIBRARY OF LOST MAPS

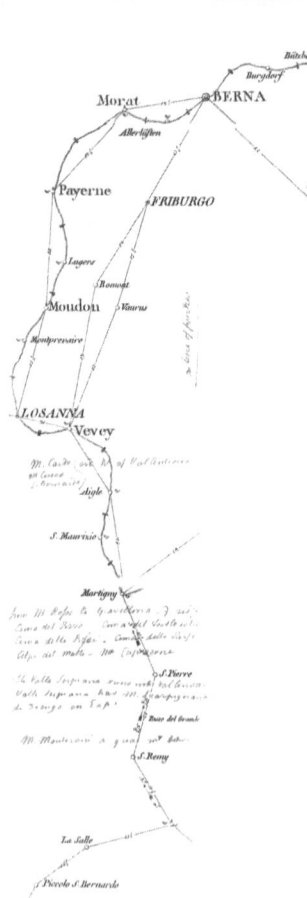

route across the Alps via Innsbruck and to Munich (here labelled Monaco, which is the Italian name for that city).

In the style of the best nineteenth-century equivalent of a satnav, the straight lines across the map show the walking times, with tiny, flecked arrows indicating the direction of travel, if one way was significantly quicker than the other.

The Mediterranean Sea makes a brief appearance to the south-west, and the Adriatic Sea to the east plays host to the most wonderful inset map of the entirety of Italy's boot, being unfurled by an eagle clutching the map in its left talon, a mace and olive branch in the other. This was a time when Italy was not united as it is today, so areas like Trieste and the peninsula of Istria were not part of the country, and Lombardy was an Austrian province. Borders were shifting towards the unification of Italy, so the Italian eagle and its regalia were there to leave no doubt about who the mapmakers thought should have sovereignty.

Looking closely, I could see in neat pencil that there were notes that started in the left margin and extended onto the map, particularly focused around the city of Milan. From what I could make out, the annotations concerned different rock formations as well as occasional tourist sights such as the Colossus of San Carlo Borromeo, a giant statue of the Archbishop of Milan in Arona, a town on the shore of Lake Maggiore. The giant copper statue was completed in 1618, so would already have been over two centuries old when these annotations were made.

Could this be a map Greenough carried with him on his travels? When searching for more information, I discovered – to my delight – that UCL's Special Collections look after a vast quantity of Greenough's papers.[26] I requested anything of Greenough's that seemed relevant to travel and I set about sifting through the ten or so boxes that came back to see if any of the material aligned with the map I had found.

There were several orange archival folders marked 'Italy' and in one I found a pocket notebook[27] with a sticker on its black marbled jacket with the title:

'from Oct. 11. 1827 – April 13. 1828.
Itinerary.'

Running to seventy-three pages, the notebook contained a list of the places that Greenough had visited on a six-month tour. With a copy of the map in front of me, I was able to make my own excursion and I spent a day reliving Greenough's journey, charting the route that he took from one map sheet to the next, ticking off the sights he'd visited and breathing life into the map. By my final count he made over sixty stopovers and visited thousands of landmarks. He was, after all, a gentleman-scientist, and so this was not an idle holiday but fieldwork to gather data about the people, the landscape and the history of Italy.

What's more, the notebook detailing the itinerary was one of several that were interleaved with additional scraps of paper with further observations scrawled on them that Greenough must have made while he was on the road. I was overjoyed to confirm that he had visited the sights noted on the map. I learned, for example, that many of the annotations were things he would have spotted from a steamboat trip on Lake Maggiore that he took on 14 April 1828; that the descent to the lake reminded him of the 'road from Kendal to Windermere' in the Lake District (see Chapter 4 for a map of this area), and that the hills above the lake evoked the Scottish landscape. He was a man who liked to anchor the unfamiliar to the world he knew.

The details on the maps and the copious notebooks made it clear that Greenough was on a mission to observe and study. The types of rock used to construct churches, and their origins, were carefully noted down; so, too, were the agricultural practices and the flora and fauna encountered. Greenough even made notes about the price of horses – 'English horses fetch immense prices in Italy. At Parma I had heard of a pair of English coach horses which had been sold there by a dealer for three hundred guineas' – and the progress of medicine: 'Vaccination is now generally practised … I happened to meet the first person who had been vaccinated – he told me it was in the year 1804.'[28] I was amazed to see that as he walked around Santa Maria della Passione, one of the most sacred buildings in Milan, Greenough copied down the words of every engraving he came across.

Greenough is giddy with excitement in Venice and is particularly impressed by the Venetians. On one scrap of paper, he has hastily scrawled: 'So handsome a race of people we have not seen in our journey – the men are dressed as well as if they were in Bond Street – the head dresses of the women a piece of muslin thrown over the head and shoulders is beautiful and the grace with which they manage to alter the folds of this drapery is peculiar to themselves.'[29]

At a time when the arts and sciences hadn't been so firmly partitioned, his connections to the English poet Samuel Taylor Coleridge have led to some referring to Greenough as a 'romantic' geologist[30] and there are occasional flashes of more evocative descriptions of landscape among the many more mundane observations. These seem to be from the road when it would have passed by only as fast as his horse and carriage would allow. It's as if he forgets himself as the 'impartial' observer and becomes infused with the beauty of the places he is experiencing, which sometimes go by too fast: he notes his regret for passing through some 'delightful' olive groves at a 'gallop'.[31]

Outside Florence he encountered a forest of 'most noble' chestnut trees, and on the Plain of Lombardy 'the fields are divided by trenches of sluggish flowing water by the side of which are poplars and vines clinging to them', and taking in the views across Lake Como, Greenough muses that the 'houses seem either to emerge from the water or to be perched in the clouds'.[32]

It was such a joy to follow Greenough along his map and get to know him in this way, as it revealed a huge amount about his character and how he saw the world. He clearly took his work as an information gatherer seriously, accumulating his lists and jottings on the scraps of paper into longer prose and deeper reflections about the places he'd visited. It's intriguing that Greenough published so few of these musings.[33] I pondered who he was writing it for: just for himself, or did he have a much grander idea in mind for a great treatise that he never got round to completing?

A couple of major exceptions to his reluctance to publish were his geological maps, the most ambitious of which were of England and Wales, and of course the map of the Indian subcontinent that I was so excited to discover.[34] In these maps Greenough was able to play to his strength as an observer, collator and synthesiser of information.

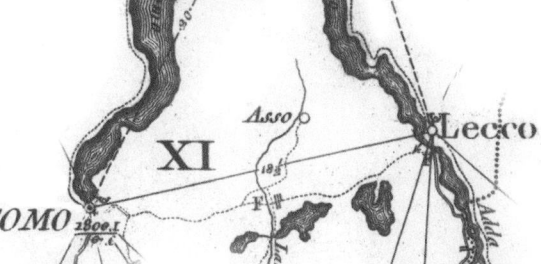

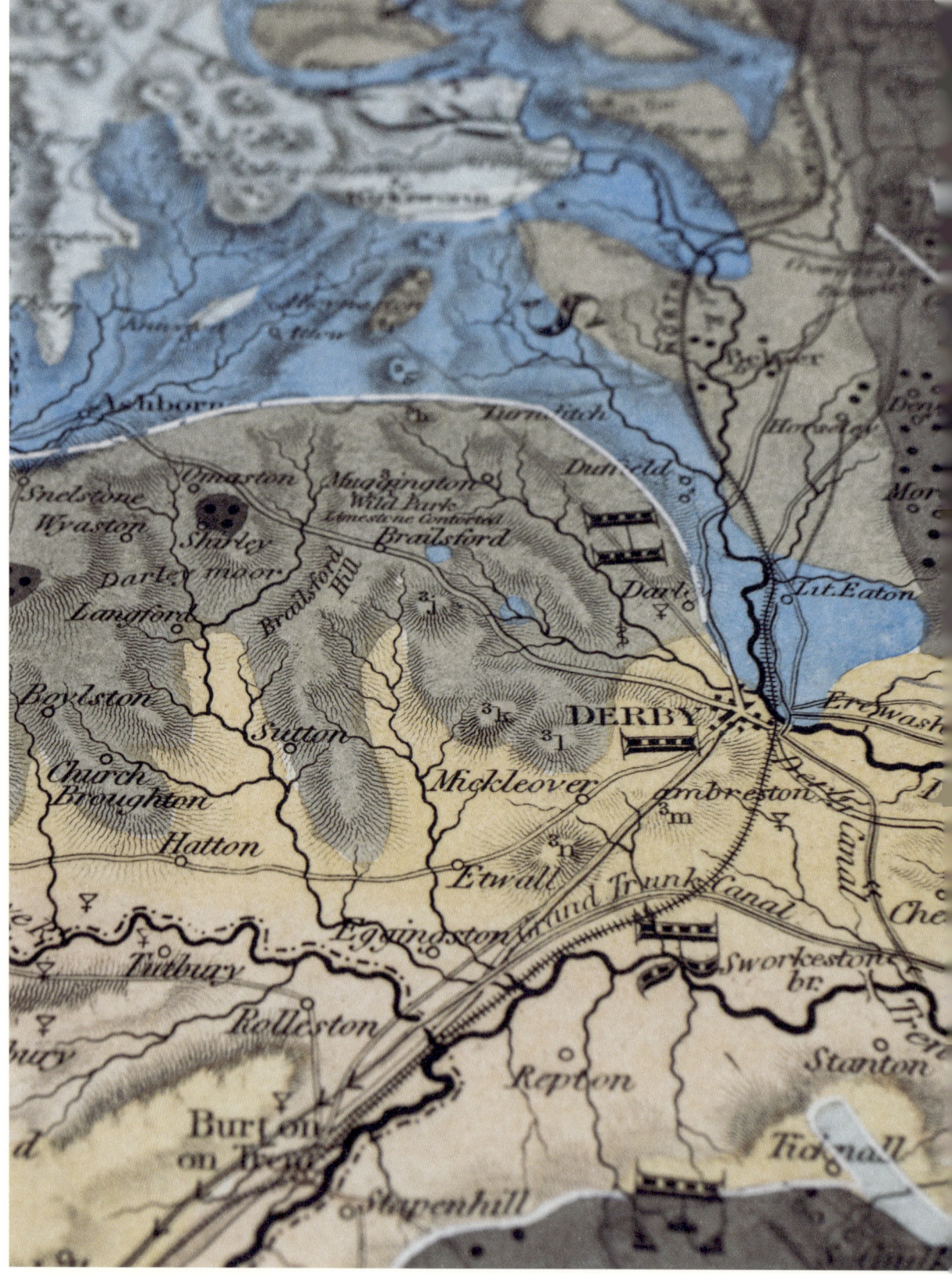

GEORGE BELLAS GREENOUGH'S REMARKABLE MAPS

Mapping the World Beneath Our Feet

In the early 1800s there was a 'little talking Geological Dinner Club'[35] that comprised an exclusive group[36] of thirteen men, who, with George Bellas Greenough among their number, resolved that a society was needed to help organise and drive progress in the nascent field of geology. So in 1807 they transitioned from the informalities of a supper club and instituted the Geological Society. This pivotal moment created a forum for many of the scientific discussions and debates that defined the Victorian era; Charles Darwin, for example, served as its secretary in 1838.

Despite his youthful age of twenty-nine, Greenough was elected in absentia the first president of the Society and served 1807–13.[37] A shrewd move, perhaps, given his huge wealth and likely longevity: he was re-elected president for 1818–20 and 1833–5 and continued to play an active role until his death in 1855.

Under Greenough's leadership, the Society concluded that geology in Britain was being held back by too much 'grand theorising'. As the Church of Scotland minister and celebrated scientist John Playfair put it in 1811, geology was considered by men 'accustomed to the more sober and cautious exertions of the understanding'[38] (such as the likes of Greenough) a 'species of mental derangement in which the patient raved continually of comets, deluges, volcanoes and earthquakes; or talked of reclaiming the great wastes of the chaos, and converting them into

Map 26 [Extract] Driven by their search for coal and other natural resources, the Victorians became masters of geological mapping. The maps they produced were often beautifully coloured by hand and drew inspiration from the hues of the rocks they were depicting. Here we see a convergence of different rock types around the East Midlands (UK) town of Derby.

A Geological Map of England and Wales: The Geological Society (1839)
160 × 188cm

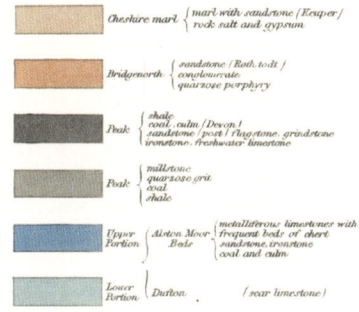

a terraqueous and habitable globe'.[39] To the young Society the cure had to be to 'purge their science of any vestige of grand theory, theological controversy and, above all, an excess of imagination'.[40]

In 1811 the Society pasted, in Latin, on the front page of its scientific journal the promise 'not to produce attractive and plausible conjectures, but certain and demonstrable knowledge'.[41]

However, 'demonstrable knowledge' in geology is a real challenge because much of what needs to be seen is underground or the result of events that occurred a very long time ago. So, rather than being able to do experiments to generate scientific observations, geologists had to content themselves with standing atop an outcrop of rock, identifying its type and then taking an educated guess as to whether it is part of the same seam that disappears beneath the surface and then might reappear tens of kilometres away.

Maps proved a successful tool for this work since they could be used to unite the fragments of different rock formations from scarce observations. By charting them out, geologists could join the dots of a rock formation from 'here' breaching the surface to similar features witnessed in a mine over 'there'. Maps also exude a confidence and sense of scientific objectivity that offered reassurances to the sceptics. But that is not to say they were uncontroversial.

How best to make a map can be the subject of disagreement, and geologists, as we'll also see in Chapter 10, are the masters of cartographic quarrels. One of the biggest arguments swirled around Greenough and another man called William Smith as they each laboured to produce the first national map of the geology of England and Wales. Both men started out around the same time, but it was Smith who published his map first in 1815, a monumental achievement. Smith had largely[42] created a map based on his extensive travels and observations across the country, which he

GEORGE BELLAS GREENOUGH'S REMARKABLE MAPS

gathered as part of his job as a self-taught surveyor and engineer.

It was to be another four and a half years before Greenough published his map, which was drawn from a compilation of data supplied by members of the Geological Society and from other sources including some of Smith's findings. Greenough's map took longer because he had commissioned a complete redrafting of the terrain (rather than use an existing map for this) and his data were more precise in some areas. But it was also seen as a direct competitor to Smith's and came under fire for its lack of acknowledgement of Smith's pioneering efforts.

Smith and Greenough disagreed about the value of the fossil record. In Smith's view, because fossils represented animals and plants that inhabited the Earth in the past, it was possible to identify the eras in which those critters existed and therefore became deposited. If Smith found the same fossil, such as a particular kind of ammonite, in two very different parts of the country – even hundreds of kilometres away – he could therefore assume the rocks they were in were of a similar age if they were in the same position in the order of the strata (the distinct units of rock). This was a liberating idea as it meant he could be much bolder when making the connections he needed between the few observations he had, thus enabling him to fill out his map.

In contrast, Greenough was very sceptical of this idea since he felt that it was just too much of an assumption to track back in time to an Earth that he could not observe. He would not bring himself to develop theories about the age of a fossil. Too much was at stake for such conjecture, not least because the maps were desperately needed to point to the coalfields that fuelled the steam engines that drove the Industrial Revolution.

Smith and Greenough therefore approached their maps in different ways. Much of Smith's view was dictated

An ammonite drawn for an 1830s magazine article about pre-historic life on earth.

Penny Magazine, October 26th 1833 (Organic Remains Restored)

by the fossils he had observed and the knowledge of where they were found. From this he could build a stratigraphy of the different layers in the bedrock of Britain. And it is these strata that could form the basis of a map, a map that might have lacked a data point for every part of the country but that could at least be filled in using Smith's theories.

Greenough, on the other hand, was not prepared to make such bold assertions about the extent of particular strata without recorded observations of them from his many data sources. This was a constraint because, in theory, he would have needed vast amounts of information if he was to be able to emphatically colour every inch of the map. So, he had to make a telling concession and copied (this was still before copyright law was enacted) from Smith's map that was published a few years before. Greenough, however, was still able to push his views that the fossil record was of little use and secure the stamp of approval from the Geological Society.

On receiving the Greenough map, Smith commented: 'This copy seemed like the ghost of my old map intruding on my business and retirement, and mocking me in the disappointments of a science with which I could scarcely be in temper. It was put out of sight.'[43]

Eventually, the burden of evidence proved the importance and value of Smith's fossil order, and Greenough had to accept his scepticism was misplaced. Smith's map had provided the key that unlocked our understanding of geological processes, but he was not given this recognition until much later in his life when, on 18 February 1831, he was the first ever recipient of the Wollaston Medal, which remains the highest honour the Geological Society can bestow.

Adam Sedgwick, the Society's president at the time, gave thanks to the man he referred to as 'The Father of English Geology':

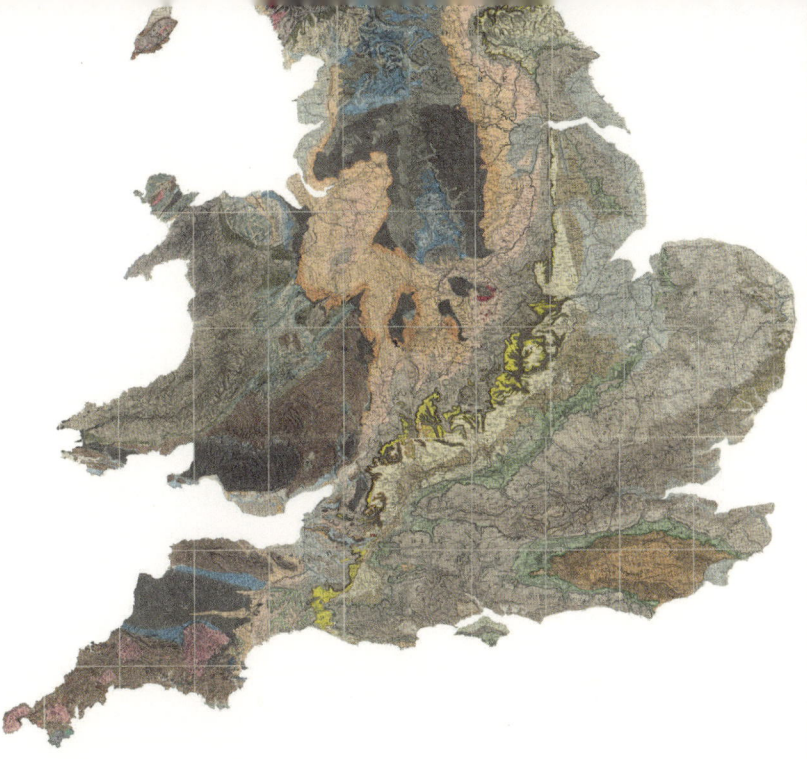

Map 26 [Extract] The second edition of Greenough's geological map is no less spectacular than his first attempt but included many refinements and additional details about the strata of England and Wales.

A Geological Map of England and Wales: The Geological Society (1839)
160 × 188cm

I for one can speak with gratitude of the practical lessons I have received from Mr Smith; it was by tracking his footsteps, with his maps in my hand, through Wiltshire and the neighbouring counties, where he had trodden nearly thirty years before, that I first learnt ... terms, which we derive from him our master, which have long become engrafted into the conventional language of England's geologists ...[44]

Despite this, Smith was never admitted as a member of the Society and it was not until twenty-four years after his death that his name was added to the Geological Society map alongside Greenough's, which was published in 1865 (Greenough had died a decade before). The impact of Smith on Greenough's thinking was perhaps more evident on the map of India I was so glad to discover in the Map Library, with its long list of fossils as well as careful listing of where information had been sourced from.

Two years after Smith was given his award, Greenough was

back as the Geological Society's president, and just as confident in his own views. For reasons he might have later regretted, he triggered a very public spat with a well-respected travel writer and illustrator named Lady Maria Callcott (better known as Maria Graham).

As with many women from her era, there is much about Callcott that has been overlooked, but a relevant characteristic here is that in her writing she'd often make the point that travellers and commentators should take proper notice of their surroundings. Callcott therefore considered herself an acute observer and built her reputation upon that. She also had substantial expertise in geology.[45]

So when, in 1822, Callcott was caught up in a major earthquake in Chile, she was well qualified first to witness and then to write about it. Given her interests in natural history, she submitted the observations to the Geological Society two years later and they were accepted to be read out on her behalf, since women could not be admitted as members. It would be another century before the Society came up with a more straightforward solution: admitting women.[46]

Callcott's observations were also published in the Society's journal,[47] making her the first woman to be so honoured. It would be a forty-year wait until the next, although, as we'll see, this is hardly surprising, given the hassle it generated for her! In her pioneering paper, Callcott (then under the surname Graham from her first marriage) states that the earthquake she witnessed had caused the land along the coast to rise out of the sea. It was an observation that at the time did not create too much of a stir. Then, in 1830, six years after its publication, it was cited by an eminent fellow of the Geological Society, Charles Lyell. He used it in his book *Principles of Geology*[48] to evidence the assertion that earthquakes could cause the elevation of land.

Greenough decided to feature in his 1834 presidential address Callcott's then decade-old account, using it as the basis to refute Lyell's theories about the changes earthquakes can make to raise the Earth's surface. Neither Lyell nor Callcott were present for this and, of course, Callcott could not have attended even if she wanted to. In his speech, Greenough attacked Callcott by doubting what she saw and, most damagingly, questioned her credibility. He implored the audience to seek a 'straight-forward description of what actually took place without the high colouring in which ignorance and terror and exaggeration are apt to indulge'.[49]

Callcott and her family were so incredulous that both her brother and husband offered to challenge Greenough to a duel. She declined their offers with a 'Be quiet, both of you! I am quite capable of fighting my own battles, and intend to do it!'[50]

Rather than have Greenough shot, Callcott's sublime retort – published in her pamphlet 'Letter to the President and Members of the Geological Society in answer to certain observations contained in Mr Greenough's Anniversary Address of 1834'[51] – was a masterclass in responding to a bungled character assassination. It was issued in the summer of 1834 and written in the third person, as, one assumes, it was to be read by a man on her behalf.

'Mrs Callcott is not prone to exaggeration' was the message substantiated over sixteen pages. The argument became international news, with copies of the exchange being sent to the great German explorer and scientist Alexander von Humboldt,[52] and Callcott's riposte being republished in full by the *American Journal of Science and Arts*.[53] Greenough's humiliation was complete when Callcott gained the endorsement of Darwin who, upon visiting the same area, noted: 'As any change of level, even in this neighbourhood, has often been disputed, I may add that I saw dead barnacles adhering to points of solid rock which were now so

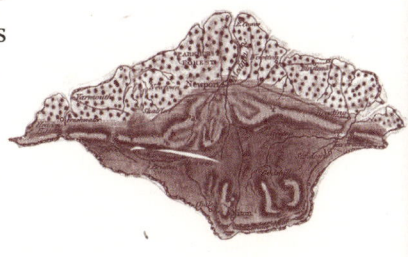

The Isle of Wight: a tiny part of Greenough's vast geological map.

much elevated, that even during gales of wind they would scarcely be wetted by the spray.'⁵⁴

So, by 1835, Greenough has made at least two very high-profile mistakes: fossils and earthquakes. Being proven wrong is often what good scholarship is about, but for all the early talk of impartiality, Greenough's behaviour towards Callcott, and perhaps to a lesser extent Smith, shows a human side to science, with all its flaws.

Greenough was known to be a 'habitual doubter of theories' and 'extremely reluctant to believe anything that was not capable of being proved'.⁵⁵ But it's clear that the burden of proof extended beyond the observation itself and to the credibility of the observer. He – with all his prejudices – would only be swayed by those he felt were worthy of being listened to and he had an almost visceral distrust of those he felt were wedded to bad ideas. A jotting in one of his notebooks reads: 'Men cling to erroneous theories when they know them to be so, as mothers have most fondness for rickety children.'⁵⁶

At the end of his life, Greenough was in reflective mood, writing an additional note on the manuscript for his *Preface to the Geological Map of England*, second edition:⁵⁷ '... in all works of much daring there must be much liability to error and those who march first are always exposed to the greater dangers, the map of England was in my hands the work of a debutant who had no previous experience ...'

Smith remarked that 'the theory of geology is in the possession of one class of men, the practice in another'.⁵⁸ After getting to know Greenough as best I could from his maps, there is no doubt he suffered the instinctive reactions of a man who spent his entire life immersed within the rigid elites of Georgian and then Victorian Britain. His lack of published work made it hard for him to be rediscovered and there were many other contemporaries, such as

Darwin, who eclipsed him. Darwin himself described Greenough as 'most obstinate but most good humoured'.[59]

I rate him based on what he started, rather than what he finished. As an independently wealthy man he could have led any life he wanted, but he chose to inject energy and organisation into the developing fields of science. He was also a man who must have been immensely conflicted by standing in direct opposition to the world that generated his vast wealth. His grandfather was mocked[60] for selling quack cures founded on no evidence, which must have seemed like ill-gotten gains for Greenough. He was a man of evidence and sought to brush aside peddlers of pseudoscience.

He was also an establishment figure keen to spread educational opportunity (in part to counter misinformation), at a time of minimal state provision, and when the idea of educating 'the masses' was subject to obstruction from many of his class. Core to Greenough's philosophy was the importance of spreading geographic literacy through use of maps to convey useful information. As part of his commitment to this he joined a group that embarked on what would become one of the most successful atlas projects of the nineteenth century. As we shall see, it was an endeavour that brought maps to the greater public and laid the foundations for their widespread use.

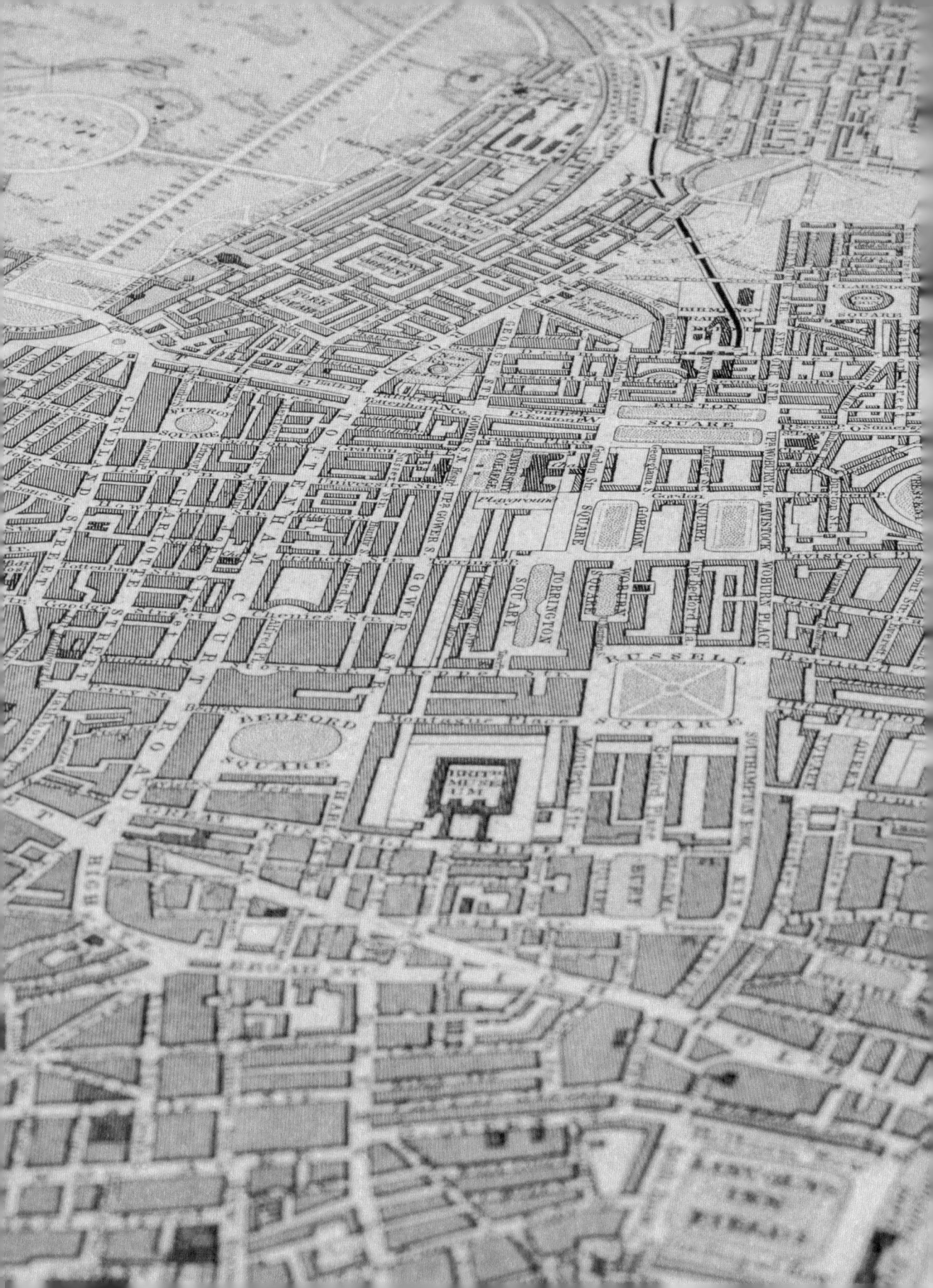

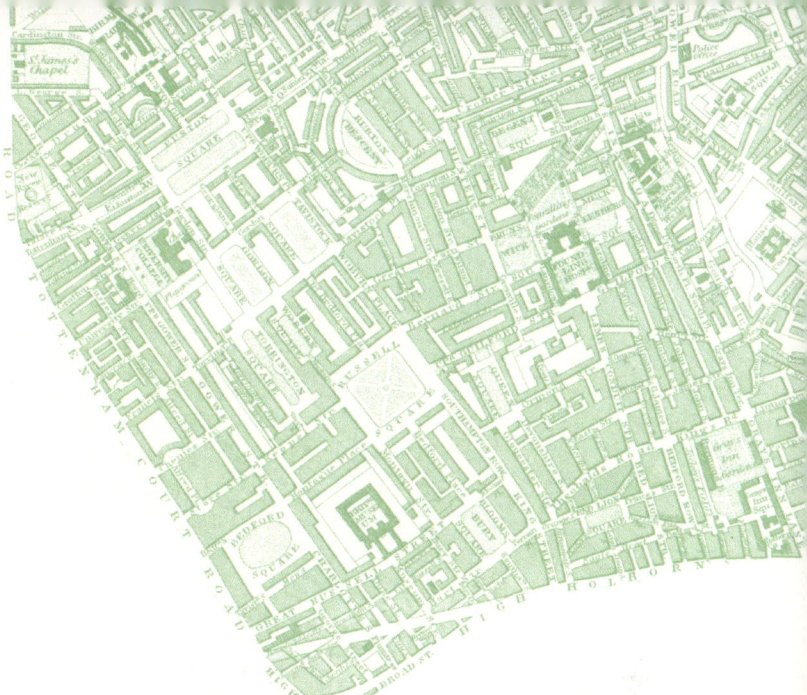

CHAPTER FOUR
Bloomsbury

'a well-drawn map is art concealing art'
Geographical Review[1]

The likes of George Bellas Greenough, and the society gents with whom he was friends, weren't just born into a world of material wealth, they were gifted the opportunity to have a broad education. Their families could pay to access schools at a time when there was no state provision: they could be financially supported through university degrees and take their 'grand tours' to experience a world far beyond the shores of Britain. And, of course, their homes would be well stocked with books (and maps) and they would have had the disposable income to grow collections of whatever trinkets took their fancy – from the finest art to the rarest fossils. In 1820s Britain the literacy rate for the general population sat at around 50 per cent, where it had been for 200 years, and there was very little incentive for the educated classes to change things.[2]

Map 27 [Extract] The neighbourhood of Bloomsbury as it was in the 1840s. Home to the Map Library today, it's a place that has long been at the heart of London's intellectual life.

London 1843: Society for the Diffusion of Useful Knowledge (1843)
63.5 x 38.5cm

However, this was a period of profound societal change, which would dramatically alter the status quo. With the Industrial Revolution going at – literally – full steam, new jobs emerged that required immense skill and knowledge, but in domains that would have been utterly unfamiliar to the Greenoughs of this world. I would not have rated their chances of servicing a steam engine or rethreading an industrial loom, for example. Other, less manual, professions emerged, too, as the demand for the bureaucracy of commerce and government grew with the population and the economy. Despite many working and living in appalling conditions, these new kinds of workers had more free time and an appetite to spend some of it bettering themselves by learning new things beyond their on-the-job skills.

Civil unrest was also bubbling up as workers were campaigning for fairer terms and more democratic representation by way of voting rights that they were being conspicuously denied. Some in the establishment feared that providing widespread education would pour fuel on the fire by equipping the masses with ideas well above their station, and, worse, encourage them to question their lot within the societal structures that had worked so well for the few. But there were some in the establishment who had formed the view that education was crucial for a functioning and successful democracy. Their leader was a prominent British statesman, lawyer and reformer named Henry Brougham, and they based themselves around the neighbourhood that is now home to the Map Library.

In 1825, Brougham made an impassioned case for the expansion of educational opportunity in a thirty-two-page pamphlet entitled *Practical Observations Upon the Education of the People Addressed to the Working Classes and their Employers*.[3] He rounds off his argument by saying:

I rejoice to think that it is not necessary to close these observations by combating objections to the diffusion of science among the working classes, arising from considerations of a political nature. Happily the time is past and gone when bigots could persuade mankind that the lights of philosophy were to be extinguished as dangerous to religion; and when tyrants could proscribe the instructors of the people as enemies to their power.[4]

And ends by first addressing the 'Upper Classes', and then the 'Working Classes':

To the Upper Classes of society, then, I would say, that the question no longer is whether or not the people shall be instructed – for that has been determined long ago, and the decision is irreversible – but whether they shall be well or ill taught – half informed or as thoroughly as their circumstances permit or their wants require ... To the Working Classes I would say, that this is the time when by a great effort they may secure for ever the inestimable blessing of knowledge.[5]

Brougham and his associates coalesced in a relatively quiet district in London, known as Bloomsbury. In her history of the neighbourhood, Rosemary Ashton sets out just how extraordinary the intellectuals of Bloomsbury were:

Motivated individuals and groups colonised the ever-developing area for their pioneering schemes: higher education for non-Anglicans and for women, education and self-help for working men and women, organised play and learning for poor and disabled children, the kindergarten system of pre-school training, schools of art and design for both men and women, progressive medical schools and specialist hospitals.[6]

Walk round Bloomsbury today and you can't miss this history – there are universities, prestigious research institutes, publishing houses, bookshops and, on its southern edge, the grandeur of the British Museum. But, unless you've spent a lot of time in London, the district may be most associated with the Bloomsbury Group of writers, intellectuals and artists who congregated there at the start of the twentieth century and counted Virginia Woolf and John Maynard Keynes among their number.[7]

Illustrious though that group are, they followed in the footsteps of Brougham and his associates who had progressive politics and educational opportunities for those outside society's elite on their agenda. An important development was the founding, in 1826, of the Map Library's home: University College London (UCL).[8] UCL opened as the only secular university in England, much to the ire of members of the conservative establishment at the time, who branded it a 'radical infidel college'. UCL was able to broaden access to university to those who cared little for religious doctrine and who had a more progressive political outlook, but it remained a relatively exclusive place of learning as fees were still charged, and you needed to have had a robust education prior to arrival.

Knowledge therefore had to spread or, as Brougham put it in his pamphlet, 'diffuse' to anyone who sought it without the inhibitions of cost. Thus, in 1826, the 'Society for the Diffusion of Useful Knowledge' (often abbreviated to SDUK) was born. Many of the same characters who founded the university took committee positions, as well as Greenough, who must have found the opportunity to join yet another society too good to miss.

The SDUK grew into a juggernaut that connected some of

the greatest and most innovative thinkers of the first half of the nineteenth century with 'the masses', in what was a very fertile time for knowledge creation. This connection was principally through a vast output of high-quality publications that included a *Penny Magazine*, *Penny Cyclopaedia* and *Library of Entertaining Knowledge*.

Some in the establishment responded with ridicule, describing the SDUK as the 'Steam Intellect Society',[9] in reference to the industries that many of SDUK's readers worked in. As the movement's leading figure, Brougham (pronounced 'broom') attracted the derision of satirists, who drew him as a literal broom, cleaning as he strode along his 'march of the intellect'. There were many such depictions, but my favourite from the time was sketched by Robert Seymour in 1828: he drew a steam-powered robot with a head made of books and wearing UCL as its crown. Brougham's head is on a broom sweeping away vicars, rectors, quack doctors and their accoutrements of obsolete laws and irrelevant acts. It was a steampunk vision of what the 'steam intellects' were up to, and one they leaned into with relish.

A satirical print by Robert Seymour. It shows a huge automaton representing the new London University (University College London) trampling over greedy clerics, doctors and lawyers.

Published c. 1828 [Extract]

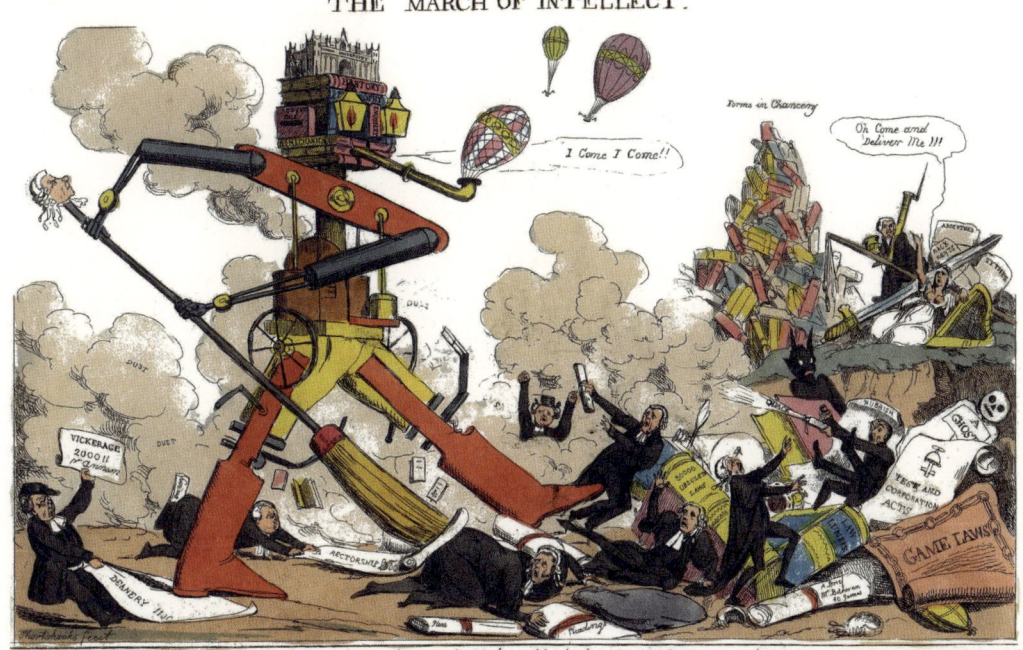

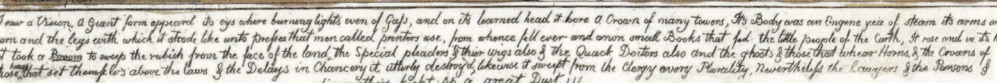

Representatives from the SDUK toured the country railing against a corrupt press and expounding the idea that, the better-informed citizens were, the less likely they were to be taken for fools by bad leaders. When he gave a talk to the members of the Windsor Public Library, Charles Knight, who published many of the SDUK's works, warned against the salacious press with 'its too commonly violent opinions, and vehement exhortations' and proffered the antidote of accessible and high-quality general knowledge which enables 'a man to reason accurately and honestly for himself', because 'the best-informed men are the least moved by the political declaimer'.[10]

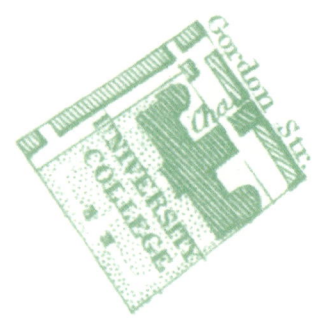

Remarkably, I can still meet one of the men who was there at the beginning, when Greenough was making his maps, and who influenced him and the likes of Brougham to set off on the March of the Intellect.[11] It was the philosopher Jeremy Bentham who popularised the phrase 'knowledge is power'. His soul may have long departed, but the physical form of Bentham – known as his 'auto-icon' – resides in the UCL Student Centre, a building I walk through on my way to the Map Library. Bentham is seated on a stool, with his favourite walking stick in hand, and wearing a hat on a waxwork replica of his head. He stares out of his glass box adjacent to one of the university's busiest thoroughfares offering an occasional greeting, when he's not burdened with students and visitors taking selfies.

I'm not superstitious, but I do feel as if I had a helping hand from the spirit of Bentham, or maybe a lurking ghost of Greenough, in my explorations of the Map Library, imploring me to find out more about both Greenough's lost maps and the events that led to the SDUK and the university I now work at.[12] In my early days rifling through piles of maps, I was completely ignorant of these characters and would not have been able to distinguish them from the other maps of a similar age and yet I kept being drawn to them.

BLOOMSBURY

Once I had identified them, then more so than any maps in the library I felt I owed it to their makers to tell their extraordinary story, not least because they are some of the most influential yet under-appreciated maps of the nineteenth century. They were published at a time of growing self-improvement and widespread education, which resulted in maps becoming a valued part of everyday life, in much the same way as they remain today.

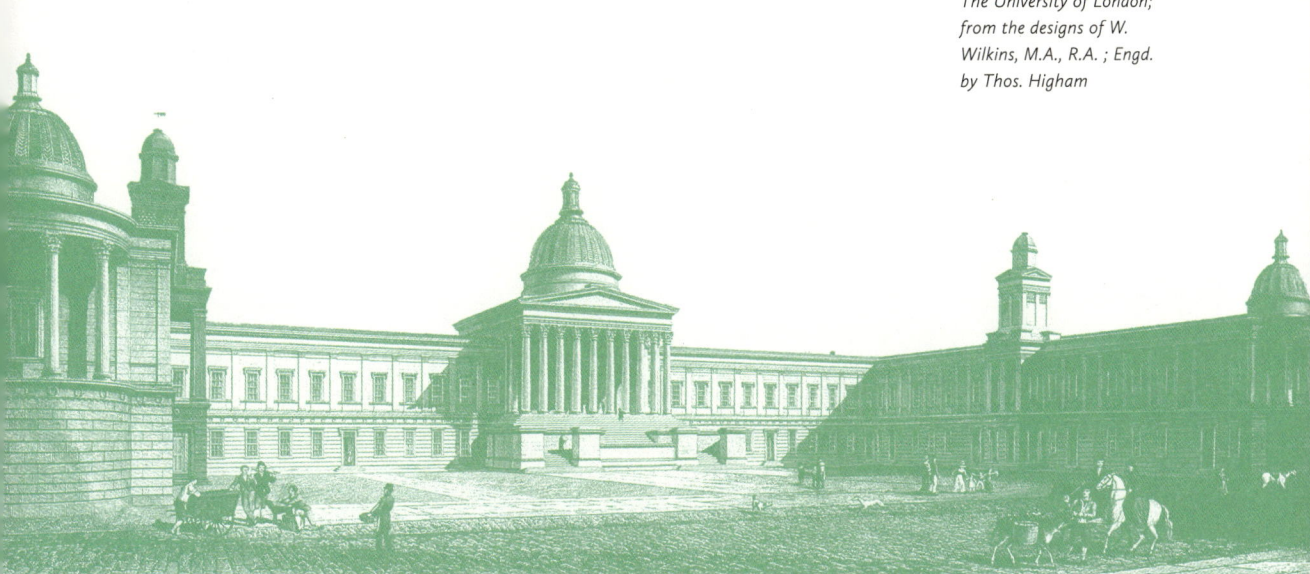

An etching of UCL based on the plans of its architect William Wilkins, who also designed Britain's National Gallery.

The University of London; from the designs of W. Wilkins, M.A., R.A. ; Engd. by Thos. Higham

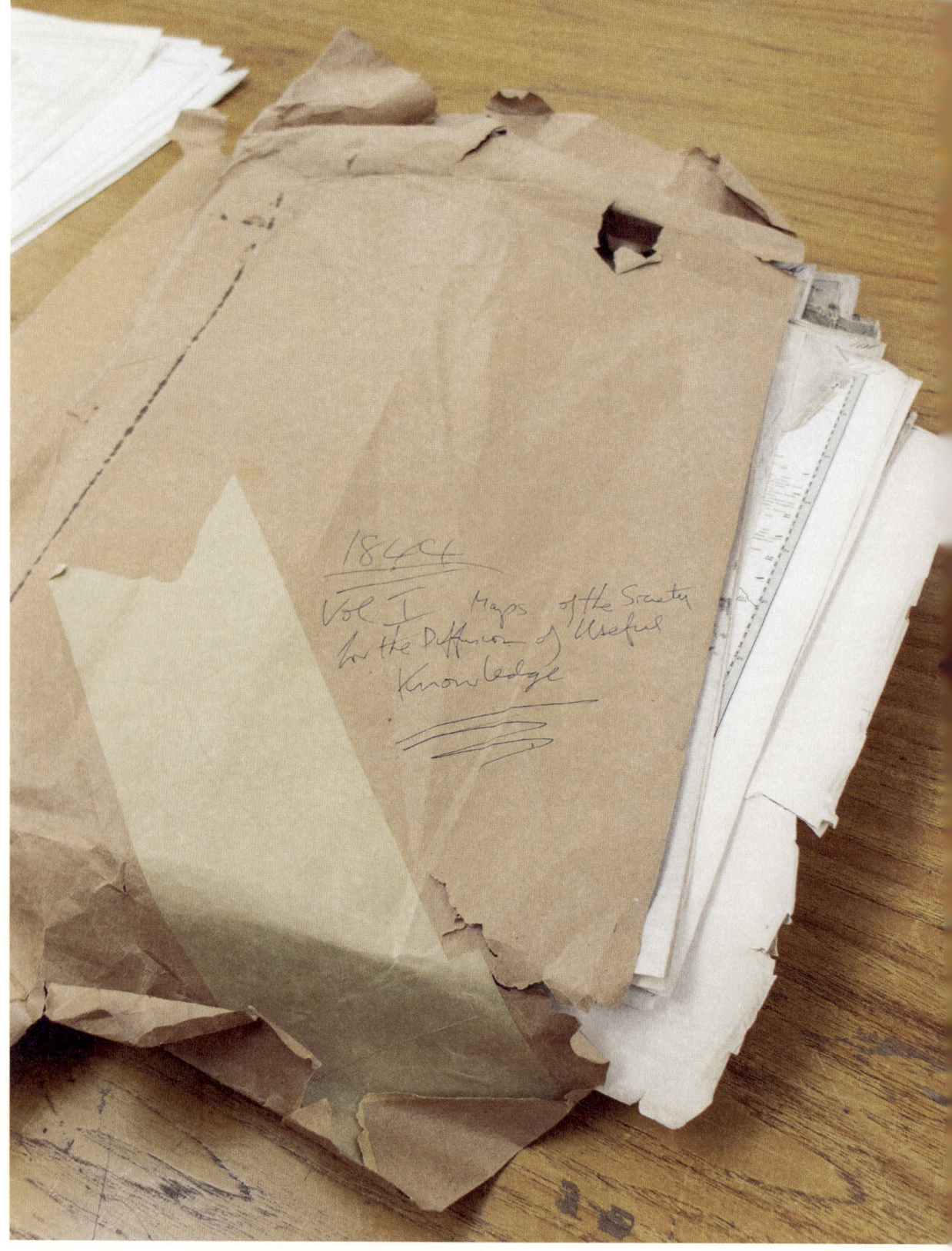

Maps to March With

If you want to set off on a long march, a map is a helpful thing to guide you, and the Society for the Diffusion of Useful Knowledge were ahead of their time in declaring to their vast audience that geographic knowledge was *useful* knowledge. In support of this they created a mass of written material that had some kind of geographic component, encompassing subjects like history, natural sciences and economics. Maps were therefore considered a necessary visual aid to accompany the written articles, particularly the Society's *Library of Useful Knowledge*, the first volume of which was published in 1827.[13]

I found a pile of the maps that SDUK produced stuffed at the back of a cabinet. It appeared as a mass of brown, crumpled paper and my first impression was that it was a nest constructed by the mice I would see in the evenings, seeking out the crumbs from the crisp packets I'd leave in the bin.[14] Liberated from its dark corner and blown free of its dust, I could see written on the wrapping was:

> 1844 Vol I Maps of the Society for the Diffusion of Useful Knowledge

Poking out from the bundle was the disintegrating shell of what seemed to be a very old book. It was shedding small pieces of yellowed and brittle paper at an alarming rate, creating drifts of sharp-edged confetti around my feet. On the top of the pile was a sheet that looked like an old treasure map with its frayed edges,

This is the photo I took of the subscription atlas from the SDUK not long after it was pulled from the back of a cupboard. It was a dishevelled pile of loose sheets, with an unpromising exterior of brown paper that disguised the quality of the maps inside.

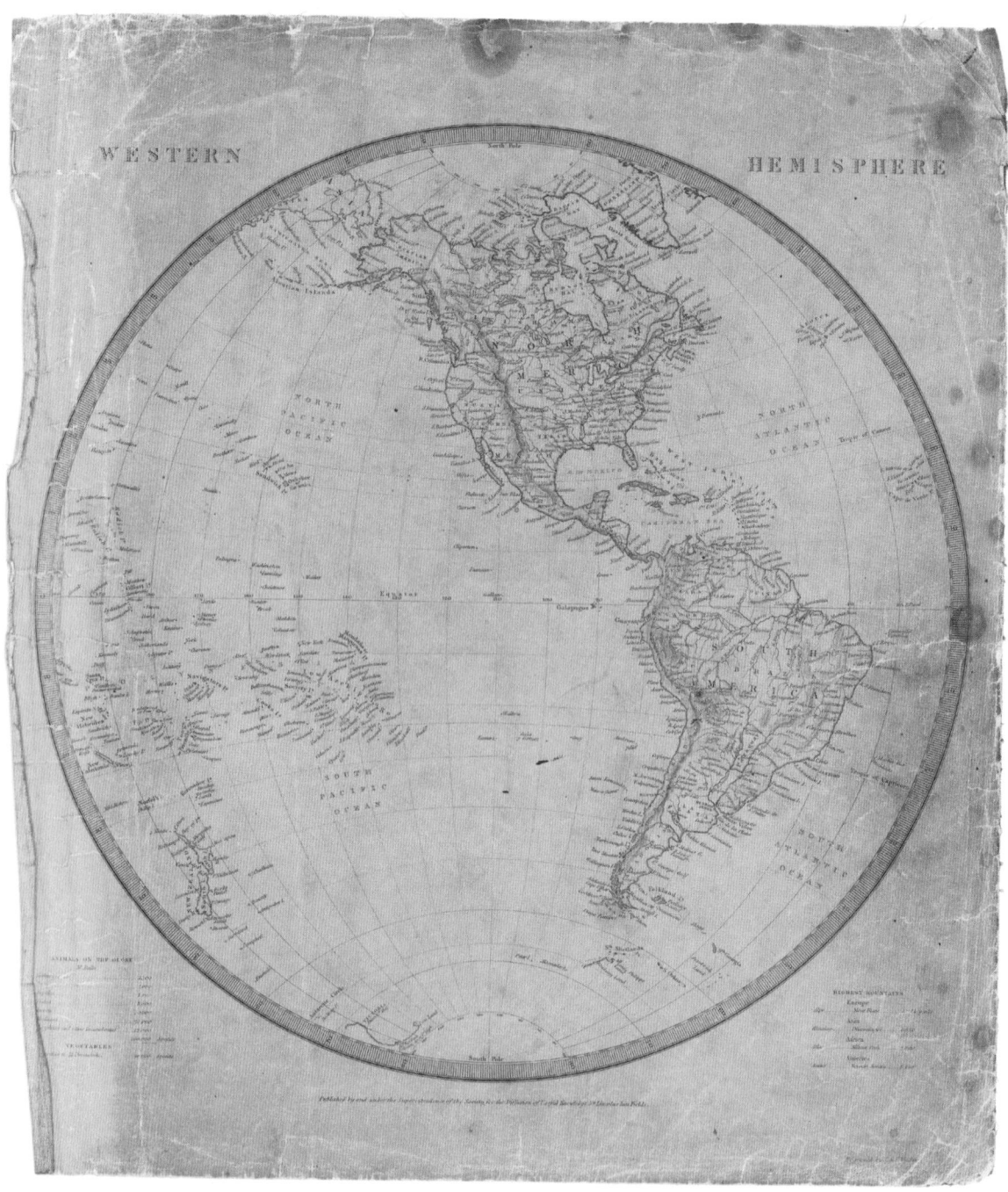

Map 28 + 29 Maps 1 & 2 of Volume 1 of the SDUK atlas. Mounted on linen, with hand-coloured borders, these show each hemisphere and have tabulated figures for the surface area of each continent, the number of inhabitants and the number of followers of major religions.

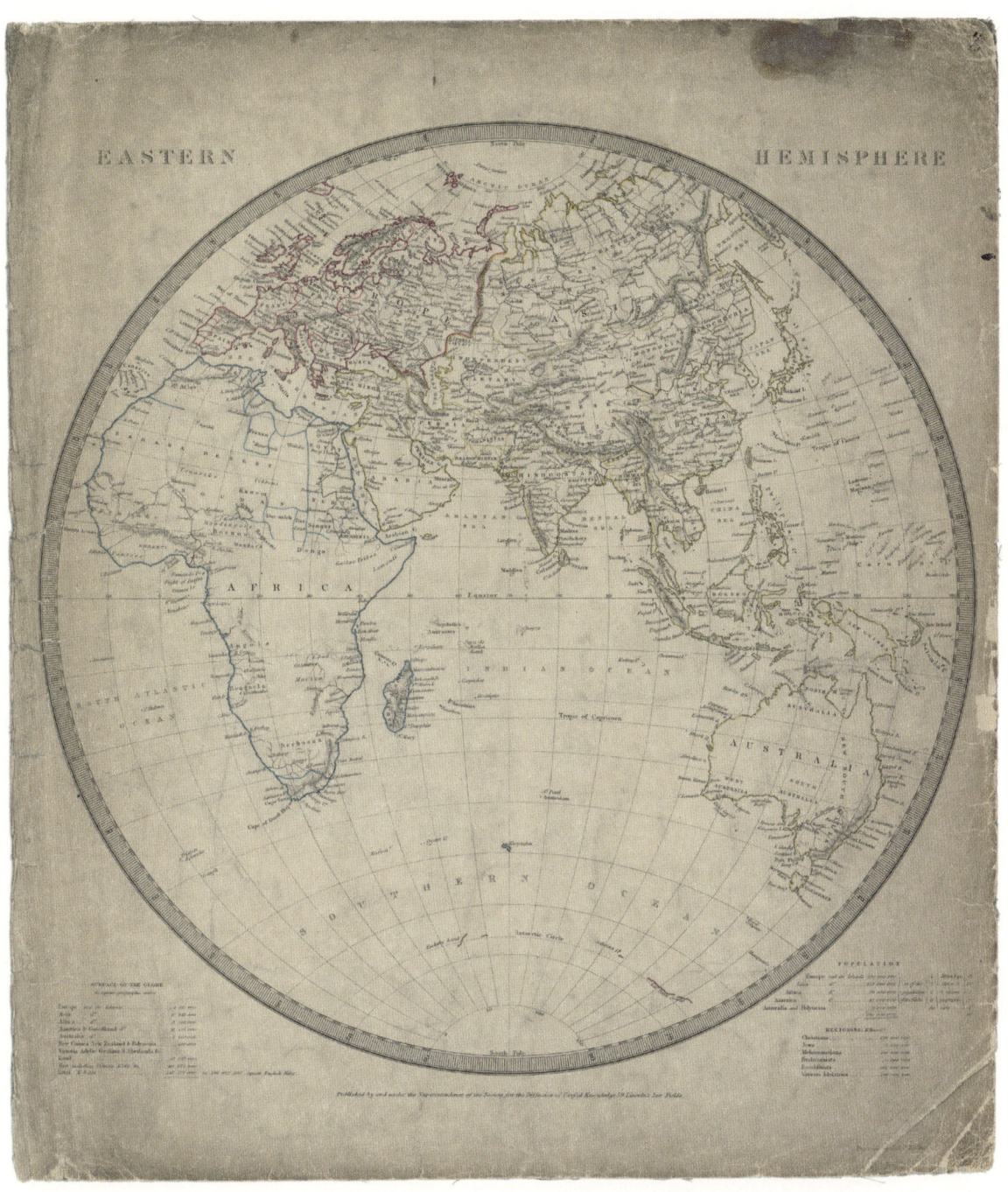

Eastern Hemisphere and Western Hemisphere: Society for the Diffusion of Useful Knowledge (1844)
Each map: 32.5 × 33cm

Map 30 [Extract] Map 180 in the SDUK series depicts the jagged ramparts of Geneva. It was published in 1841 and engraved by a B.R. Davies of Euston Square, which – pleasingly – is an address that was only a few metres from my office at UCL.

Geneva: Society for the Diffusion of Useful Knowledge (1841)
37.5 × 28.5cm

oily blotches and a layer of grime so thick it rendered the beautiful outline of the 'Western Hemisphere' almost illegible.

To the gentle patter of paper shards hitting the table from the disintegrating package, I gritted my teeth and turned the map over to see how the next in the pile had fared. This was the partner of the first map: the Eastern Hemisphere struggling to be seen against the gloom of two centuries of accumulated London filth. The next two maps showed the world using a different map projection (Mercator's) followed by a jump in the number on the top right corner from 5 to 12 (suggesting maps were missing) to a map of Europe and then another jump from 13 to 31 to a zoom in on 'The Netherlands and Belgium'.

I pressed on, despairing as the pages continued to fracture and crackle. But there were reasons to be optimistic: with every turn the loose map sheets, backed on linen, were becoming cleaner and brighter.

It was as if the sun was emerging from behind dark clouds with each new page. By number 58 ('Italy, General, with Sicily, Malta, Sardinia and Corsica') the clouds had completely evaporated, and the maps soon appeared in almost mint condition. They seemed to get more and more beautiful, reaching a climax in their magnificence and quality towards the bottom of the pile, which comprised a selection of city maps.

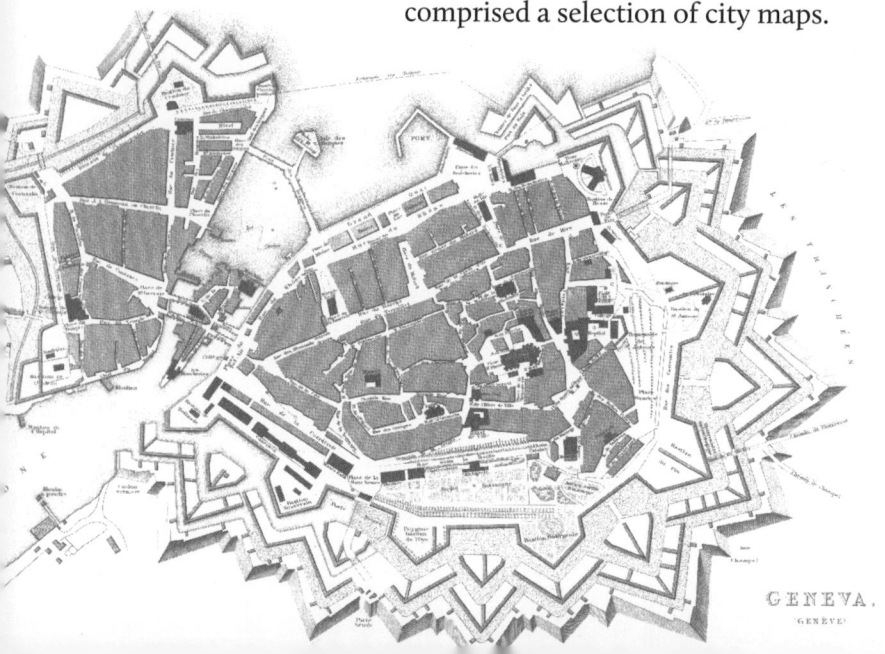

From Geneva (Switzerland) to Moscow (Russia), Calcutta (now Kolkata, India) to Constantinople (now Istanbul, Türkiye (Turkey)), each one was depicted in the same detail and accompanied by exquisite etchings of its most impressive architecture. These cities were frozen in time, just at the moment before their populations would outgrow their defensive lines and spill out over their walls. Had the maps been made a decade or two later many would have become more amoeba-like without the abrupt delineations of their walls.

New York is rotated so that north is to the right, but it is still recognisable from its distinctive grid of streets, many of which have been laid out but not yet filled in with buildings. Kolkata (rotated with north to the left) by comparison appears a very different city from the others in terms of its urban form, its fort appearing like a rosette buffered from the encroaching city by a large park and race ground (still there today), with depictions of both the people and the animals that inhabit the city.

Map 31 [Extract]
Published in November 1842, the SDUK's choice to include a map of Calcutta (now Kolkata) reflected Britain's growing colonial interests in India.

Calcutta: Society for the Diffusion of Useful Knowledge (1842)
40.5 × 31cm

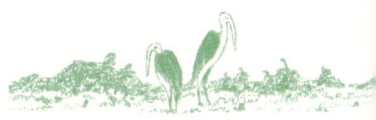

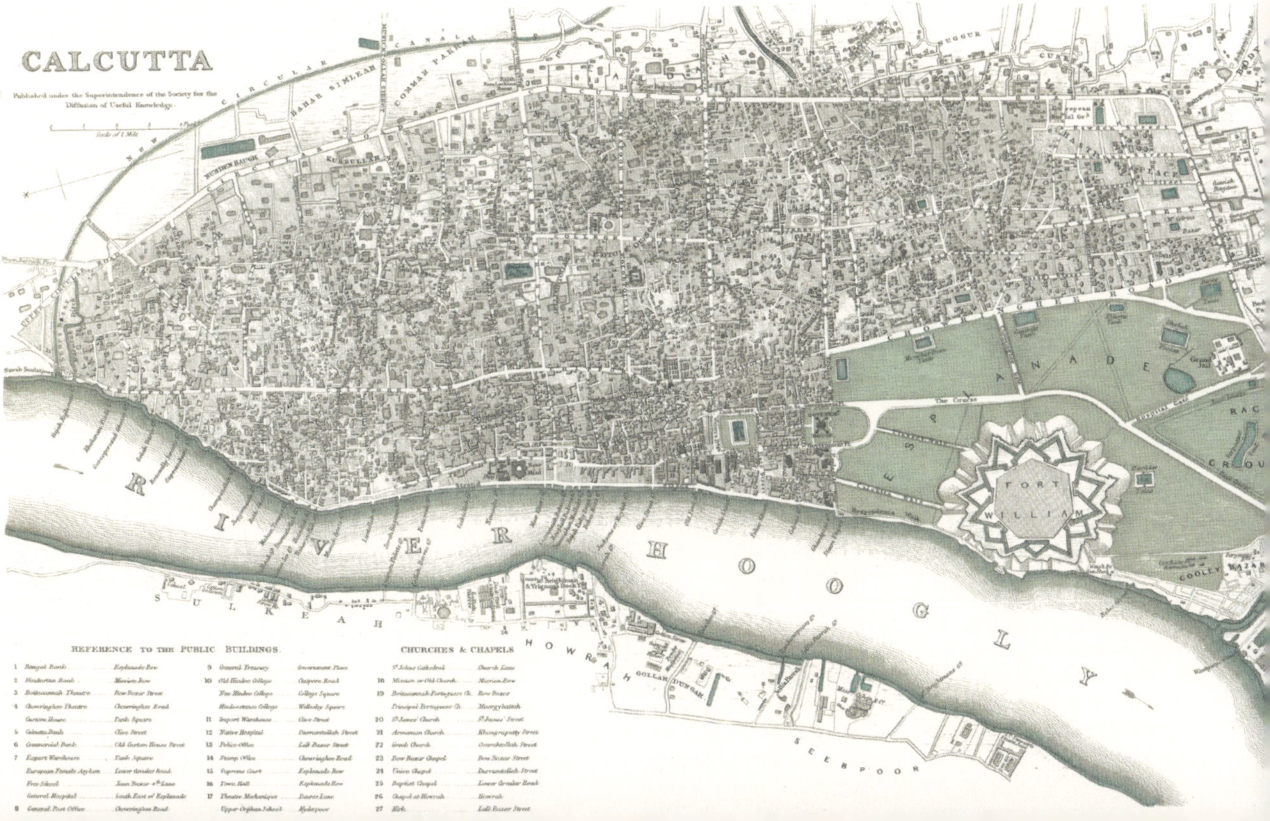

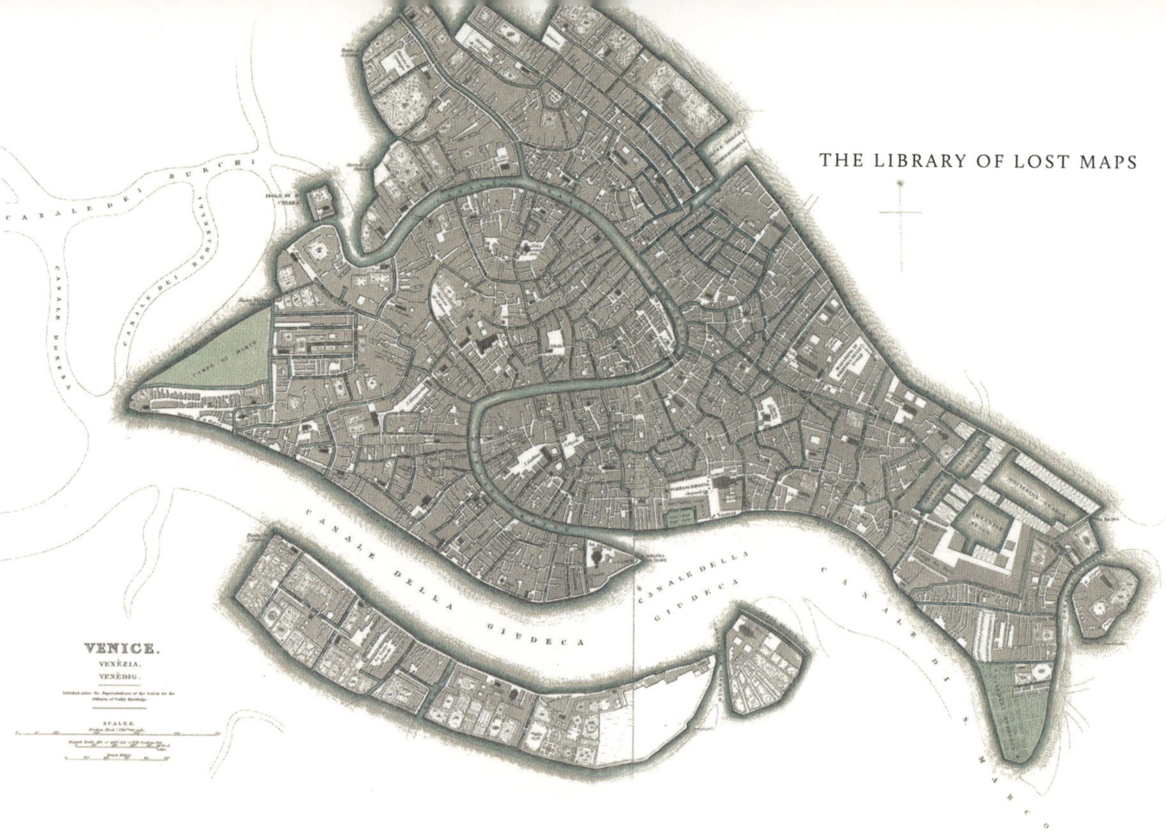

THE LIBRARY OF LOST MAPS

Map 32 [Extract] The SDUK published this glorious map of Venice in 1838. It is accompanied by delicate etchings of the city's famous architecture along its bottom edge.

Venice: Society for the Diffusion of Useful Knowledge (1838)
40.5 × 31cm

Goethe, the great German literary figure, remarked that Venice can only be compared to itself, and after seeing the city map it's hard to disagree. The Grand Canal sweeps through the centre of a city that appears like a giant iceberg breaking up as it drifts into the Adriatic. I thought back to George Bellas Greenough's travel diaries and I could see him stepping off the gondola, notebook in hand, and into the throng of well-dressed Venetians.

After getting lost in the streets of these enticing cities, I took a step back and pondered the pile of maps as a whole. Despite the unpromising start wrapped in the tatty brown paper, I was startled that the final tally had reached 190 across two volumes, a number confirmed by a table of contents sheet at the bottom, where each of the maps had been ticked off by whoever assembled the collection (thirty-four or so were missing). As I read down this list, my mind jumped from one place to the next, but as it did so it occurred to me that this atlas was not doing the one thing you'd expect it to help with, which is to make it easy to find a place on the map.

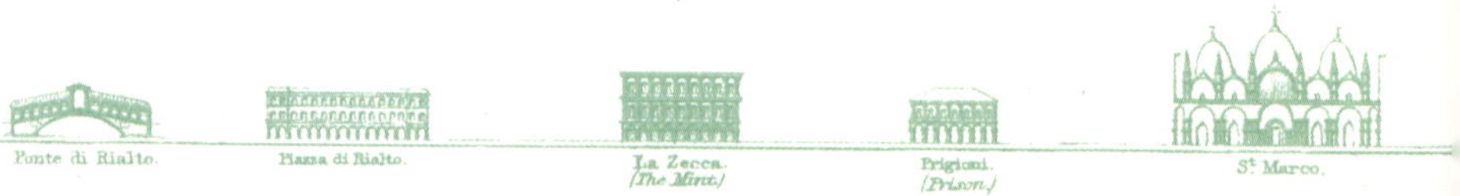

BLOOMSBURY

If you know the name of the place you are looking for, you should be able to reach for an atlas and turn to the index. To find Timbuktu in Mali[15] you'd scan down the places in alphabetical order and it wouldn't matter that the neighbours in the list are Timaru (New Zealand) and Timiskaming (in Quebec), which are thousands of kilometres in opposite directions. You just care about the page reference and then the grid reference to guide you to the precise location on the map you have turned to.

If, however, you're planning a road trip across Europe you might expect to start with Spain and then, as you turn the pages, track from one country to the next in a geographical order. Alphabetising the order of the maps makes no sense, for you'd find yourself jumping from Spain to Sri Lanka over 8,000 km away, rather than to the neighbouring countries of France or Portugal.

It struck me that the Society's atlas was an odd mix of alphabetical maps – so turning the pages you'd jump from Birmingham (UK) to Bordeaux (France) and then Boston (USA) – and a system of country groupings that made slightly more sense, for example, the 'Islands in the Atlantic' providing a stepping stone from 'Africa' to 'America, North'. But India is covered by twelve maps (which were the maps used as the basis of Greenough's geological map of India), while only two depict all of China. What's more, the list of cities seemed slightly eclectic. I wondered why, for example, there was a map of Syracuse in the USA but not one of Washington. I also pondered why the maps were loose in this way. And why was there no detailed index or system used to find the places? The answers to these questions, as we shall see, revealed themselves when I researched how the atlas was created.

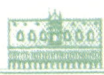
The Palace.

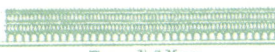
Piazza di S. Marco.

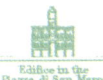
Edifice in the
Piazza di San Marco.

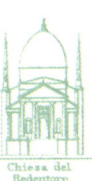
Chiesa del
Redentore.

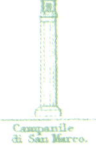
Campanile
di San Marco.

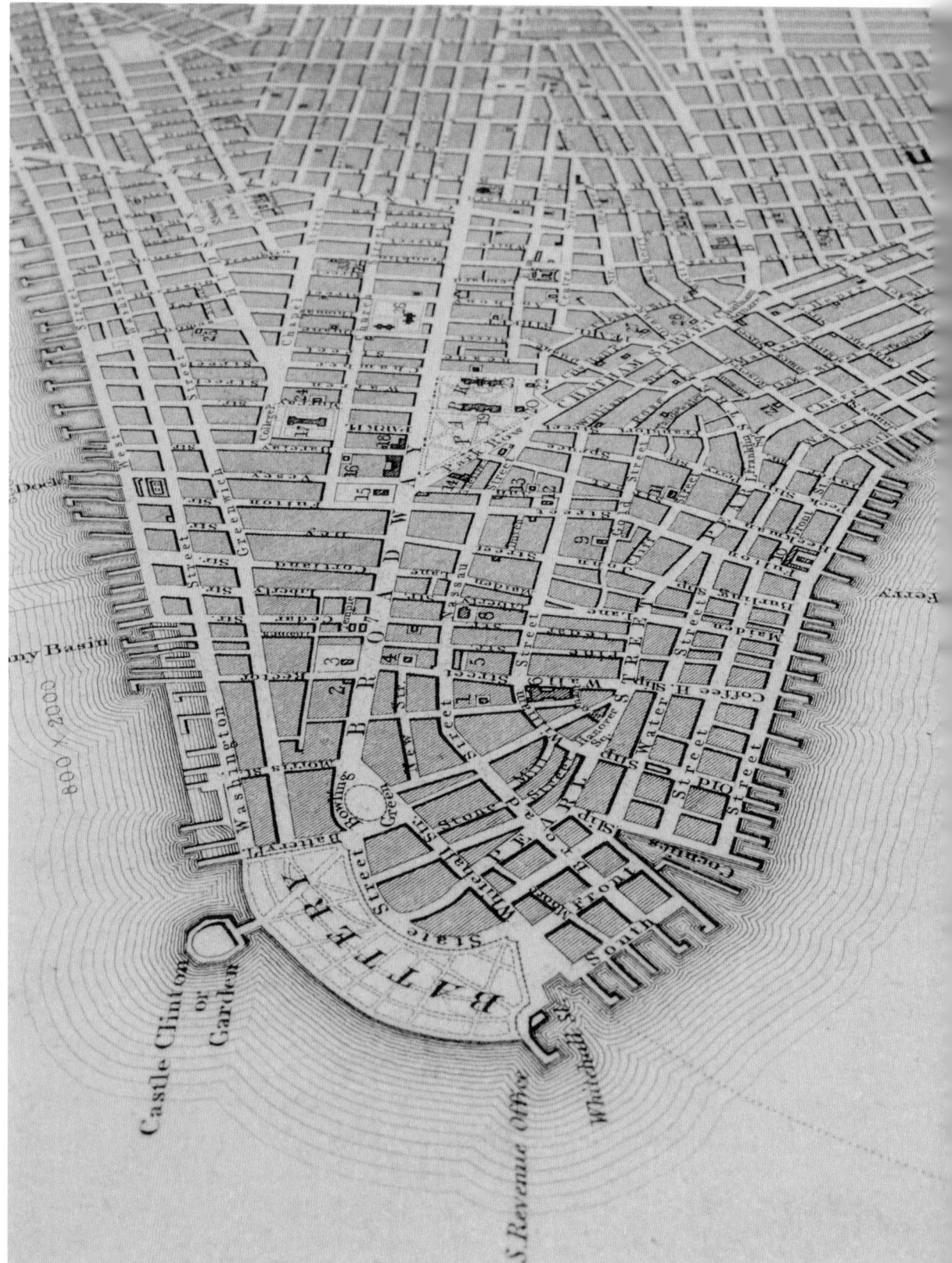

An Idiosyncratic Atlas

For much of the first half of the nineteenth century, maps were a luxury well beyond the reach of most readers. They were expensive to produce because of the labour required to print them, not least with the manually operated presses and the need for hand-colouring. What's more they appeared increasingly out of date in the light of the rapid developments of the age, and the market for them remained relatively small, reduced by both cost and the illiteracy of the population. Things began to change thanks to steam-powered printing presses, an increasingly educated population (with some disposable income) and the funding of comprehensive national surveys to keep the maps up to date.[16]

The Society for the Diffusion of Useful Knowledge became a major contributor to these developments by setting themselves the seemingly impossible task of creating maps of the highest quality at the lowest price. To keep the upfront cost low they adopted a monthly subscription model, where readers would be sent a couple of maps at a time. To ensure quality, a top-class group (which included George Bellas Greenough) was assembled to serve on its Map Committee. The driving force behind the project was Francis Beaufort, an accomplished surveyor and cartographer, as well as hydrographer to the navy, which gave him access to lots of additional sources of information. Sailors will know him as the originator of the 'Beaufort Scale', which is used to describe wind strength. He was unwavering in his commitment to the project, providing important consistency as committee members

Map 33 [Extract] The SDUK map of New York shows that the first streets on the tip of Manhattan developed the more organic layout of European cities. It was the Commissioners' Plan of 1811 that established the gridiron layout of the streets that is so famous today across the rest of Manhattan and that was bedding in when this map was published.

New York: Society for the Diffusion of Useful Knowledge (1838/1840) 37 × 30cm

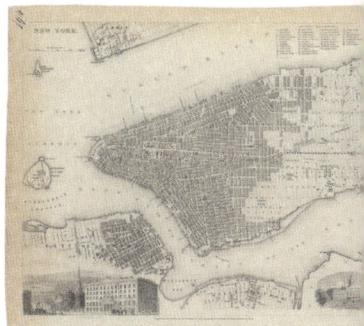

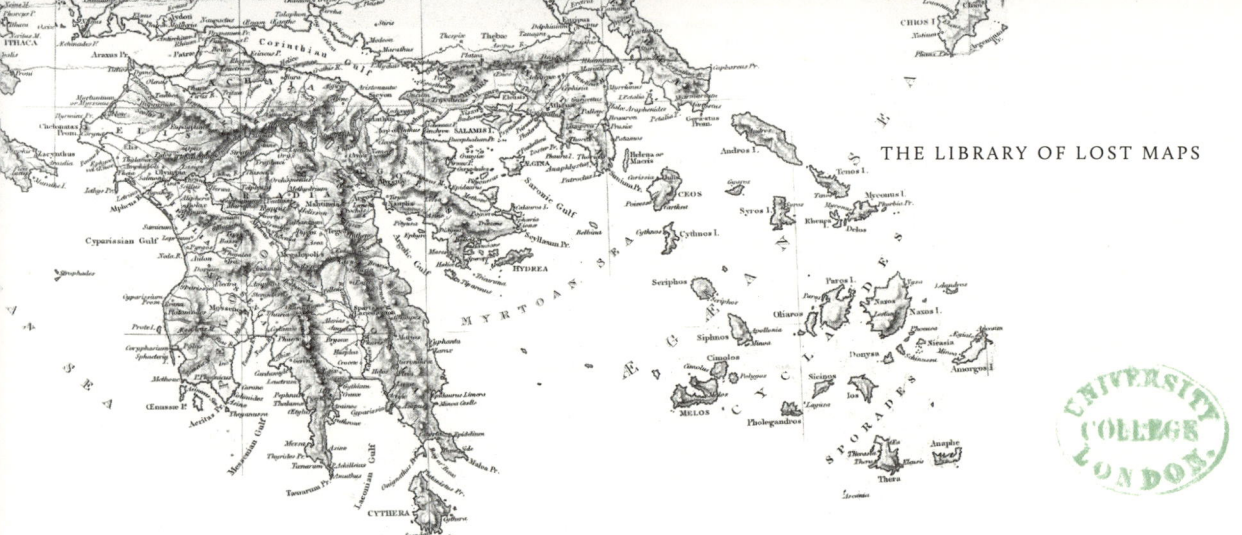

Map 34 [Extract] This map of Ancient Greece was the first the SDUK published (on 1 September 1829). The Society encouraged its subscribers to use maps for 'exercise of the intellect' by combining history as well as geography and to be observant of 'the changes that have occurred in a country or countries [a map] delineates. The gradual change from wild wastes to densely populated districts; the growth of towns; the spread of civilisation; the march of conquest or of commerce.' 3

Ancient Greece Southern Part: Society for the Diffusion of Useful Knowledge (1829)
37 × 30cm

drifted in and out. Attendance at their meetings was variable – to say the least – with the minutes for 9 June 1831 stating simply: 'The Committee was summoned for this day but Capt. Beaufort alone attended.'[17]

Beaufort accepted a low payment as he viewed the maps more as a leisure pursuit, despite it being reported that he worked on them between 5 and 6 a.m. every day. In a letter[18] to Henry Bellenden Kerr, who had the idea for the atlas and had shrewdly requested Beaufort's assistance with it, Beaufort sketched out his ambition:

> It will convey a sufficiently clear Idea of the general distribution of land and water and we must have recourse to 4 or 5 Maps of the great divisions of the world in order to shew the relative situations of the different Countries. Of these there cannot be well fewer than 16, which, with the former 4 or 5 and the projection of the Globe in separate Hemispheres will swell our Atlas to 22 or 23 plates. Even such an atlas would be a prodigious acquisition to a very large portion of the Community, but might not we attempt still higher game ... [and aspire to having a] copy find its way into every house in the Empire & would not this be diffusing real tangible knowledge?[19]

To limit the price, the Map Committee had to work hard to drive production costs down, but were under instructions not to compromise on the printing quality as it was deemed 'inexpedient to sacrifice any of the beauty of the Maps'.[20] For example, the decision was taken to produce the maps from steel printing plates, which were more expensive and more challenging for engravers to work with. But they were also hard-wearing and could therefore be used for the huge print runs anticipated, as well as offering superior detail.

Ultimately, the Society realised that most of the expense in map production occurs before the first sheet is printed. Considerable amounts of time (and therefore money) were needed to compile information, draft, edit and redraft the maps and then engrave them onto the printing plates. So they reasoned that if they could generate a higher volume of sales at a lower unit cost, they would still recoup these costs, even with a smaller profit margin on each sale.

In September 1829 the first two maps rolled off the printing presses, showing 'Ancient Greece, the Southern Part' and 'Part of Turkey, containing Southern Greece and Candia'.

They came loosely stitched into a thin paper wrapper, which set out the ambition of the SDUK atlas to subscribers:

> Moderate in size, yet capable of distinctly showing every place of interest; of unexampled cheapness, yet finished in the best manner; and the accuracy of which may safely be relied upon, from the arrangements made for their composition and execution. They will be engraved on Steel; the size will be 11 inches by 14 [28 × 36cm]; and Two of them will be delivered in a Wrapper for One Shilling, or with the Outlines coloured for One Shilling and Sixpence.

> The Series will consist of at least Fifty Plates; and a Number will appear at intervals of Two Months, or more frequently, if they can be more speedily completed.[21]

The promise of 'at least' fifty maps was more than met, the final tally comfortably exceeding four times that number at 224 (and even included six maps of the stars). This extraordinary total, however, was accomplished in fits and starts and would only have been enjoyed by the most patient (and perhaps younger) subscribers. To be the first to get each map in the complete series you would have needed to sign up in 1829 and paid your dues until 1844. It was a fifteen-year commitment that was not to be undertaken lightly, and concerns about not surviving long enough to enjoy a complete atlas were set out in the letters of complaint.

For example, a Mr Cassin of Liverpool in December 1842[22] pleaded: 'I have now above 90 Nos. by me, the accumulation of years & in their present state absolutely useless – my patience is nearly exhausted, in fact I shall grow old waiting for the completion of a work which I once expected would have been useful in my younger days, & now I have the prospect of bequeathing it to the next generation, & am not sure of its being finished in their days.'

The policy of sending maps in the order in which they were ready became especially eccentric for the atlas's structure and completeness because the pairs of maps at the moment of issue did not even relate to one another. Subscribers would be left frustrated waiting to complete countries, let alone continents.

The maps were dispatched at wildly different rates, too. In 1833 subscribers enjoyed twenty-six maps (issued in thirteen instalments), but only six maps (three instalments) in 1839. There was also little explanation provided about why the maps were chosen, generating further complaints, even about the wondrous city maps.[23]

The SDUK's maps would be sent out in pairs, loosely bound into paper wrappers, but, to the annoyance of subscribers, the pairings were rarely related to one another. In this case a map entitled 'Scotland, III, Orkneys, Shetland, and Hebrides' arrives, not with a map covering the rest of Scotland, but with 'The Principal Rivers of the World' (shown on page 136).

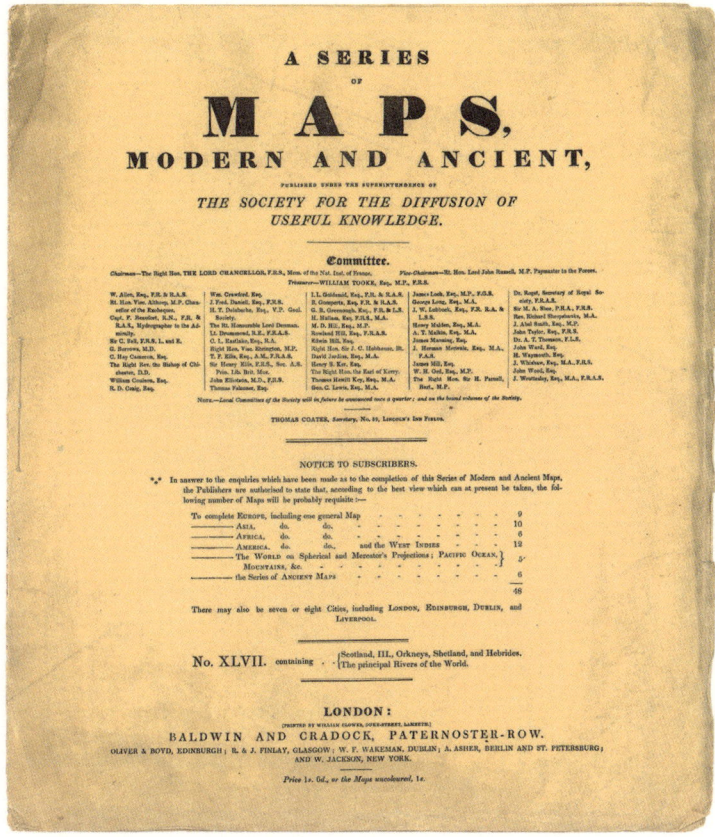

As I have been taught that nothing sublimary [sic] is eternal I presume that even the Series of Maps published by your society will <u>one day</u> have an end; but as present appearances induce me to think that such an event must be witnessed by one of my remote posterity … I find that the Society in its Zeal for the promotion of Geography has most acutely discovered that this world contains cities as well as countries … Now really a plan of Bam-boo-Zlee-Kiana in China is not at all interesting to me … (a Mr E. Mullineaux of Manchester)

Concerns, too, were also raised about the usefulness of the maps without an index and so a Revd James Mickleburgh was tasked with doing this. Given his vocation you might think the reverend would have had the patience of a saint, but this was sorely tested:

> Even you cannot estimate the slavery of the concern. I have neglected my friends, my parishes, and myself for the sake of this work, and its preposterous issue. You will naturally call to mind the 'Nil Sine Magnus'[24] – of the Roman poet – which is deeply verified in the present instance. It is of no avail to hint neglect on my part – you cannot have impossibilities.[25]

It must have been a thankless task to carefully read each label off the maps, noting its location and then assembling a list of all such places in alphabetical order. But Beaufort was unwavering in this approach, even threatening to resign, if quality was compromised, as a letter to a disgruntled subscriber makes clear:

> [The Committee] prefer this course to a more speedy publication; because they know that increase of speed will involve imperfection in the Maps, a neglect of the most recent authorities in their compilation, thus making of them a mere copy of Maps already published ... Even if the Committee were to satisfy themselves with a publication imperfect & inaccurate for the sake of obtaining a more speedy sale; the condition upon which they must obtain that result will undoubtedly be the loss of their present editor.[26]

The maps therefore took as long as they needed and required Beaufort's sign-off before he let them anywhere near a printing press. Thanks to this approach they were truly maps for the masses who could now own something of a level of quality that would previously have been out of reach.²⁷

They weren't perfect, not least with the appearance of the mythical 'Kong Mountains' (see Chapter 1) on the SDUK maps of the African continent, and Beaufort did sail a bit close to the wind at times by copying from other maps. In 1843 the committee of the SDUK wrote to Beaufort seeking permission to credit him on the maps and to ask that he disclose his sources, to which he replied: 'The early maps were ... mere copies of such materials as were accessible to us without any dangerous breach of copyright.'²⁸

Elsewhere in Africa the SDUK maps offer some wonderfully vague but enticing labels including 'High Acacia trees' just to the north of 'Herds of Elephants' and the 'Loose Sand' marking a transition to 'Herds of Gazelles'.²⁹

Quibbles aside, there is no doubt that Beaufort wanted the maps to be the best they could possibly be for the era. He could have compromised and produced the maps as cheap copies at bargain basement prices, justifying his efforts as still giving the working classes something they would not ordinarily have, or perhaps that such a class could not appreciate the artistic or scientific qualities imbued in the maps.

Would a coal miner or textile worker really notice or complain that their map is lacking the latest border or a recently named peak or river? It was a sign of respect to these workers that they were not offered compromises. It was also an indication of how seriously the SDUK took geography and enabling everyone to see the world beyond their horizons.

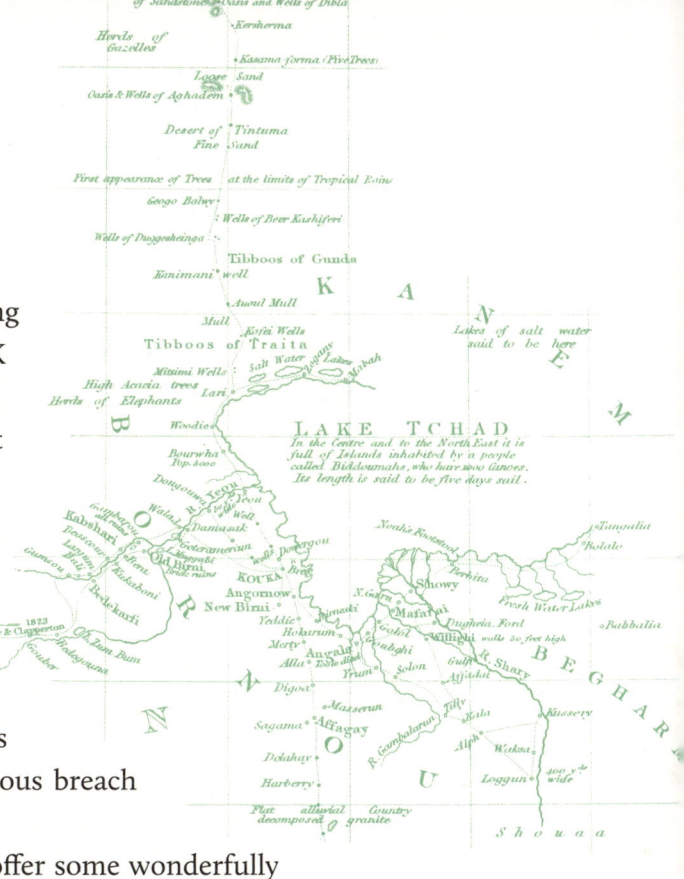

Map 35 [Extract]

West Africa II: Society for the Diffusion of Useful Knowledge (1839)
37.5 × 29.5cm

THE LIBRARY OF LOST MAPS

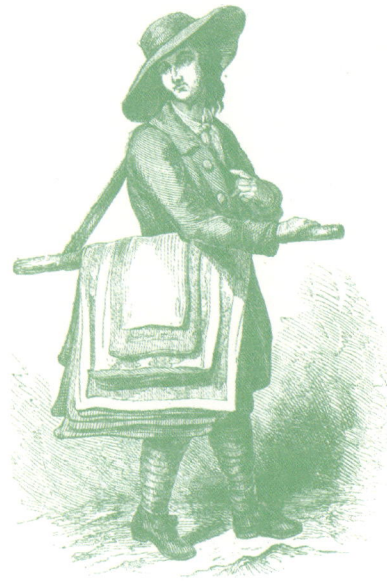

[German Itinerant Map-Seller.]

The Penny Magazine 31 October 1840: Society for the Diffusion of Useful Knowledge (1840)

The mission to expand the use of maps, and thus create a more vibrant market for them, was set out clearly in the SDUK *Penny Magazine* on 31 October 1840. It bemoans the lack of appreciation of, and therefore demand for, quality maps in England, contrasting it with Germany, where the SDUK perceive higher demand fuelled by better geography education. The article is illustrated with a man in a large, brimmed hat, neck scarf and buttoned jacket. Slung across his body is a strap that is bearing the weight of a wooden pole draped with maps, sorted from smallest to largest. It is captioned 'German Itinerant Map-Seller'.

'We wish we could have given an "English" itinerant map-seller in our cut, but it will be long, we fear, before such a character will be found amongst us as the representative of a class ... A desire for maps must be created before the hawker will find it profitable to include them in his little stock ...'[30]

To the SDUK the Germans had a higher standard of education, which generated the demand for maps among its geographically literate population. The article also included a quotation from Greenough's presidential address to the Royal Geographical Society given in May 1840: 'Words following words in long succession, however ably selected those words may be, can never convey so distinct an idea of the visible forms of the earth as the first glance of a good Map.'[31]

Despite their concerns about the lukewarm desire for maps in Britain, by 1844 the Society had sold more than three million.[32] It's a staggering number that is testament to the dogged determination of Beaufort and his committee not to use price, or indeed the intended audience, as a reason to compromise on quality. The SDUK atlas remains one of the most successful mapping projects in history.

The maps remained profitable, but financial pressures due to a fall in sales across the SDUK's other publications meant that the SDUK was wound up on 7 June 1848 over dinner at the Piazza Coffee House in Covent Garden, on the site of which is now the Royal Opera House.[33] But the maps continued to be in print in some form or another for the next fifty years as the hard-wearing steel printing plates were bought and sold by a succession of publishers.

And the Society's influence did not end there. If you are lucky enough to visit the Geography and Maps Division of the Library of Congress in Washington DC, you will be greeted by an old globe. The caption alongside it reads:

Theodore Roosevelt – Globe of the Vice President

This globe, published in 1882 under 'the Superintendence of the Society for the Diffusion of Useful Knowledge', was in the office of Theodore Roosevelt while he served as Vice President of the United States in 1901.

When I read this, I did a double take. The date of 1882 is thirty-four years after the SDUK wrapped up, so perhaps the Library of Congress was mistaken? I walked round the fragile globe and, sure enough, the SDUK are credited alongside the maker 'Malby Globes', a name I had seen in the minutes of the Map Committee. It appears that Malby continued to make globes based on the maps of the SDUK decades after the Society's final dinner.

Knowing how overlooked the SDUK maps are, and their deep connection to Bloomsbury and the university I work in, I felt an odd sense of pride seeing the globe there, thousands of miles from home, but celebrated for being in the office of one of the most powerful men in the world. I just could not believe that over seventy years after the first of a series of maps, conceived for the masses of Victorian Britain, had rolled off the printing press they were in the office of a future president of the USA, shaping his worldview as he glanced over at it countless times each day.

Another publisher who bought the plates was Edward Stanford, whose map shop Stanfords was founded in 1853[34] and continues to trade from Covent Garden today. When the SDUK wrapped up, the steel plates changed hands a couple of times before Stanford purchased them for his shop in 1856.[35] The previous year he had acquired the rights to publish Greenough's India geology map,[36] and, thanks to the likes of the SDUK, there was now a growing public interest in affordable maps. In fact, Stanfords kept the SDUK brand and its maps alive well into the 1860s. These gave the business – still a relatively new map shop and publisher – a head start with some global maps that could be updated and sold as individual sheets (plain for 6d or lightly coloured for 9d). They were also packaged up as a couple of new atlases:[37] *The Harrow Atlas of Modern Geography* for use at Harrow School (not entirely aligned with the egalitarian values of SDUK as Harrow is an elite private school) and more significantly, in the accessible format of the cosy sounding *Family Atlas*, issued in twenty parts and published between 1863 and 1865. It received some enthusiastic reviews, among them: '[A] splendid series of maps ... the latest discoveries, the most recent improvements, the submarine telegraphs, for example – are all introduced ...'[38] and 'exceedingly valuable, while its cheapness makes it very largely available for the purpose of reference ... In the present publication, we have not merely the new arrangement of the kingdom of Italy, but we find the eastern portion of the continent of Africa filled up much more completely than we have been accustomed to see it ...'[39]

The reviews are a reminder of the changes taking place in the world in the latter half of the nineteenth century that mapmakers had to keep pace with, such as the creation of new countries, more detailed surveys and technological developments speeding up communications. The reputation of its maps meant that Stanfords' products were in greater demand and the firm expanded further, moving to a new address in Charing Cross as well as opening its print works on Long Acre, an address it would occupy for 118 years (the shop moved to a new premises in 2019), passed by millions of people as they walked from Leicester Square to Covent Garden.

As well as the exploration and colonisation that was the focus of Stanfords' patrons in the 1850s, throughout the remainder of the nineteenth century Victorians developed an appreciation of the value of maps to explain a fast-changing world, and to help address pressing societal issues. For example, the revolutionary nurse Florence Nightingale was a customer, and Charles Booth chose a Stanfords *Library Map of London*, which was known for its quality and clarity, as the basis of his pioneering maps of London Poverty.

That said, not all Victorians wanted to see the industrial wonders (and challenges) of the age appearing on a map. Writing to 'Messrs Stanford & Co' in February 1887, the Victorian polymath John Ruskin had the following request:

> Gentleman,
> Have you a school atlas or any other sort of atlas on sale at present without railroads in its maps? Of all the entirely odd stupidities of modern education, railroads in maps are the infinitely oddest to my mind.
> Ever your faithful servant and victim
> J. Ruskin[40]

Ruskin, who was a professor at the Slade School of Fine Art (now part of UCL), was waging a campaign against the rapid expansion of railways. A year before his letter to Stanfords, he wrote about the consequences of expanding the railways into the Lake District to increase access to one of Britain's most cherished landscapes:

> But the stupid herds of modern tourists let themselves be emptied, like coals from a sack, at Windermere and Keswick. Having got there, what the new railway has to do is to shovel those who have come to Keswick, to Windermere – and to shovel those who have come to Windermere, to Keswick. And what then?[41]

Map 36 [Extract] This map of the Lake District comprises an Ordnance Survey map that Stanfords made more appealing to the Victorian tourists who flocked to the region by mounting it on folded linen and adding some hand colouring. It does, of course, feature the railways (see previous page), so Ruskin would not have approved!

Stanford's English Lakes (Hachures) No CII.S.W. (Appleby): Ordnance Survey and Edward Stanford (1872) 94 × 91cm

However, it was the 'stupid herds of modern tourists' who were becoming an important class of customer for Stanfords.[42] Thanks to more leisure time and disposable income, and the founding of the pioneering Thomas Cook Tours, Victorians needed maps and guidebooks to take with them on holiday. Maps were often recoloured and repurposed from official sources such as Ordnance Survey to appeal to a range of interests from customers who were keen not just to plan excursions but to inform themselves about the landscape they were passing through.

Therefore, in the first half of the nineteenth century, maps had become firmly established as a truthful and informative resource for all: as important for learning as for leisure. Maps were also becoming appreciated as a credible way of articulating the complexities of the world that were being uncovered by the Victorians in their quest for scientific advancement.

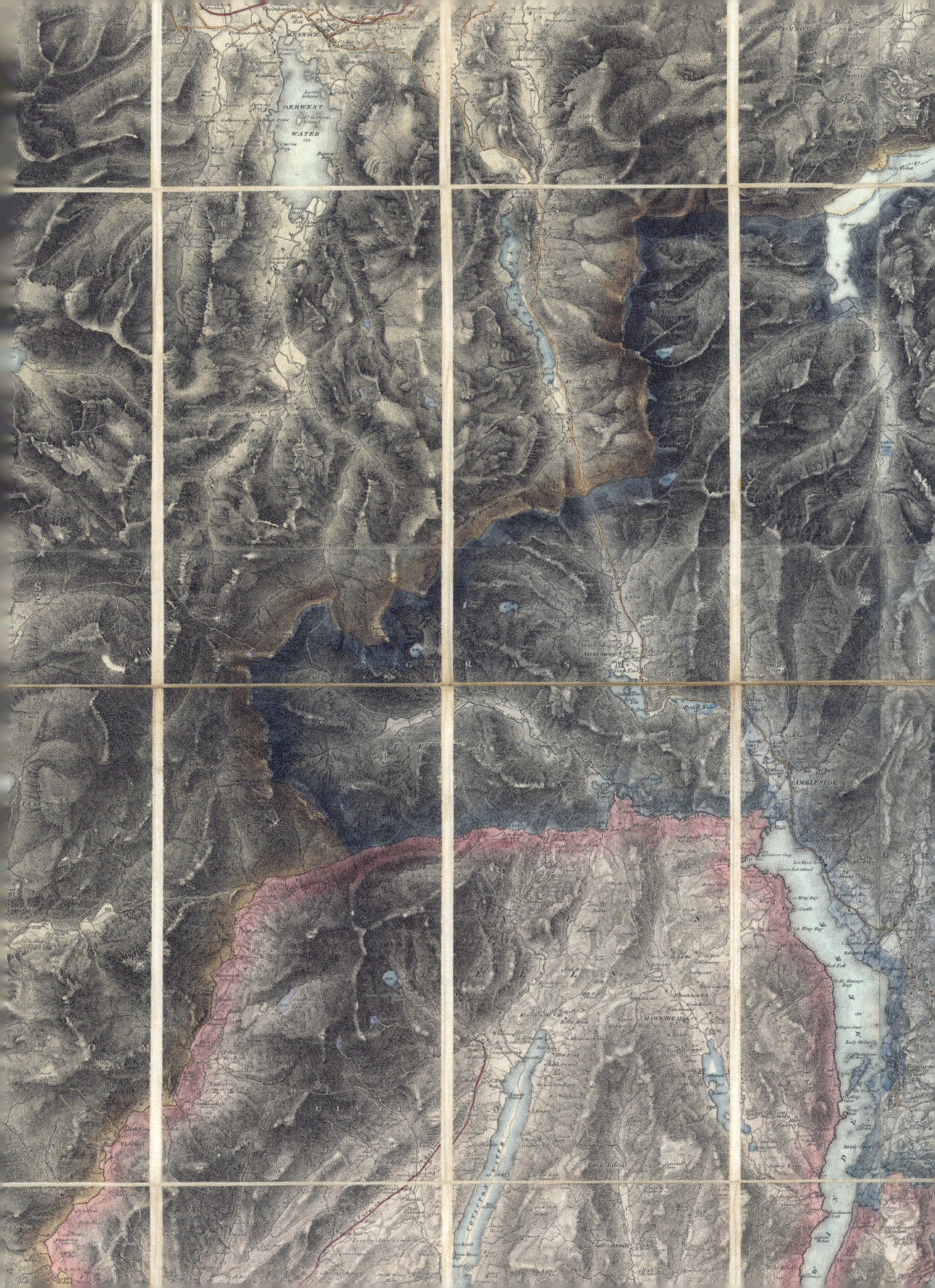

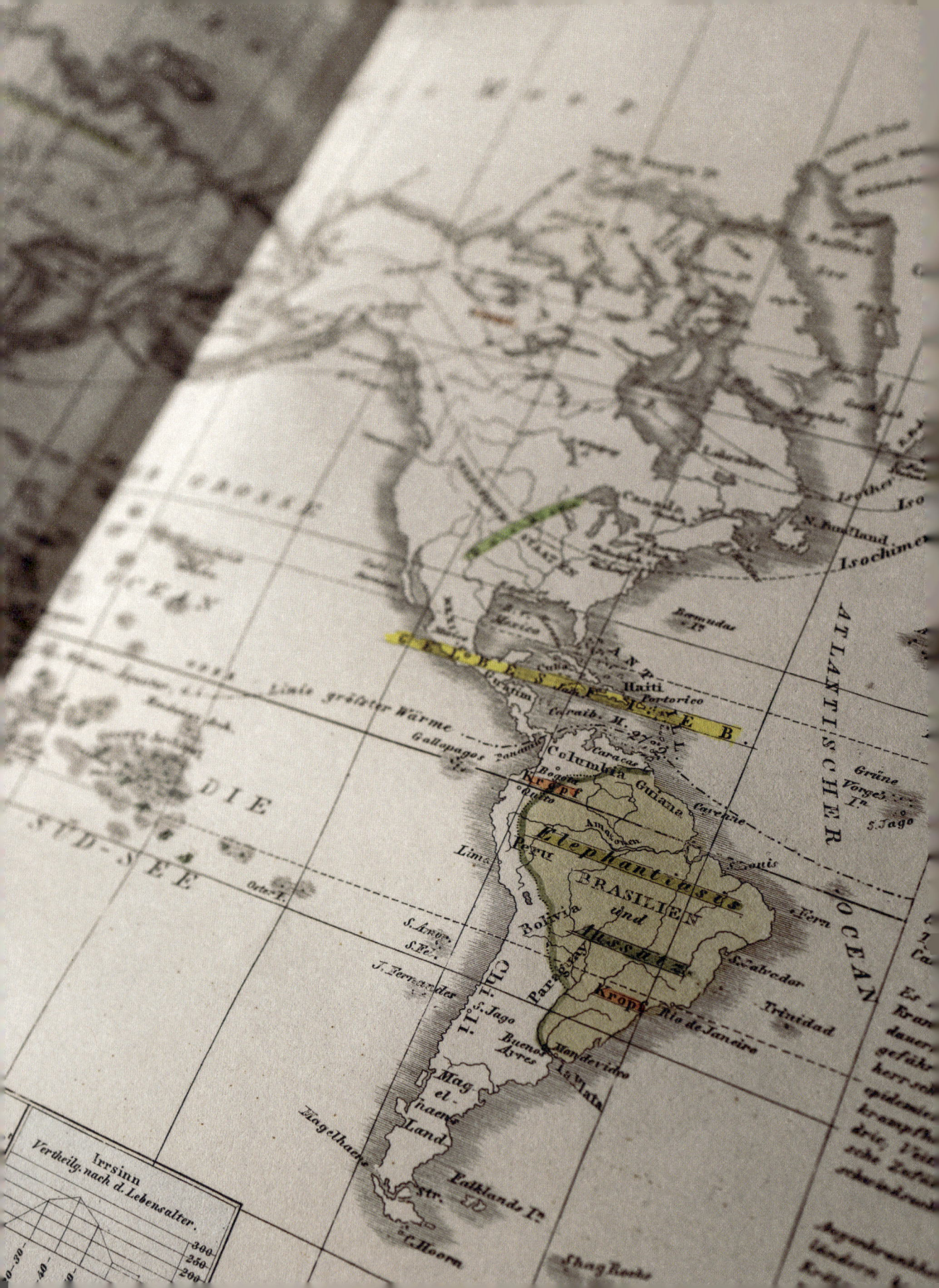

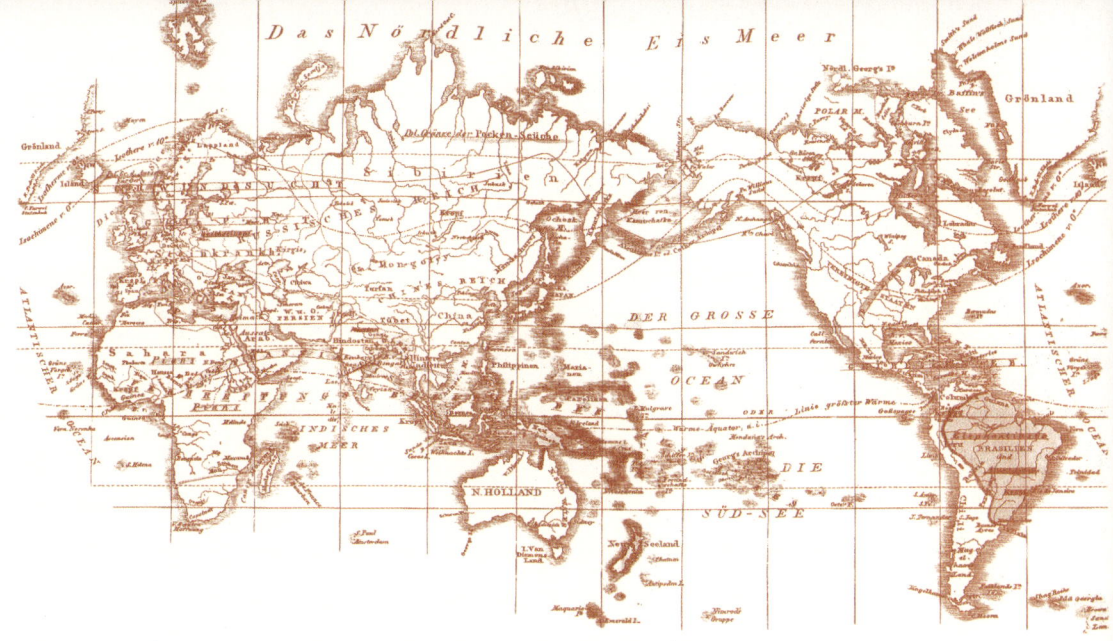

CHAPTER FIVE

Knowledge Is Power

'Roll up that map. 'Twill not be needed now
These ten years! Realms, laws, peoples, dynasties,
Are churning to a pulp within the maw
Of empire-making Lust and personal Gain!'
Thomas Hardy[1]

While the focus of the Society for the Diffusion of Useful Knowledge (SDUK) was on topographical mapping, to extend people's horizons beyond their own town or village, they also began to develop topic- rather than place-based maps. For example, they published a geological map of Britain (which George Bellas Greenough was meant to oversee but this instead fell to Roderick Murchison) as well as a rather wonderful comparative map showing the lengths of the world's rivers as if flowing outwards from the same source. This innovation away from maps as largely topographic endeavours gathered pace towards the middle of the nineteenth century.

Map 37 [Extract] This is one of the first global maps of disease to be published. It was created in the 1840s for Heinrich Berghaus's *Physikalischer Atlas* to detail all manner of ailments, and where they might be found. At the time cholera was ravaging India, tuberculosis (*Schwindsucht*) was sweeping across more northern latitudes and – as can be seen here – elephantiasis (a condition spread by mosquitos) was troubling South America.

Physikalischer Atlas (Planiglob zur Übersicht der geographischen Verbreitung der vornehmsten Krankheiten, denen der Mensch auf der ganzen Erde ausgesetzt ist): Justus Perthes (1845–1848) 35.5 × 28.5cm

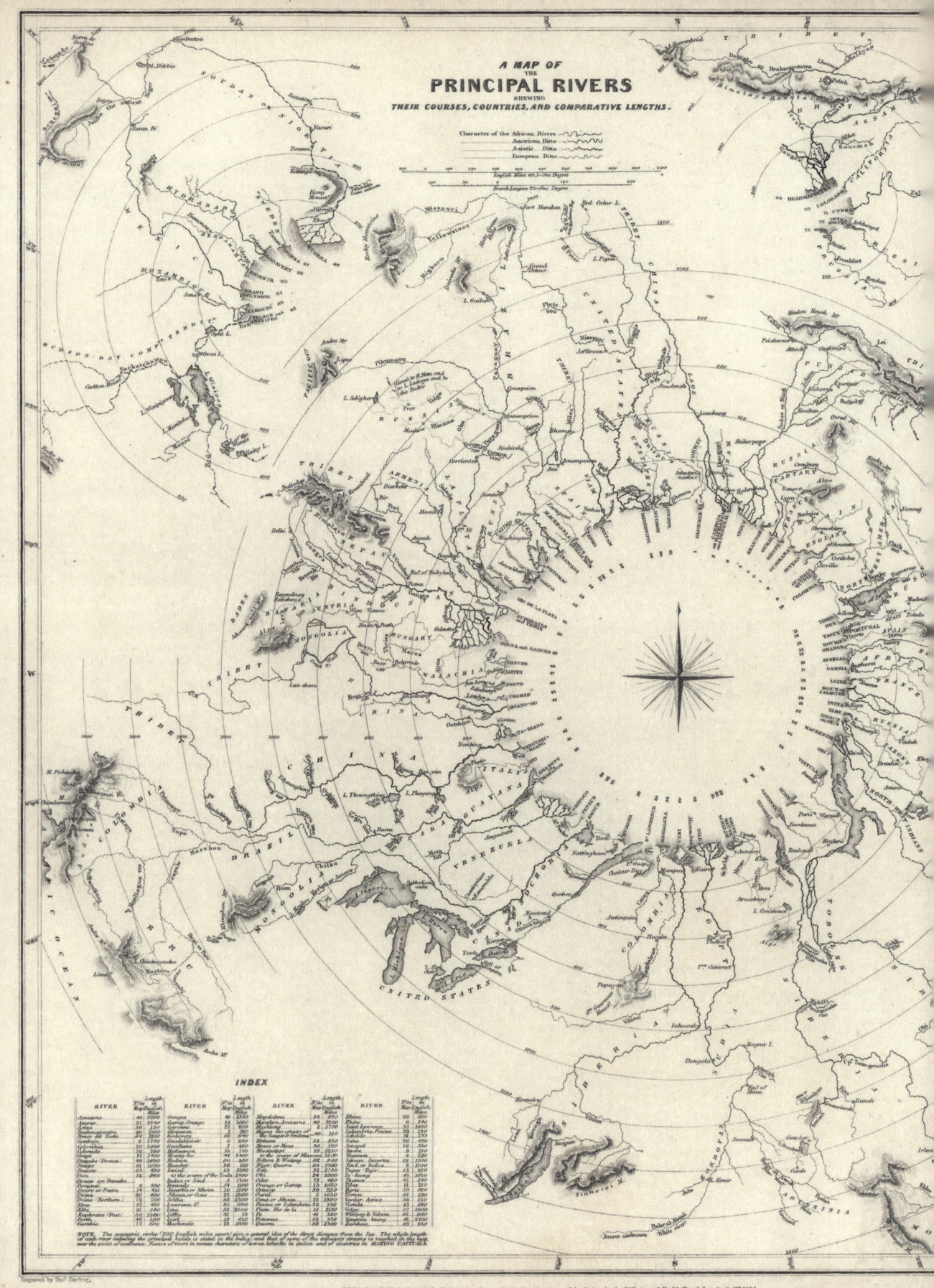

Greenough himself was doing a lot of thinking about what other topics maps could be used for, and among his notes[2] I found a few sheets of paper upon which he explored the idea of a category of maps called 'monographic maps'. Nowadays these would be called 'thematic maps' as they are maps that focus on a particular theme or topic, rather than simply showing roads, rivers, mountains and so on in situ, as topographic maps do so well. I couldn't discern a precise date for this note, but many of the first versions of the kinds of maps that Greenough listed appeared after his death. He was clearly ahead of his time.

His list is a long one and appears to be written as a stream of consciousness since there is no clear order to it (Greenough liked to alphabetise things, so anything not in alphabetical order either needed a good reason not to be, or was an earlier draft of some thoughts he'd later organise). Towards the top of the list there are 'Military' and 'Territorial' maps and then 'Statistical' maps which include 'Population', 'Education', 'Condition of the People'. 'Medical' maps, including those of disease etc., come a little further down and then 'Ethnological' appears, after a run of 'Orographical', 'Geological', 'Hydrological' and 'Botanical', and just before 'Historical' (which includes 'conquests') and then 'Meteorological'. Many of these were developing fields of science in their own right and Greenough clearly saw the value of maps in both informing scientific discoveries and communicating them.

The name that Greenough had pencilled against a number of these types of maps was that of Heinrich Berghaus. Berghaus was an extraordinary

Map 38 The SDUK stretch the idea of a 'map' here with this fabulous illustration of the world's major rivers. They have been arranged around an imaginary pole and are flowing outwards towards the margins. The rivers are roughly ordered by the longitude of their source, and their lengths are indicated by concentric circles radiating outwards. It's a map that takes a moment to become oriented to but once you get your eye in there is so much to see (though remember that in 1834 many of these rivers had not yet been surveyed in detail).

A Map of the Principal Rivers Shewing Their Course, Countries, and Comparative Lengths: Society for the Diffusion of Useful Knowledge (1834)
30.5 × 38cm

mapmaker who created the first atlas devoted to thematic maps. The *Physikalischer Atlas* (Physical Atlas) is the atlas that all cartographers would love to own as it is regarded as one of the most pioneering and inspirational collections of maps ever published. I had only seen photographs of it and had imagined a grand tome, bound in thick leather. Convinced it would have passed the Map Library by, I was astounded to see both volumes of the first edition on a shelf among a cluster of German world atlases. Rather than the showy exteriors I had expected, each volume was rather more understated and slightly smaller in comparison to the other atlases propping them up on the shelf.

They were bound in a fine blue cloth, embossed with a gold border and gold lettering. The interior pages of both volumes had sustained water damage, although the maps were still clear, and there seemed to be a German shelf mark label, suggesting they had been picked up as part of the spoils of war in 1945.

On the inside cover is printed praise from German and British press and periodicals. The *Athenaeum* offers a fabulous summary:

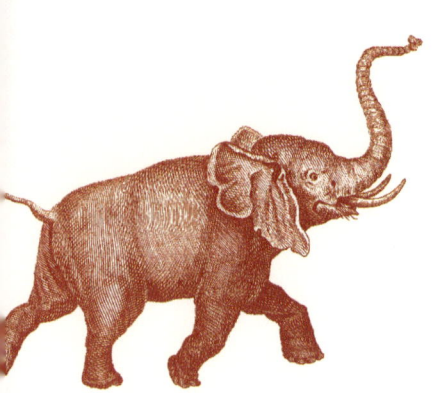

Physikalischer Atlas (Geographische Verbreitung: Dickhäuter): Justus Perthes (1845–1848)

All possible relations of time and space, heat and cold, wet and dry, frost and snow, volcano and storm, current and tide, plant and beast, race and religion, attraction and repulsion, glacier and avalanche, fossil and mammoth, river and mountain, mine and forest, air and cloud, and sea and sky – all in the earth, and under the earth, and on the earth, and above the earth, that the heart of man has conceived or his head understood – are brought together by a marvellous microcosm, and planted on these little sheets of paper.

The *Scotsman* was enthusiastic but not as breathless, instead endorsing the scientific merit of the atlas and the degree of effort that must have been expended, highlighting that 'some single maps in the series are the product of two or three years' continued research, by a person well prepared ...' It goes on to say that the 'beauty of the Atlas renders it appropriate for the drawing room table', but that it is not just for decoration, as 'in families where there are young ladies, it will be found a good substitute for the lecture room'.

The first part of the *Physikalischer Atlas* was published in 1838. Further sections were released as they were finished, with the final instalment available in 1848, comprising ninety maps in all. Heinrich Berghaus edited the first two editions himself, while his nephew Hermann edited the third edition. Some maps of the first edition also carry the signature 'drawn and engraved in the "Geographische Kunstschule" at Potsdam'. This Geographic Art School was run by Berghaus, and he paid out of his own pocket for a few of his star students to attend. As compensation the pupils were put to work on his atlas projects, including the *Physikalischer Atlas*.

The atlas contained many unprecedented maps, from the first maps of disease to world maps of the distribution of animals and plants, so there are many reasons for Berghaus's name to have appeared on Greenough's list of thematic maps.

While Berghaus and his students may have been its cartographers, the origins of this remarkable atlas lie with the Prussian polymath, geographer, naturalist and explorer Alexander von Humboldt. Humboldt is perhaps best known for his extensive travels and explorations in Latin America, where he had direct experience of the region's ecosystems, climates and geology.[3] He was also a great gatherer and synthesiser of information from secondary sources via a huge network of correspondence. This gave him a global perspective that enabled him to create his hugely influential book *Cosmos*, where he aimed to take all his acquired data and to unify diverse branches of scientific knowledge in

THE LIBRARY OF LOST MAPS

Map 39 The first temperature maps showed the entirety of the Earth with isolines running from west to east, rippling along the lines of latitude. In his atlas, Berghaus included this view of the northern hemisphere, where the isolines resemble the wavelets on a pond with a stone dropped at the North Pole. The number of temperature readings underpinning these first maps was quite small, but as the science developed the isolines on maps became more undulating as they were tensioned around additional data points.

Physikalischer Atlas (Die Isothermkurven der Nördlichen Halbkugel): Justus Perthes (1845–1848) 27.5 × 27.5cm

order to present a holistic view of the universe. During his life he was extremely well known, but there is recognition today that Humboldt's ideas and arguments have never been more relevant. For example, he was one of the first to stress the interconnectedness of nature. In 1844 he wrote: 'Nature … [is] a harmony blending together all created things, however dissimilar in form and attributes; one great whole animated by the breath of life.'[4]

In addition to *Cosmos*, Humboldt knew it would take an atlas to help conjure this grand vision and so turned to Berghaus to create one. Such an atlas had to include 'maps for the world-wide distribution of plants and animals, for rivers and oceans, for the distribution of active volcanoes, for magnetic declination and inclination, intensity of magnetic energy, for the ebb and flow of ocean currents, air currents, the course of mountains, deserts and plains, for the distribution of human races, as well as for the representation of mountain heights, river lengths etc'.[5]

Berghaus lived life on the edge of bankruptcy because of the time it took him to complete projects and his investment in his own cartography school.[6] So, Humboldt needed to finance their collaboration as well as offer creative inputs. It was a partnership that resulted in new ways of visualising the world through maps that did more than simply describing natural phenomena: they allowed maps to demonstrate the way that the planet operates as the interconnected systems that Humboldt describes.

There are plenty of maps in the Map Library that have been influenced by this thinking. For example, climatologists and cartographers have Humboldt to thank for an innovative approach to mapping temperature over the Earth's surface. In 1817 he developed the concept of 'isolines' to show lines of constant temperature across the Earth.[7] At this time it was thought that temperature was purely a function of latitude, thus getting warmer towards the equator and cooler towards the poles at a constant

constant rate. If this were true, Humboldt would have drawn his isolines as equally spaced and parallel to the lines of latitude. But his lines gave a very different impression of the temperature on the Earth's surface by kinking away from the poles in relatively cooler regions and away from the equator where the climate was warmer. Berghaus devoted the first three double pages in the *Physikalischer Atlas* to these innovative isoline maps, the first looking at the entire world using Mercator's projection, the second from the North Pole and the third map just covering Europe.

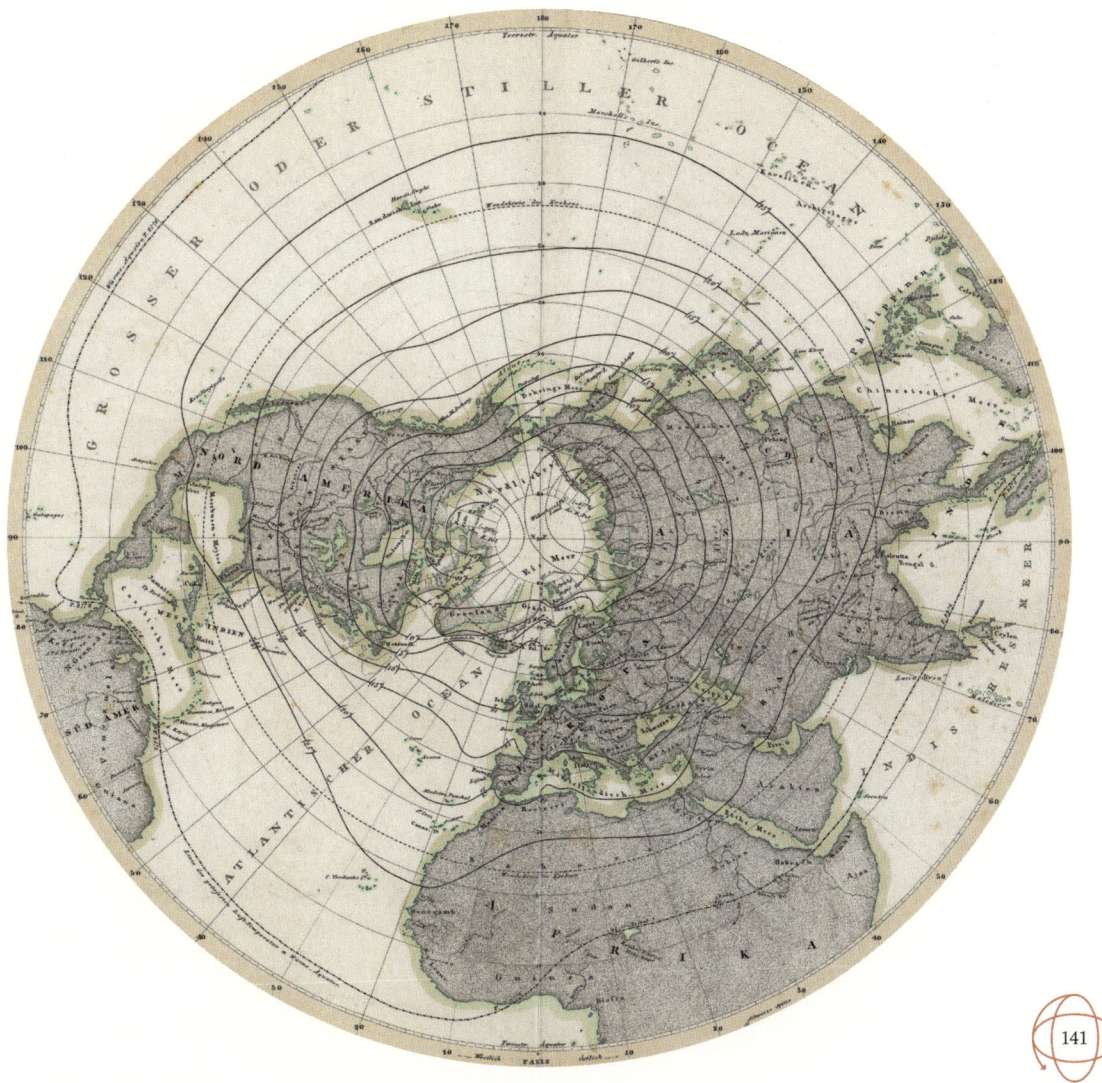

THE LIBRARY OF LOST MAPS

Once a map shows where the temperature differences occur, it begs the question: why are they there? Through modern eyes we can now appreciate the impacts of the North Atlantic Current, channelling warmer waters and therefore warmer temperatures between Iceland and Norway, causing the isolines to kink up from the south. While the elbow around the Alps showed the impact that altitude has on temperature, so these isoline maps were hugely influential, not just in driving scientific study, but in providing insights upon which people made real-life decisions. In North America, for example, they became part of campaigns to tempt people out west, with the maps in some cases being doctored to suggest more appealing climates (and therefore better prospects of making a living off the land) than the reality people found when they arrived.[8] And they were used in the peace negotiations at the end of the First World War, as we shall see in Chapter 6.

Humboldt's collaboration with Berghaus helped confirm the status of maps as a level-headed scientific tool. The *Physikalischer Atlas* therefore set the standard for what an atlas could be by articulating an approach to science that welcomed maps on a range of different aspects of the natural and human world. Maps had become data-driven and were to be trusted.

In this context, the genre that cropped up again and again in the Map Library was 'ethnographic' maps. These begin to appear from the 1840s, growing in number for the next century. After 1945 they abruptly stop, with only occasional examples from the second half of the twentieth century. Seeking to find out why this might be the case set me off on a journey of discovery that unearthed uncomfortable truths about the role that maps and their makers have played, not just in witnessing, but also in fuelling, some of the most horrific periods of history.

The maps that assigned specific racial/ethnic categories to populations only really developed towards the middle of the nineteenth century. Berghaus's atlas was, perhaps unsurprisingly, among the first to include ethnographic maps. He featured four maps on 'Anthropographie' and nineteen on 'Ethnographie' that came towards the end of the second volume. This included four double-page spreads that could be pulled from the atlas and pieced together into a giant map of Europe's ethnic groups. It is more beautiful than it is accurate, with oddities and omissions that disappeared from later maps, such as distinguishing those living along the English coast from those inland, for example, but that is beside the point. This is the only topic for which Berghaus did a special pull-out map, suggesting something of significant interest to his readers. By putting ethnography[9] in a 'physical' atlas, mapmakers were seeking to give it a greater and more scientific status – at a time when the sciences were becoming more specialised – as a pursuit akin to geology or climatology.

Around the time of Berghaus's atlas, James Cowles Prichard, an English physician, is credited with creating the first atlas devoted exclusively to ethnography (he also features on Greenough's list of pioneering thematic maps and their makers). It comprised a map for each of the continents (plus a view of the Pacific, with surrounding landmasses) and was published in London and New York in 1843 to support Prichard's seminal book, *The Natural History of Man*.[10] Prichard was a scientist from the Greenough mould of dealing only in what he saw as observable facts, and, indeed, the two men knew each other well.

As with the early years of the Geological Society, the formation of the appropriate society to support the science, in this case the 'Ethnological Society of London', was shaped by debates arising from doubts about the role of God in the creation of Earth and the prevalence of conflicting high-profile theories about how the world came to be. It was founded in 1843 as a spinoff of the Aborigines' Protection Society (APS), which was more of a humanitarian

(Overleaf) Map 40

The inclusion of a four-part ethnographic map of Europe was another major innovation in Berghaus's magnificent atlas. The map is extraordinarily detailed in the way it shows the terrain and other features, but the novelty here is the inclusion of the ethnographic regions, as they were understood at the time, painted on in delicate watercolours.

Physikalischer Atlas (Ethnographische Karte von Europa): Justus Perthes (1845–1848)
40.4 × 33cm (each of the four)

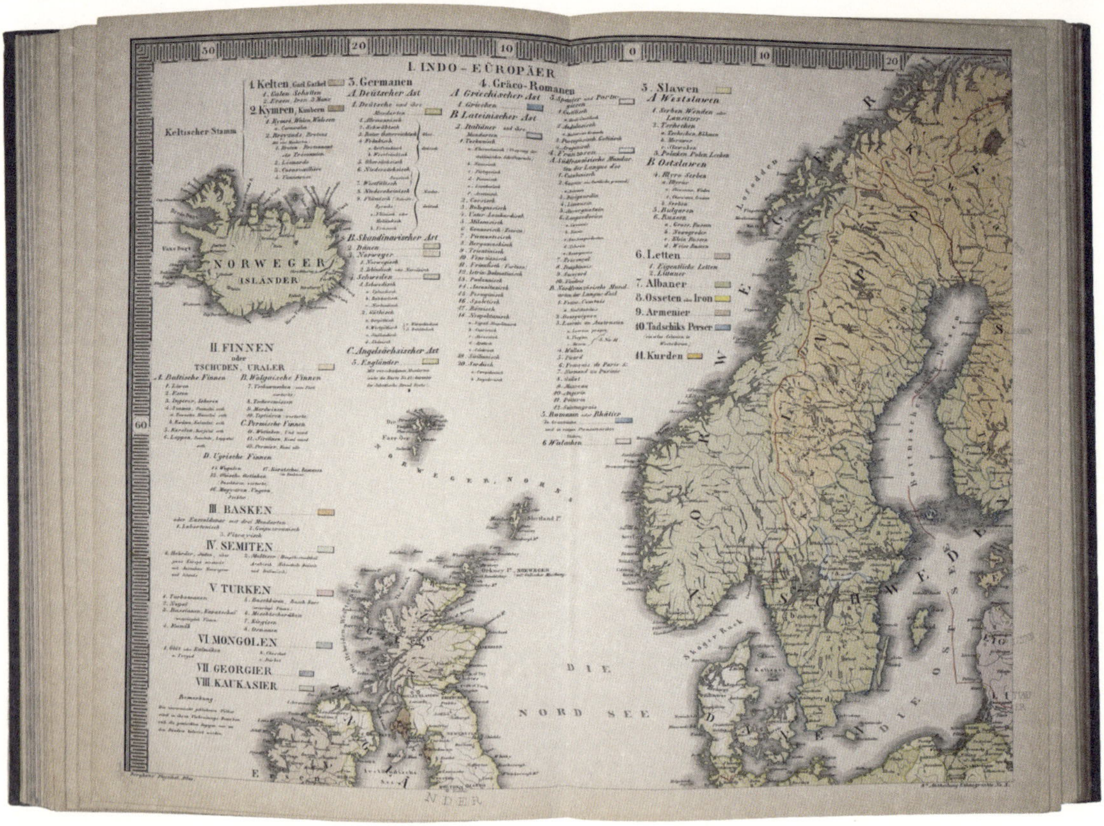

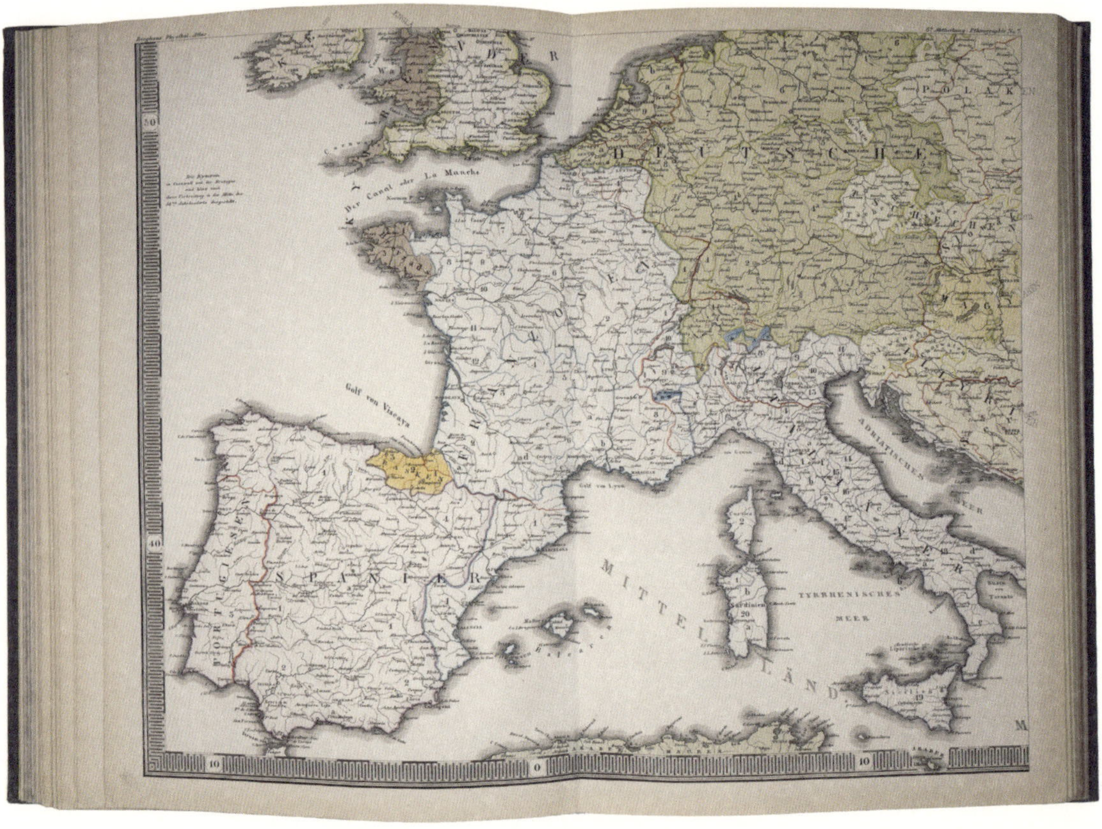

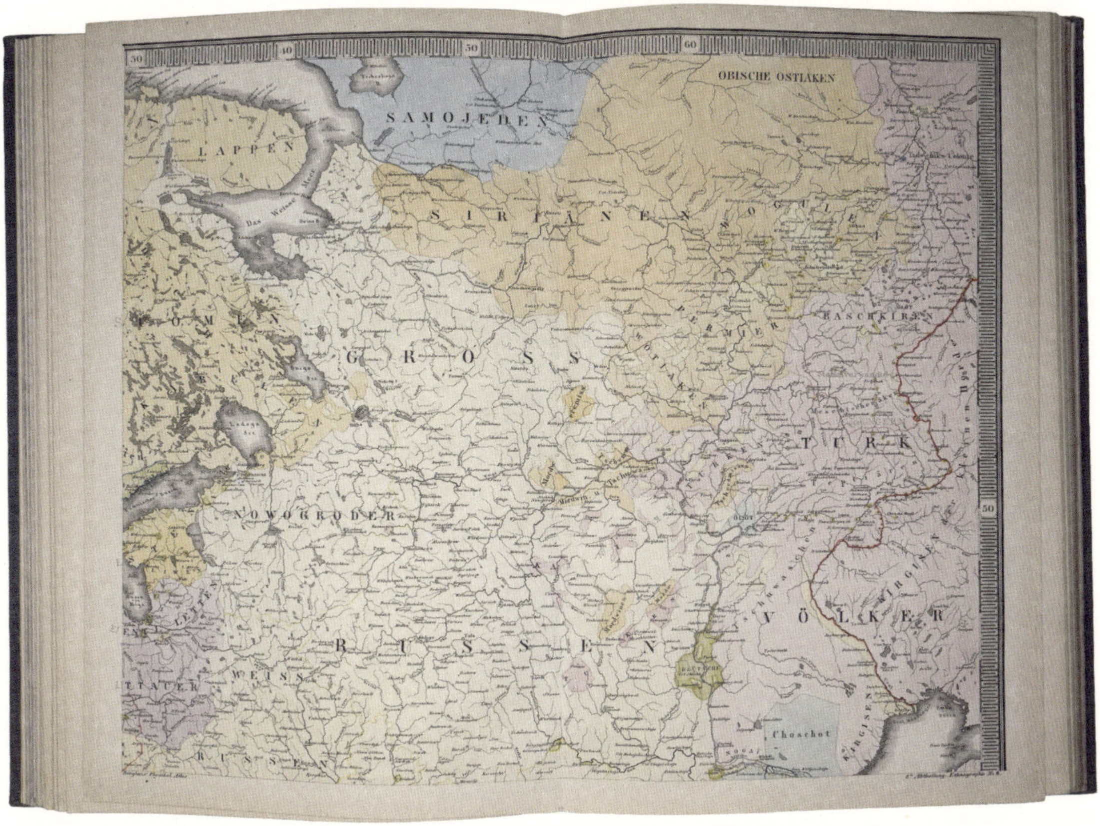

Ethnographische Karte von Europa

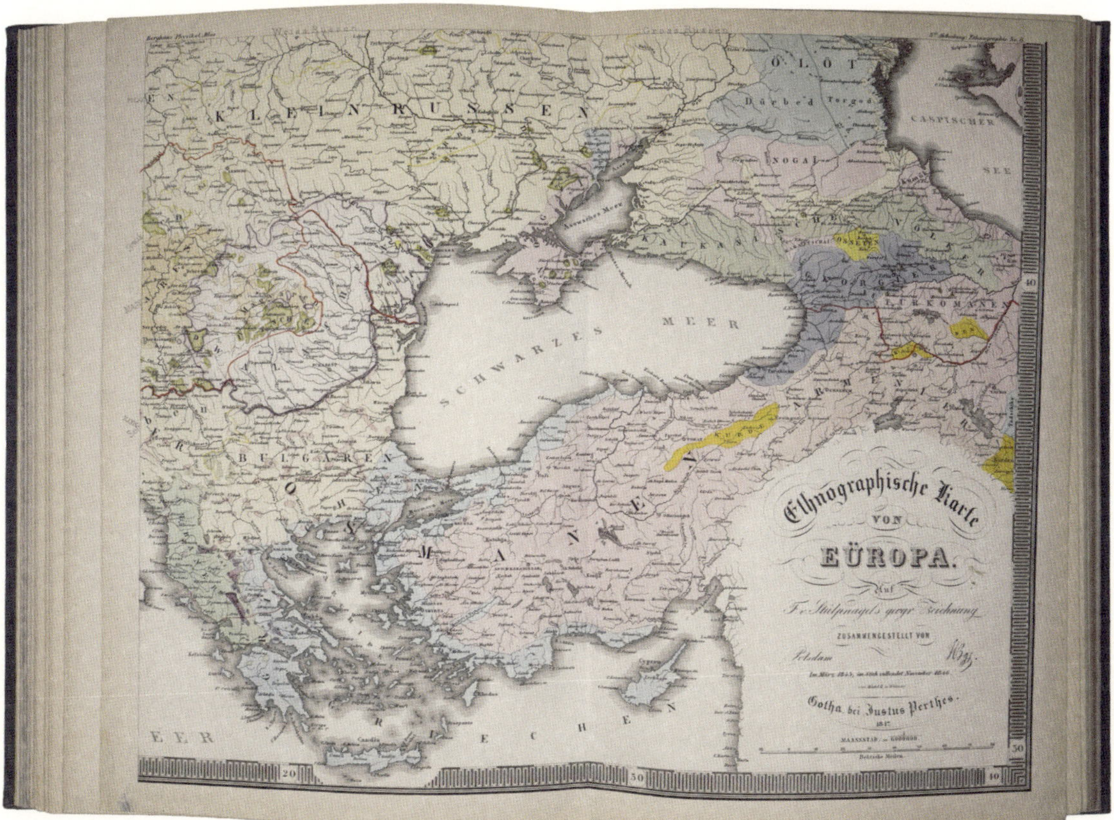

organisation founded by Evangelical and Quaker philanthropists whose aim was to protect and advance the rights of indigenous peoples in the British Empire and beyond. In the 1830s, the APS's charitable efforts began to develop into ethnographic and anthropological interests, spearheaded by academics like Prichard and his more scientific study of human races and cultures.

This became the predominant focus for a critical mass of members who moved to create the Ethnological Society of London. Its purpose was to be 'a centre and depository for the collection and systematisation of all observations made on human races' and of course Greenough became an enthusiastic member, serving on its council and as vice-president only a couple of years before his death (1855).

Into the 1860s there was the rise of the Anthropological Society of London (ASL), the proponents of which had a keen interest in physical anthropology and the classification of human races. The ASL had some especially abhorrent views (even in the context of the time), driven by white supremacy and the conviction that there were multiple human races with no common ancestors between them. These opinions were at odds with the Ethnographical Society's scientific theories about evolution.[11] Arguments between these factions continued throughout the 1860s before an eventual merger in 1871 created the Anthropological Institute of Great Britain and Ireland, with a general acceptance of Darwinian theory.[12] This institute still exists today, now with the 'Royal' prefix, which was granted in 1907.

With the formation of such a society and the development of the theories of what distinguished different races, ethnographic maps assumed the status of a scientific product. In the minds of the intellectual classes of the nineteenth century, they crystallised ideas about the groups of people who could live side by side and those who were destined for conflict. To their eyes the maps

Men from the Tyrol region (today's Austria), Paris (France), and Central Asia, each dressed distinctively and used to illustrate a map of the different clothing styles and fabrics used around the world.

Physikalischer Atlas (PLANIGLOB zur Übersicht der verschiedenen BEKLEIDUNG's Weise der Bewohner des ganzen ERDBODENS): Justus Perthes (1845–1848)

offered unequivocal contrasts between 'civilised' and 'uncivilised' areas and, as we shall see, underpinned the contrasting fortunes of those in power, and in possession of the maps, and those who were subject to them.

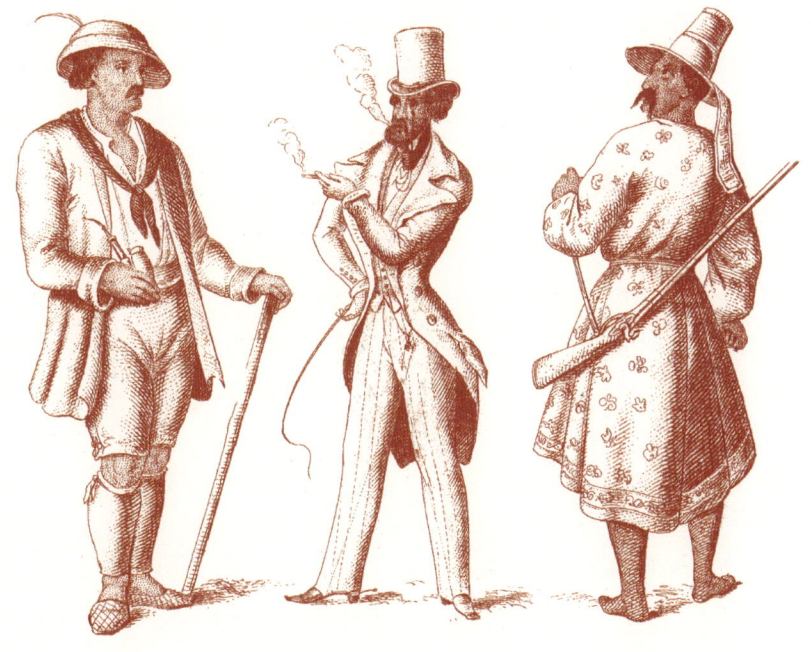

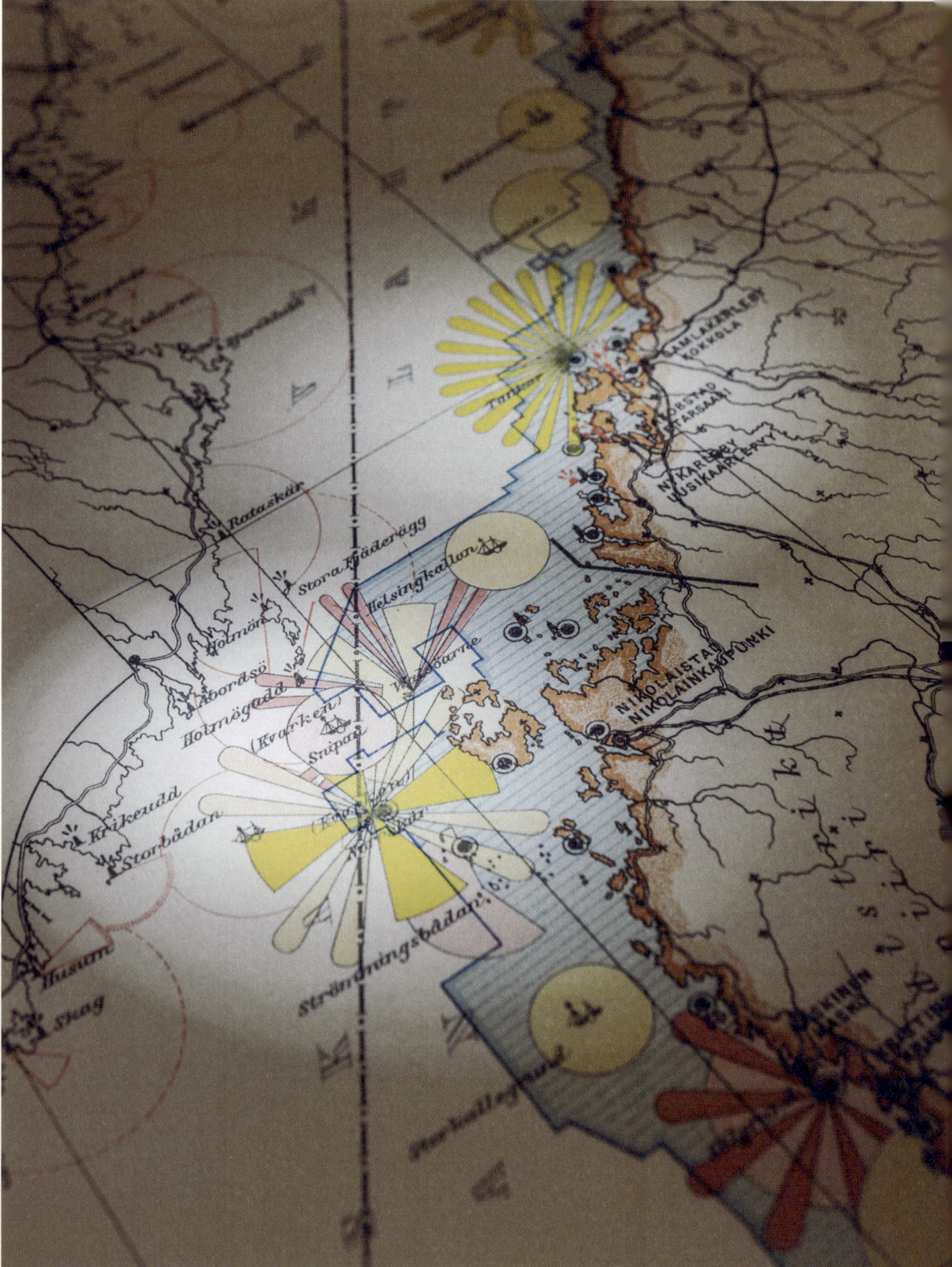

KNOWLEDGE IS POWER

Maps and Nationality

The development of ethnographic maps brought a new wave of evidence to the table for politicians looking to stake out their territory. They could for the first time compare their borders to the people that were encapsulated by them, and if they felt an ethnic or cultural group appeared to be dissected they could craft an argument to extend the border into a neighbouring state.

To understand how, we first need to immerse ourselves in the map of Europe as it was in the 1870s, and I found a map that provides the perfect portal to get us there. This was created by one of Heinrich Berghaus's star pupils (and adopted son), and the man George Bellas Greenough commissioned to engrave his geology of India map: Augustus Petermann. Petermann was one of the most prolific and famous cartographers of the nineteenth century. He moved to Britain to work on the English translation of the *Physikalischer Atlas* before setting up his own studio, working his way up to the status of 'Physical Geographer to the Queen' before moving back to Germany to take over the map publisher Justus Perthes (the firm that I mentioned in Chapter 2 was raided by the OSS nearly a century later).

The map's full title is *Karte von Europa und dem Mittelländischen Meere* (Map of Europe and the Mediterranean Sea) and it was published in Gotha (Germany) by Justus Perthes in 1871. Mounted on fine linen and with delicate loops stitched to its top corners from which it could be hung, its title tells us that it is the result of Petermann's enlargement and

Map 41 [Extract]
My favourite page from the *Atlas de Finlande*, which was an atlas that helped to define the nation as it was beginning to break away from the Russian Empire. All aspects of Finnish life and culture were mapped, including the architecture of its many lighthouses and their sequence of flashes. Here we see the lighthouses on the Finnish side of the border that bisects the Gulf of Bothnia (the northernmost part of the Baltic Sea).

Atlas de Finlande: Société de Géographie de Finlande (1899)
49.5 × 39cm

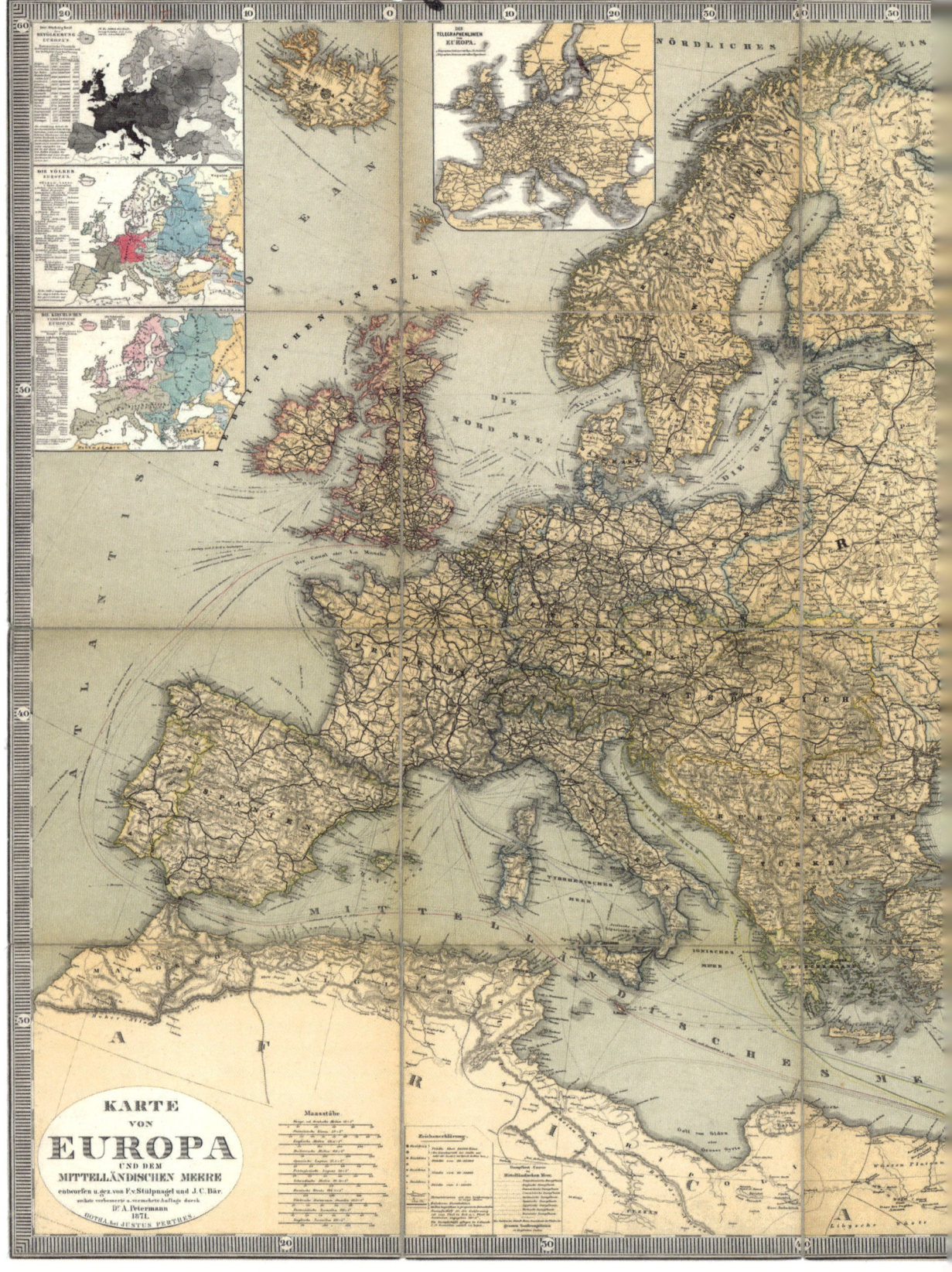

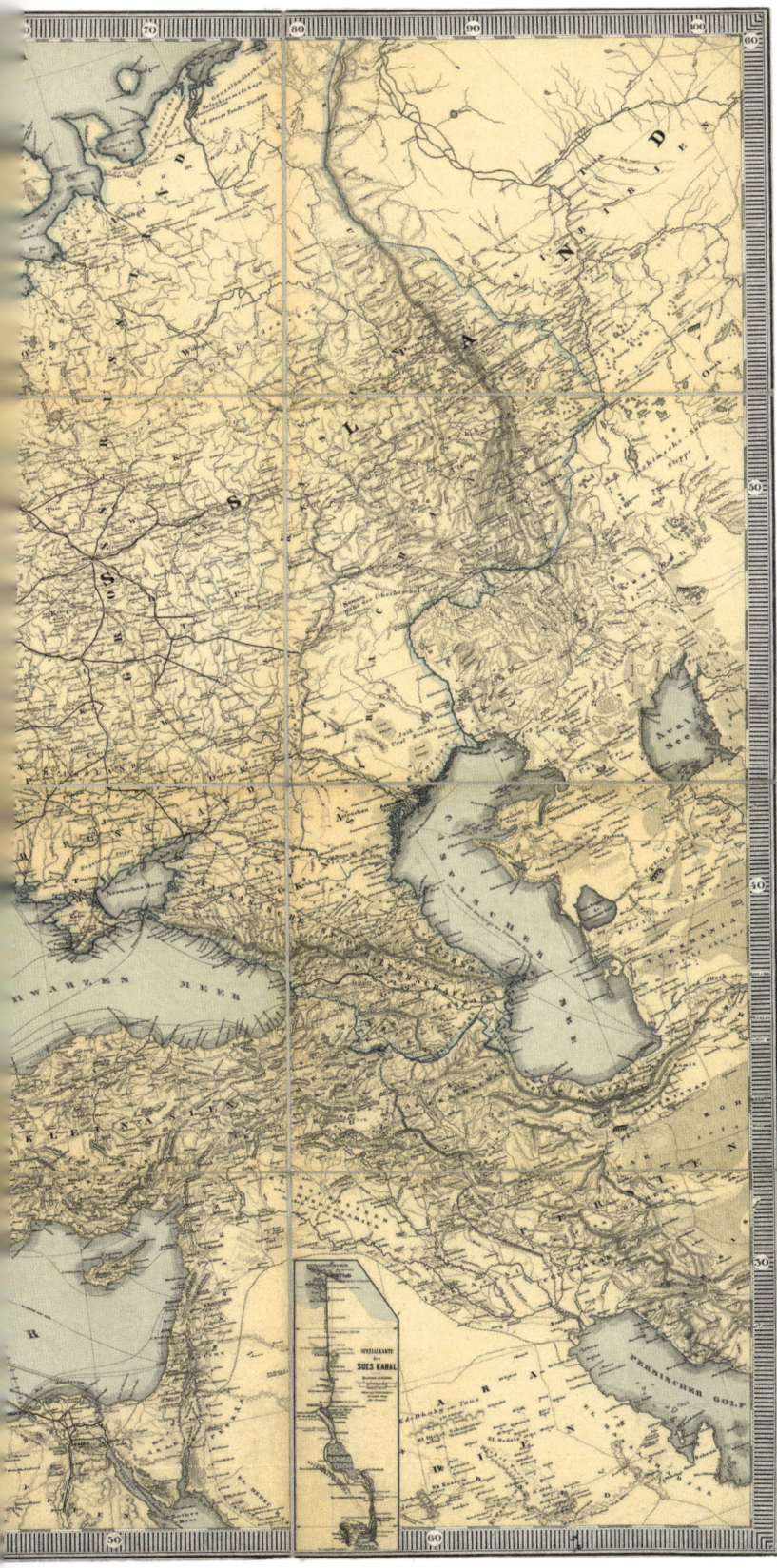

Map 42 One of Augustus Petermann's maps of Europe. It is a fabulous example of the high-quality cartographic outputs he produced during his time with the publisher Justus Perthes. You may have spotted that the decorative border around the edge of the map is the same as the one used on Berghaus's ethnographic map, which is a reminder of his time as an apprentice at Berghaus's school. The map also captures an important moment in European history, as states developed and became better connected, reducing the friction of distance but increasing geopolitical tensions.

Karte von Europa und dem Mittelländischen Meere:
Justus Perthes (1871)
108 × 86.5cm

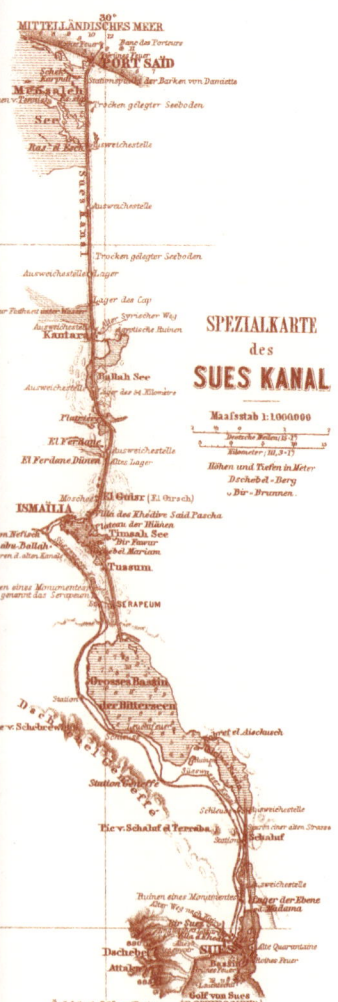

improvement of a previous map by F. V. Stülpnagel and J. C. Bär. The land is a pale straw and the sea a delicate blue. Country borders are distinguishable by thin colours hand-painted in watercolour. The thickest black lines are reserved for the growing railway network and insets are given for telegraph lines that criss-cross the continent as well as the recently opened Suez Canal.

Before researching the map, I got the sense that this was a hasty update of a previous edition: the Suez Canal inset looked like a last-minute addition, tacked on in the bottom right and overlapping the border of the map (known as the neatline) in contrast to all the other neatly placed insets. And, sure enough, this is a later edition of the original published in 1867,[13] without the canal that had only opened in 1869. In the few years between the two versions, Russia had also added a sixth radial railway line from Moscow and connected its southerly railways to the Black Sea.

In the top left corner, Petermann has included three thematic maps about Europe's growing population. The population density map of the continent – shaded in greys and blacks and updated to 1870 – shows a more populated core that spans the diagonal between Italy and the British Isles, with lower densities the further eastwards and westwards you go. Petermann then invites us to compare that pattern to the ethnic groupings on the continent, with 'Deutsche' in vibrant pink, 'Russe' in turquoise and other groupings in more muted tones. He also gives estimates of the size of these populations. The final map, again with an accompanying table of data, shows the religious groupings, dominated by Christianity, although reaching Islam on the far south and east.

When taken as a whole, it is a map that at once shows the static, rooted aspects of culture and religion against a world experiencing population growth and almost unlimited transport possibilities. Today's trains are faster, and steamships have been replaced by aeroplanes for most long-distance hops across the sea,

KNOWLEDGE IS POWER

but the European transport and communications networks were already extensive when this map was created 150 years ago.

At the time of Petermann's map, however, the idea of a 'nation' as we think of it today hadn't yet crystallised. It was nonetheless developing as those in power pushed a fairly loose collection of cultural groupings that affiliated to particular kingdoms/empires (such as the Kingdom of Prussia, the Austrian Empire or the Kingdom of the Two Sicilies) towards more centralised control. It is therefore likely that the inclusion of the Suez Canal and additional railways was a secondary consideration to showing the numerous shifts in Europe's borders that had occurred since 1867.

Given the German-speaking audience, it would also have been important for Petermann to show the success of Otto von Bismarck's ambition to build the Second German Reich. Bismarck proved an effective strategist who had just dealt Napoleon III a serious defeat in the Franco-Prussian War. The peace negotiations that followed led to the Treaty of Frankfurt, which was signed in May 1871. It determined that Alsace and part of Lorraine would go to the German Reich, which had itself only just been created by the unification of the

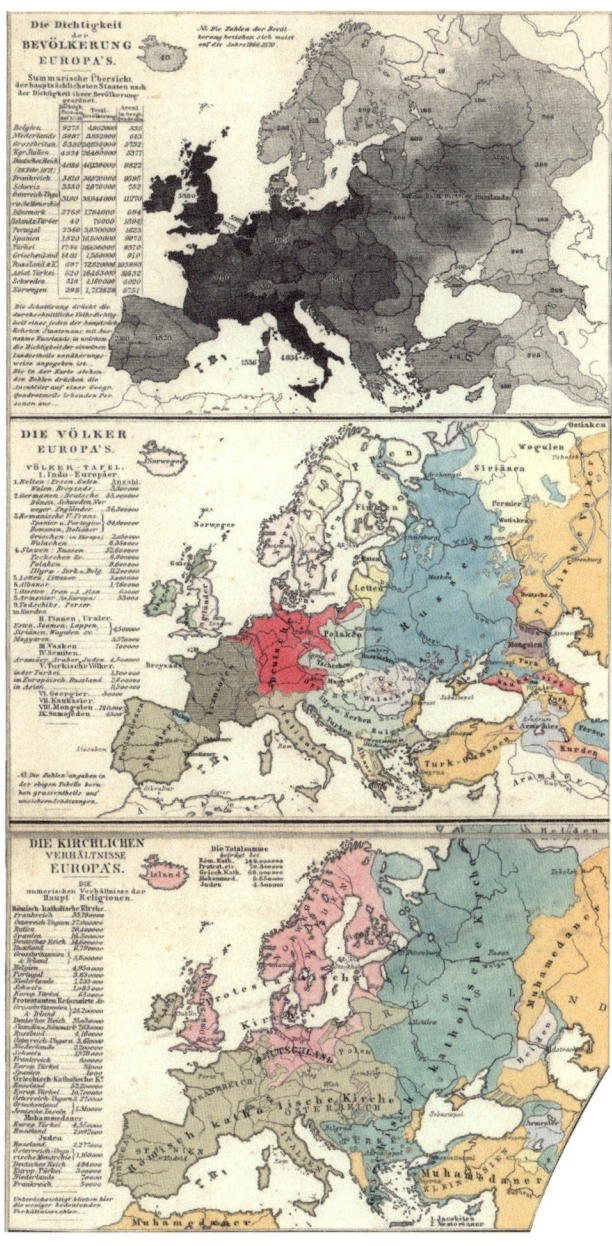

Map 42 [Extract] The population and ethnographic inset of Petermann's grand map of Europe. Maps like these would come to define the next century.

collection of smaller Germanic states that fell outside of the Austro-Hungarian Empire.

Interestingly, at the same time as Petermann was updating his *Europa und dem Mittelländischen Meere* map in response to changes in the size and shape of states, another of his maps was being used to inform the new boundaries as they were negotiated. A linguistic map[14] of the Alsace region (in today's France) that he had produced was used by the victorious Bismarck to support the claim that the territory should be part of Germany.[15] It was a map that showed the national borders as they were in 1870 and then the 'linguistic' borders between the German- and French-speaking populations that were not well aligned to the states.

It was an early example of the way that a thematic map showing ethnographical or cultural information could be used to leverage negotiations in tandem with the usual strategic and punitive considerations that accompanied peace talks. In this case, Bismarck was driven by the desire for a defensive buffer against France. The map gave him the knowledge, and therefore the power, to make both a militaristic and a cultural argument.

This was an important moment in Europe's history, as Karl Marx reflected at the time: 'History will measure its retribution not on the extension of square miles ripped from France, but from the scale of the crimes created by the policy of conquest in the second half of the 19th century.'[16] The fate of these regions continued to hang in the balance until the end of the Second World War and ethnographic maps became seen as the legitimate basis to territorial conquests.[17]

Combined with the desire to demarcate nation states based on ethnographic groups, there was a growing sense that, for a nation to survive, it needed to outcompete its neighbours for resources and influence. The extraordinary growth in transport connectivity shown on Petermann's map meant that goods and people (and armies) could be moved more rapidly over greater distances. Politicians saw a more distant horizon to reach for and over which to exert their influence. As well as expanding and securing borders, there was also a heightened concern about the increasing resource demands within them, as populations grew and industrialised.

Thanks to cartographers like Petermann, maps could now provide an image of the geographic distributions of people, resources and economic outputs that, with Bismarck's use of the language map for political ends, demonstrated the power of cartography to change history and build nations. The realisation that maps could show the many different dimensions of a state, such as its outlines, its physical and natural resources, population characteristics, industrial output and so on led to the natural conclusion that such maps could be bound into a 'national atlas'.

National atlases were very much in the mould of what Berghaus achieved with his *Physikalischer Atlas*, but rather than offering a global view, national atlases could be bespoke to the country in question. In an increasingly competitive and hostile world, they also had the advantage of controlling the narrative because they were created (and funded) by the state for the state. Borders could be drawn where the nation believed they should be, regardless of what their neighbours thought. As we saw with the Barbie movie (see Chapter 1), this attitude still plays out today, and widespread censorship and prohibition of the 'incorrect' borders within some countries can manifest itself in surprising ways.

For example, India, Pakistan and China are currently involved in a border dispute over the Aksai Chin region of Kashmir, and I spotted that this is even evident in the most peaceful setting: the globe Christmas baubles I hang from my tree each year. On the Indian-made decorations the Indian border has a kink that prevents Pakistan from touching China. On the Chinese baubles we see Pakistan extending to China and the kink is gone. Of course, these differences will usually go unnoticed, but if I were to take the Indian globe to China I could be reprimanded by the authorities. Travel to India and you will have to sign a customs form declaring that your luggage does not contain narcotics, drugs and psychotropic substances, wildlife products and 'maps and literature where Indian external boundaries have been shown incorrectly'.[18] And it's not just baubles: I've seen the same border discrepancies on globes and map prints in homeware stores that would see you in trouble if you were to display them in the wrong country.

In 1899, the *Atlas de Finlande* started this trend with a bang, by creating what is considered to be the first national atlas and asserting the Grand Duchy of Finland's independence from Russia.[19] At the time Finland was an autonomous region within the Russian Empire. It therefore stands to reason that the atlas would have shown Finland's internal border with Russia as a thin line, setting it apart from the thicker lines that defined her international boundaries with Sweden and Norway. Finland, however, took the provocative decision to use that same thick international border line to separate herself from Russia. The atlas was also written in French, and therefore for an international audience, which Finland knew it would need to support its claims, in much the same way it did when the frontier was back in the spotlight with Finland's accession to NATO in April 2023.

Leafing through the copy I discovered on a shelf alongside various Scandinavian atlases, I did not immediately appreciate the political statement the *Atlas de Finlande* was making. Instead, I became distracted by the sheer range of topics it featured, covering everything from the trade of sawn wood to the location and architectural style of lighthouses; from the extent of pine forests to the prevailing wind direction. These were all things that asserted a sense of the soon-to-be nation's independence and

KNOWLEDGE IS POWER

uniqueness. The maps were a good accompaniment to images of rural, lake and forest scenes and wilderness areas, which were also deployed to evoke a sense of national identity.[20] Finland showed that to be a serious country you therefore needed a serious atlas: clearly many other countries agreed and we have atlases in the Map Library from Mexico to Indonesia and many states (current and former) in between.

National atlases required their makers to step back and assume a god's eye view to define what makes a nation. They were considered by their creators and most of their readers as books for reference rather than rhetoric, even if they were treated very differently by those who disagreed with their contents. This tension is one that would only build as the nineteenth century progressed into the twentieth.

In the first half of the twentieth century, atlases became statements of national identity and strength. This atlas of Mexico, published in 1946, was a low-budget production compared to some of its European counterparts since it is printed on very thin, fragile paper, but this frontispiece is the grandest example I saw in the Map Library.

Atlas Geographico Estados Unidos Mexicanos (1946)
48.5 × 37.5cm

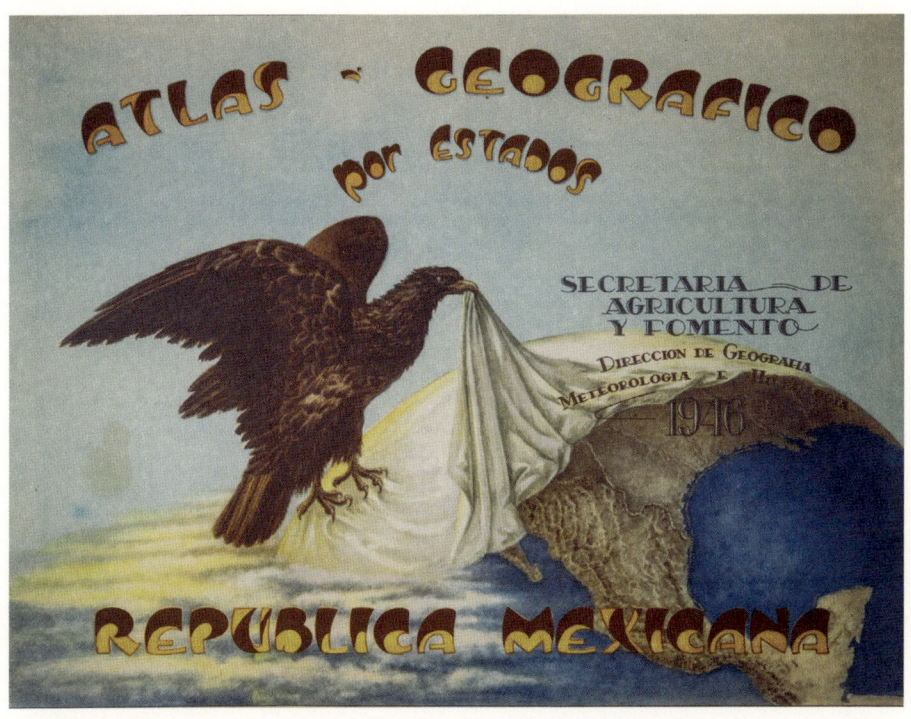

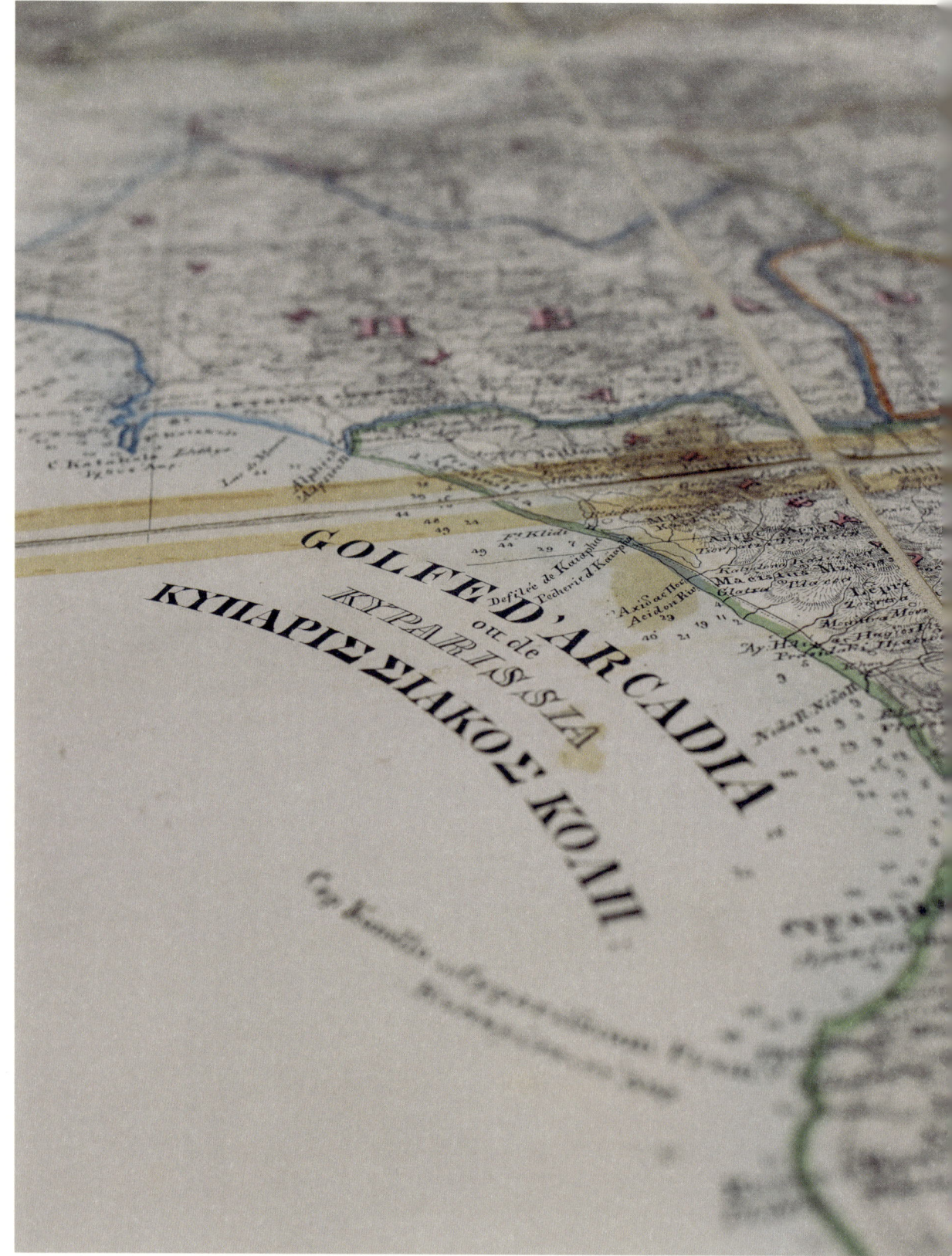

KNOWLEDGE IS POWER

The World as a Chess Board

Throughout the nineteenth century there was a growing appreciation that scientific advances were offering important explanations for some of the biggest questions about life on Earth and enabling industrial progress at an unprecedented rate. The more scientifically 'advanced' nations, equipped with their recently acquired maps, began to consider themselves world-builders as they gave themselves precedence over those who they considered were lagging.

Maps meant that these nations were no longer abstract ideas, but something more tangible and politically manageable.[21] To some leaders, their maps resembled a board game of opportunities and threats. There had to be winners and losers: it was 'survival of the fittest'.[22] Indeed, the territorial rivalry in Central Asia between the British and Russian empires became known as the 'Great Game', and maps became the playing surface for such rivalries.

One figure offering commentary on how the game might be played was the British geographer and politician Halford John Mackinder. In January 1904 he read what would become an iconic paper (to a far from sellout audience[23]) at the Royal Geographical Society in London titled 'The Geographical Pivot of History'. He set out his idea that there was a 'Heartland' that lies at the centre of the 'World-Island', stretching from the Volga to the Yangtze and from the Arctic to the Himalayas. The key to global dominance was the control of this central area of Eurasia, due to its geographic and strategic advantages. Mackinder summarised his theory thus:

Map 43 [Extract] This is a small detail of an extremely rare map of the Kingdom of Greece that was created in 1838, six years after its establishment. Published by the Royal Lithographic Printing House in Athens, and dedicated to King Otto, this map makes a statement about the sovereignty of the young country. It is one of the largest maps in the collection and extremely well executed with labels not just in Greek, but also in French (French surveyors had a long history of mapping the region). The map includes areas (such as Crete and Macedonia) outside of the kingdom, suggesting aspirations for territorial expansion beyond its shores.[4]

Χάρτης του Βασιλείου της Ελλάδας/*Carte du Royaume de la Grèce*: ΒΑΣΙΛΙΚΗ ΛΙΘΟΓΡΑΦΙΑ/*Lithographie Royale* (1838)
169 × 111cm

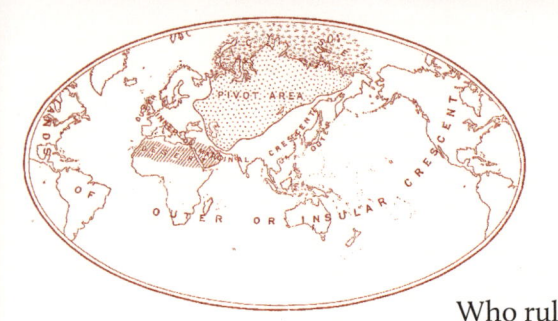

Map 44 The simple map Mackinder used to illustrate his concept of a geographical 'pivot area' centred over Eurasia.

The Geographical Pivot of History: The Geographical Journal (1904)
13 × 18.5cm

Who rules East Europe commands the Heartland;
Who rules the Heartland commands the World-Island;
Who rules the World-Island commands the World.[24]

He sounds a bit like a sports pundit making predictions for a World Cup tournament as he sets out the dependencies and vulnerabilities of each of the players and how they might perform in the game of geopolitics:

> Russia replaces the Mongol Empire. Her pressure on Finland, on Scandinavia, on Poland, on Turkey, on Persia, on India, and on China, replaces the centrifugal raids of the steppemen. In the world at large she occupies the central strategical position held by Germany in Europe. She can strike on all sides and be struck from all sides, save the north.[25]

After Mackinder had read his paper there was a discussion led by an academic and writer named Spencer Wilkinson, who was clear about the way that the world should be seen:

> Whereas only half a century ago statesmen played on a few squares of a chess-board of which the remainder was vacant, in the present day the world is an enclosed chess-board, and every movement of the statesman must take account of all the squares in it. I myself can only wish that we had ministers who would give more time to studying their policy from the point of view that you cannot move any one piece without considering all the squares on the board.[26]

Mackinder had many supporters, but there was nonetheless philosophical disagreement about the nature of geopolitics among intellectuals at the time. An influential German geographer named Friedrich Ratzel had looked at a world map and, rather than seeing static countries seeking control over a particular piece of territory, he saw moving amoeba-like organisms that needed to sustain themselves by outcompeting their neighbours in true Darwinian fashion. For Ratzel, who would have been in his twenties when Germany was unified in 1871, the creation of nation states was 'the greatest achievement of man on earth' and the 'climax of all phenomena connected to the spread of life'.[27] In his view, the strategic objective of a nation once it had formed should be to acquire enough space for its population to survive.

It is this thinking that led to the now infamous concept of *Lebensraum*, defined as 'the geographical surface area required to support a living species at its current population size and mode of existence'.[28] According to Ratzel: '[Every] new form of life needs space in order to come into existence, and yet more space to establish and pass on its characteristics.'[29] Through Ratzel's eyes, borders become dynamic and contested as states jostle for space. He, like many thinkers at the time, also put significant weight on the way that natural and physical characteristics of a state (its climate, terrain, etc.) can dictate how successful it is.

In addition to Mackinder and Ratzel, there were other theorists[30] such as France's Paul Vidal de La Blache, whose *Atlas général Vidal-Lablache* the Map Library seems to have acquired from the Bodleian Library in 1946.[31] His view was more nuanced than Ratzel's in the sense that he placed more emphasis on a two-way relationship between humans and their environment. Vidal de La Blache developed the concept of 'possibilism' which gives societies more agency in the way that they engage with their surroundings, thus reducing the constraints that the physical environment may have on them. As inhabitants of each region developed adaptations to its environment, such as specific architecture and agriculture, they established their own unique characteristics that formed cultural landscapes.

Understanding how such landscapes developed over time and how they might differ between regions, Vidal de La Blache argued, would reveal answers to the creation of political entities. He was keen to stress this in the preface to his atlas, where he encourages readers not to look at any map of a country or region in isolation but to look across the multiple maps he provides to get a sense of the interrelationships between the phenomena. By expounding gentle evolution rather than the constant sense of competition and opportunism, this was a much less boisterous perspective than Ratzel's.[32]

The use of maps to demarcate the field of play for the game of geopolitics was an important moment in global history because maps gave statesmen the perspective of a world maker, not a world taker, and they had the support of their leading intellectuals. It was a potent combination that played out on the ground as those in power mistook the maps before them for the territory itself.

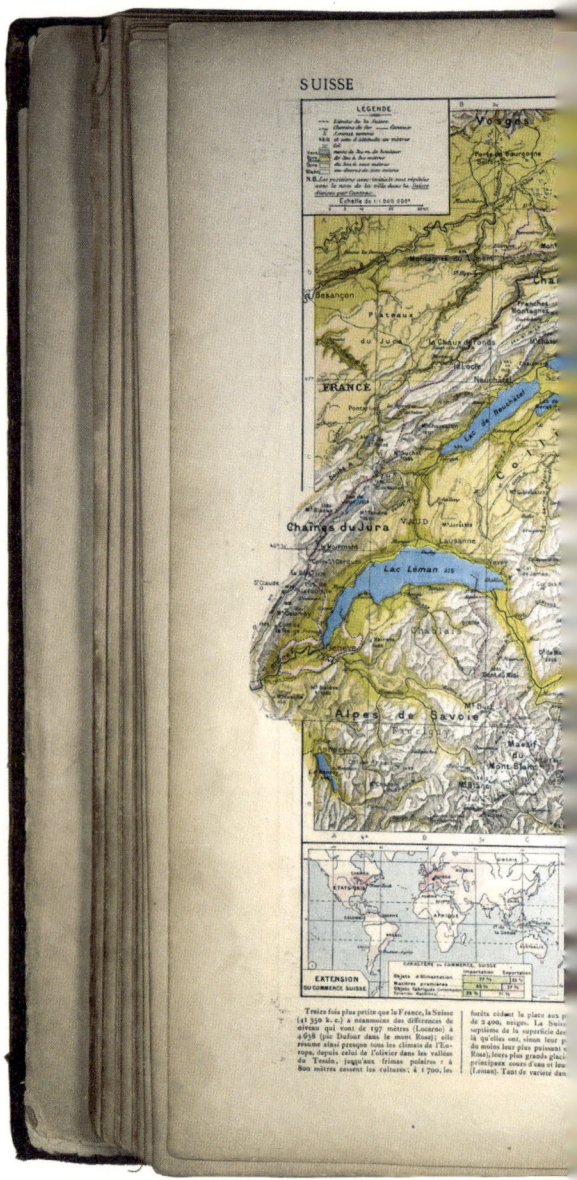

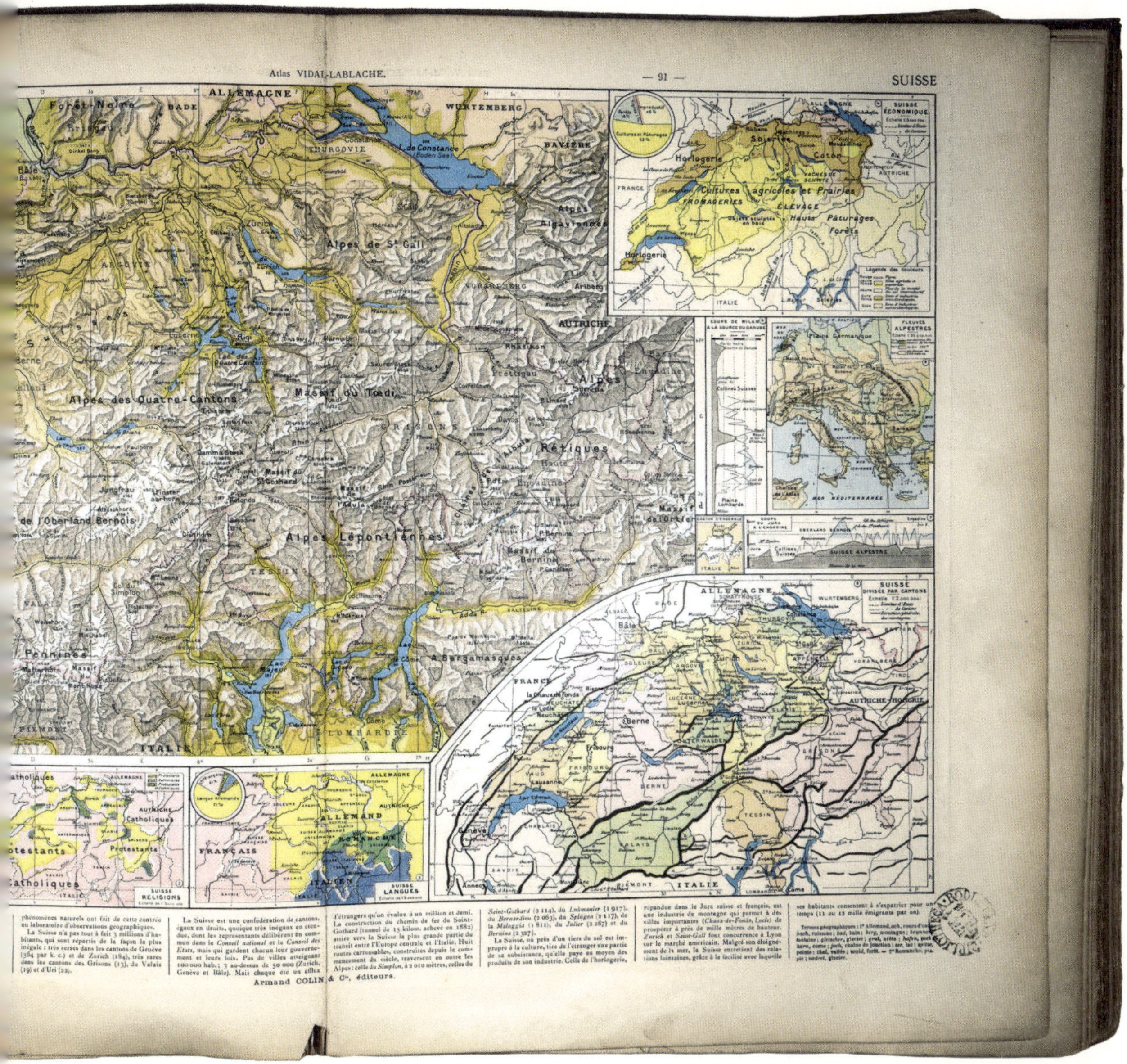

Map 45 In the preface to his atlas, Paul Vidal de la Blache encourages readers to look at the multiple dimensions of the territories depicted. For Switzerland, it being a relatively small country, he packs several panels of data onto a single double-page spread, first showing its mountainous terrain, then also economic outputs and ethnic and religious groups.

Atlas général Vidal-Lablache (Suisse):
Armand Colin & Co (1894)
43.5 × 28.5cm

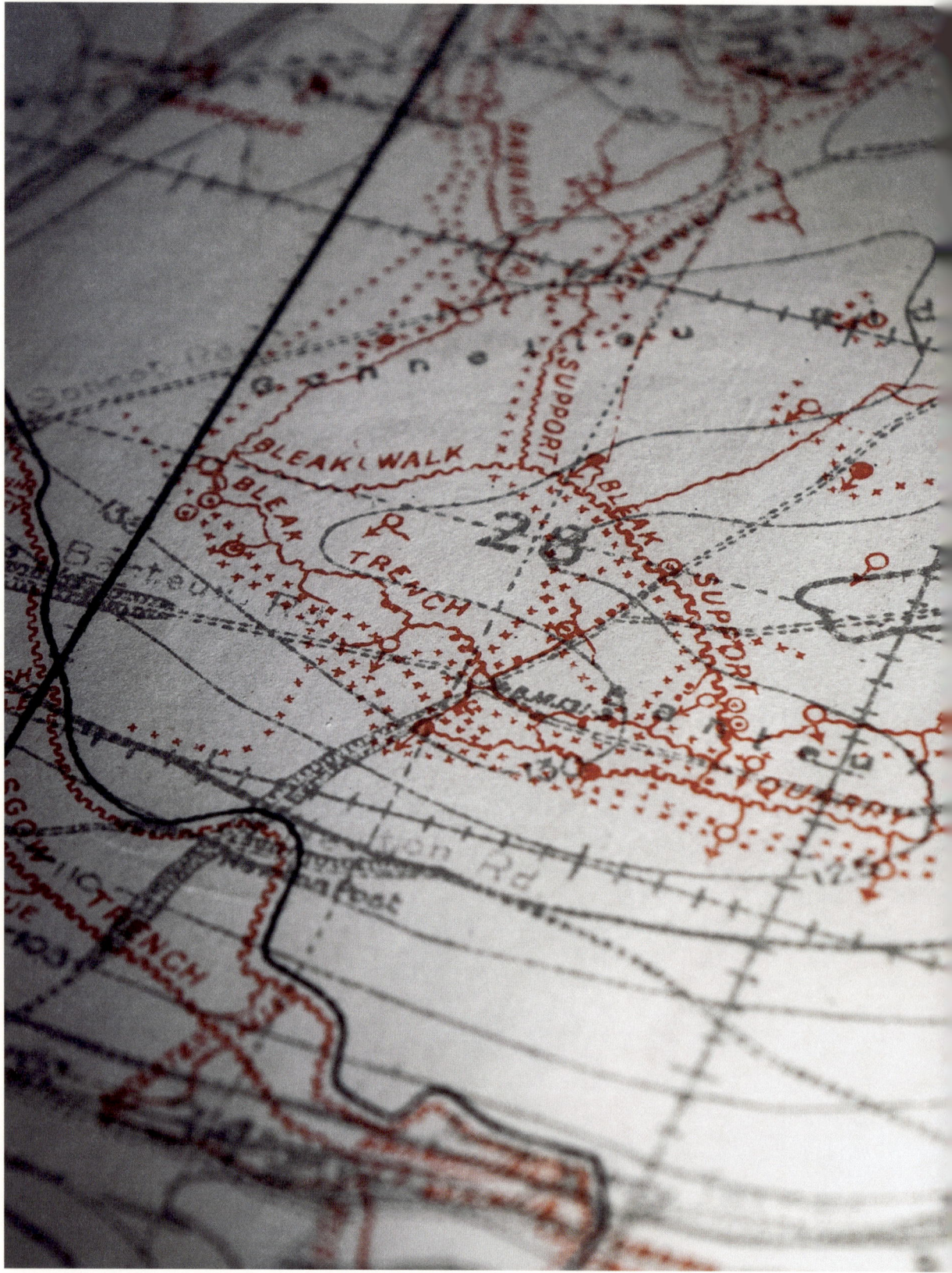

KNOWLEDGE IS POWER

A Game with Consequences

While these geopolitical debates were being rehearsed among academics, the world outside of universities and learned societies was accelerating towards the first global war as the complex web of treaties and alliances that had held Europe together began to stretch. They reached breaking point when a Bosnian Serb nationalist assassinated Archduke Franz Ferdinand, heir to the Austro-Hungarian throne, on 28 June 1914. Convinced that Serbian nationalism and Russian Balkan ambitions were disintegrating its empire, Austria-Hungary hoped to trigger a limited war against Serbia, and that strong German support would force Russia to keep out of the war and weaken its influence in the Balkans. Things then spiralled and Britain declared war on Germany on 4 August 1914, ostensibly to protect Belgian neutrality, which had been compromised by Germany's invasion of the country en route to France (with which Germany was at war due to France's allegiance to Russia), but also in fear that victory for Germany would leave Britain vulnerable without any allies. Perversely, as we shall see, rather than prove the folly of some of the more combative ideas of geopolitics, these events bolstered them.

My time in the Map Library has left me in no doubt that maps can be incredibly self-serving. They can make the argument for war in the first place and then assume even greater importance than the weapons used to fight it, before becoming the images used to communicate its consequences to a public away from the front lines and hungry for news. This was the case in the First World War,

Map 46 [Extract] In the relative tranquillity of the Map Library, it is simply impossible to imagine what some maps bore witness to. First World War maps volunteer glimpses of the horrors in the names that the soldiers assigned to some of the trenches. I invite you to pause and consider what a night spent along 'Bleak Walk' must have been like for those stationed there.

Positions Map 30-10-17 Gouzeaucourt: Geographical Section General Staff (1917) 77.5 × 52.5cm

THE LIBRARY OF LOST MAPS

Map 47 'Charing Cross yesterday resembled the Tower of Babel. Nearly every nationality was represented in the crowds that besieged the booking-offices for tickets to the Continent ... Men who, in a few days, may be trying their hardest to kill each other laughed and joked as if war were only a pastime.'

First Signs of War in London: Foreign Reservists leave in Hundreds: Daily Sketch (1914) 27.5 × 20cm

where maps were created to enable new types of warfare (long-range artillery, for example) but also to provide updates to a huge newspaper readership.[33]

There is one map that I purchased a few years ago that, despite being incredibly simple in its construction, captures the naivety of what was to come as Europe barrelled towards war. Pulled from the *Daily Sketch* newspaper, it is titled:

FIRST SIGNS OF WAR IN LONDON:
FOREIGN RESERVISTS LEAVE IN HUNDREDS.

It is little more than an outline of London, with Bloomsbury at its centre, and the major parks and the River Thames there as a minimum to help readers orientate themselves. Alongside the place names are the counts of foreign residents by nationality; for example, we see that Hampstead was home to 6,000 Germans and Soho to 3,000 French, Germans and Italians.

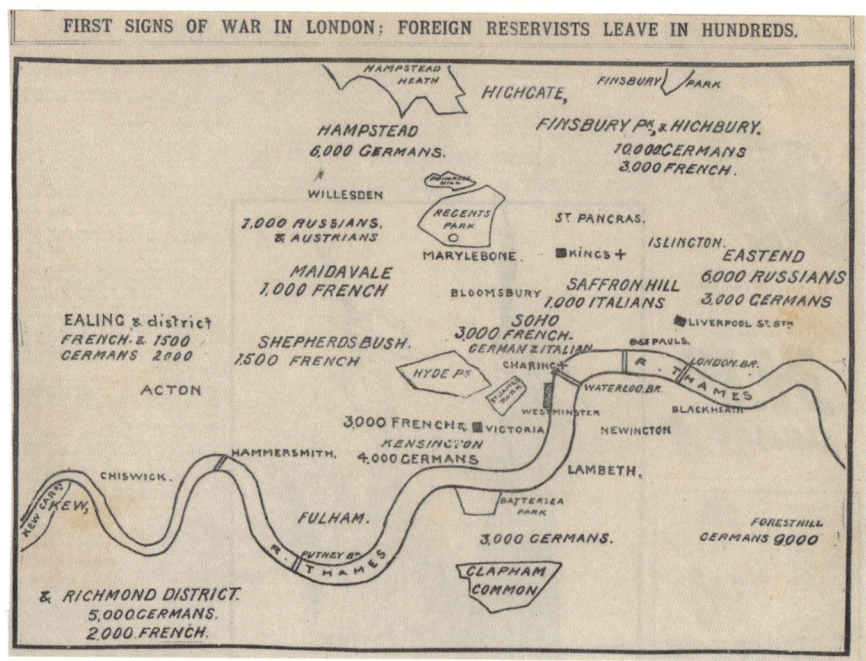

For something so simple it successfully evokes a cosmopolitan city with many nationalities living side by side. But it also conveys an ignorance of what the war would mean, compounded on the next page of the paper, which features an illustration of two women sporting enormous feathers in their hats under the headline 'Fashion for Cowes and After: By Mrs Gossip'.

The Great War started with the futility of cavalry charges and ended with the terror of gas attacks, tanks and submarines, the apocalyptic technologies that are perhaps best known. However, the innovations in map-making were just as consequential for the way that the First World War (and subsequent wars) was fought.[34]

A cautious estimate tallies 800 million map sheets printed by Germany, 60 million by Austria-Hungary and 40 million by France.[35] The British military printed about 34 million maps for the Western Front alone.[36] They were produced at various scales for different purposes such as general maps for administration, more precise maps that were developed for the artillery to pinpoint their targets and the highly detailed maps to document the networks of trenches that marked the front line. Most of the British maps showed only the German trenches for fear they might fall into enemy hands, although from September 1915 onwards 'secret' editions were created to show the British side.[37]

While it had printed huge numbers of maps by the end of the conflict, the British mapping operation got off to an inauspicious start. At the onset of the war, surveyors arrived in France with the belief that traditional mapping approaches, using on-the-ground surveys, would suffice. The War Office had provided medium-scale topographic maps of Belgium and north-eastern France and anticipated a trickle of updates from surveyors in the field. General Hague, commander of the British Expeditionary Force, endorsed this approach, commenting, 'I hope none of you gentlemen is so

(Overleaf)

Map 48 [Extract] The Map Library has only a few maps from the First World War, but I was pleased to find a fascinating map, in two parts, that showed the Russian railway network and its daily capacity during the harsh winter of 1916–17. This would have been essential for planning the logistics of war, but also to assess Russia's capacity to transport food at a time of serious shortages, which contributed to the unrest that led to the overthrow of the Russian monarchy in February 1917.

Railway Administration Map (Russian Railways Sheet 1 & 2): Ordnance Survey (1918) 125 × 89cm

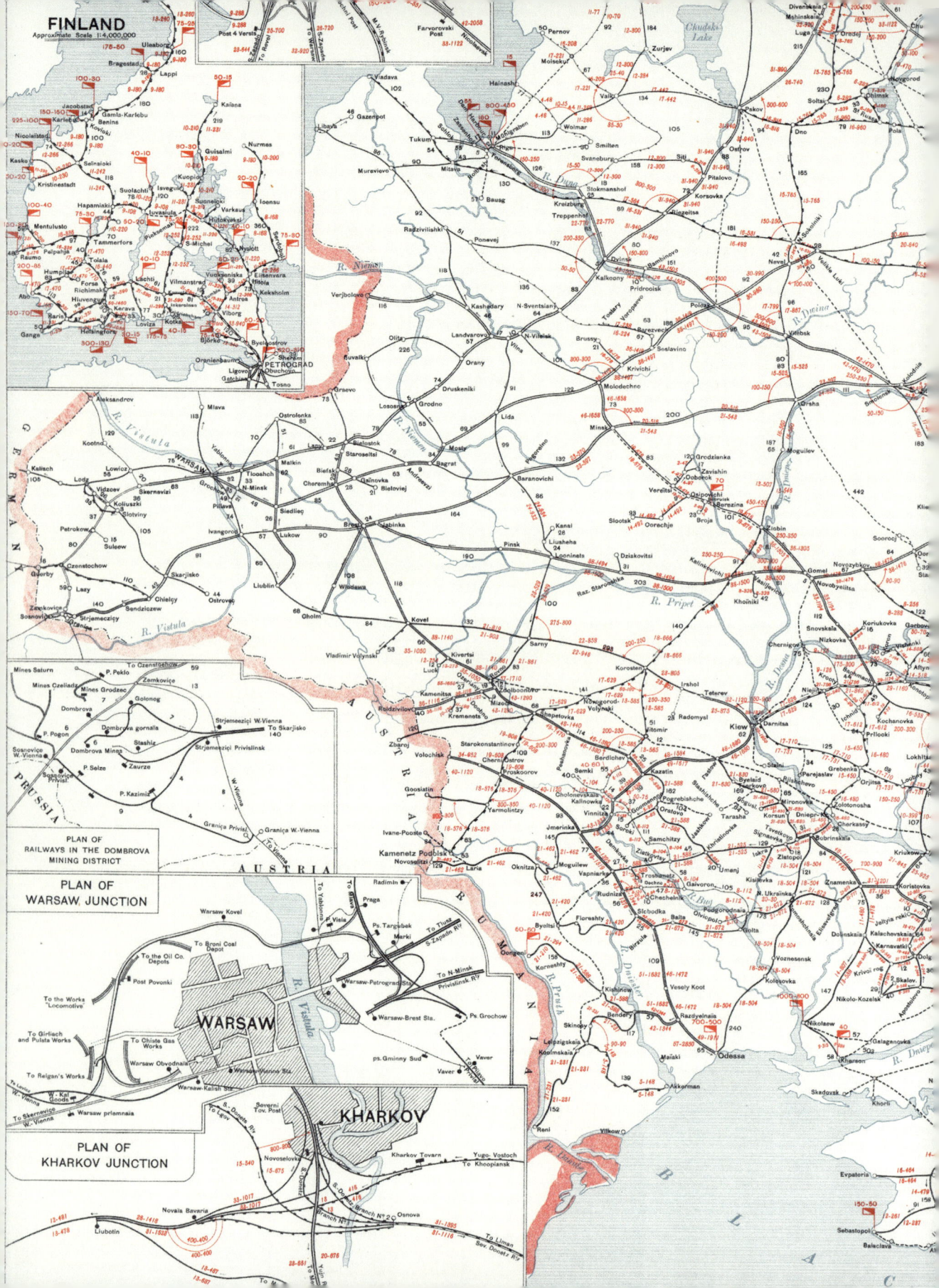

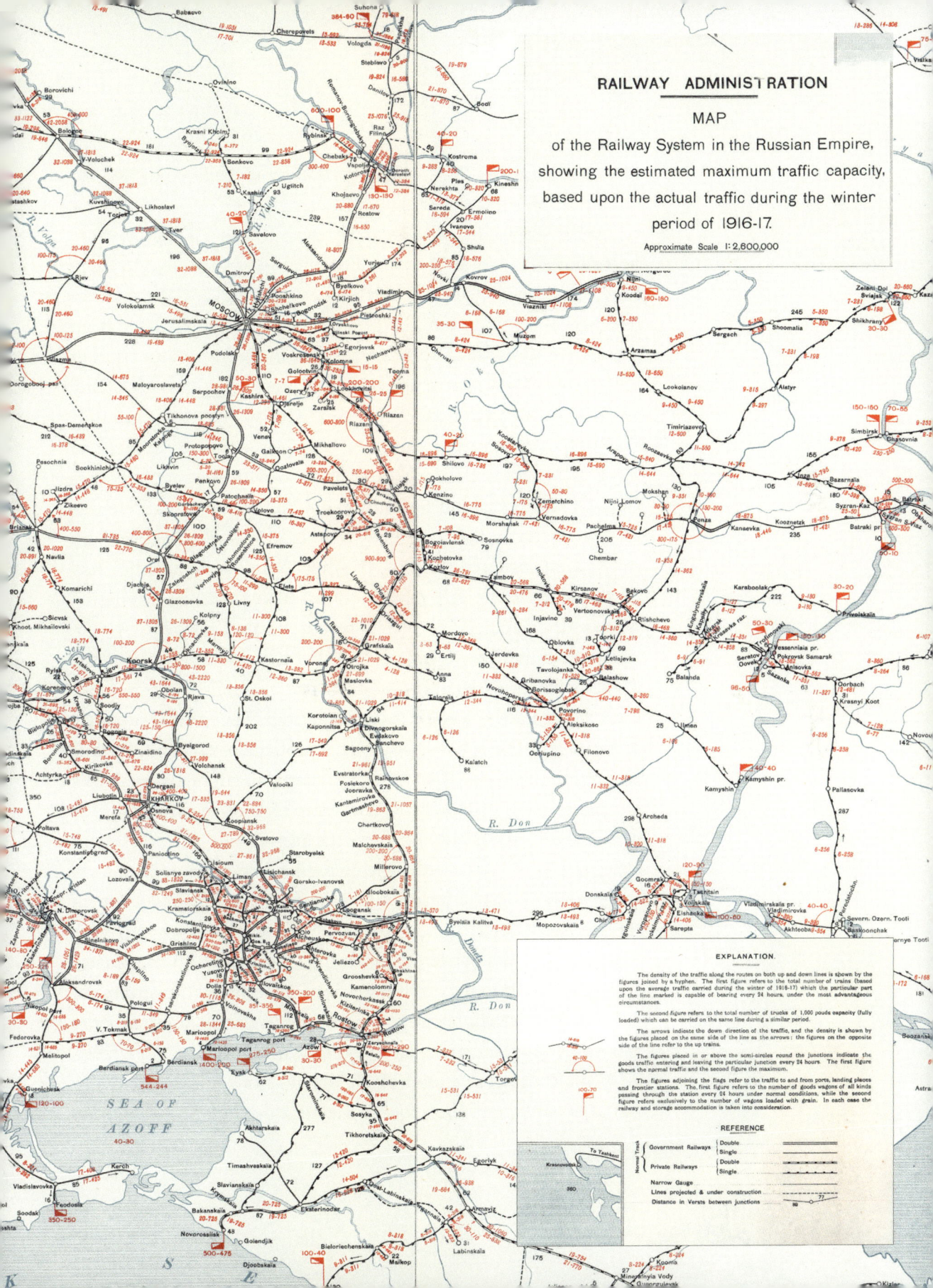

foolish as to think that aeroplanes will be usefully employed for reconnaissance purposes in war.'[38]

With hindsight, however, it's clear that such tactics were never going to work. They required teams of surveyors to carry heavy equipment, such as theodolites, tripods, drawing boards, telescopes and long lengths of steel tape.[39] It would have been impossible to move at speed, especially in the notorious mud of the Western Front. Surveyors became sitting ducks for both enemy and friendly fire. One was even arrested for spying on the grounds that only the Germans would be carrying survey equipment in the area in which he was picked up.[40] So with the methods that General Hague had put his faith in proving inadequate, there was no option but to bring in the Royal Flying Corps to photograph the battlefields from the sky. This new approach, once they had mastered it, was hugely effective and enabled the rapid updating of maps that changed battlefield reconnaissance forever.

Gathering the information needed was only one part of the process. The updated maps had to be printed and sent to the front lines. The military had field printing stations they could set up close to battle, but these, like the survey equipment, were cumbersome and required specialist staffing. For example, engravers were still needed to etch into the heavy lithographic printing stones which were used to press the ink onto the paper, and once this work was completed the infield printing press could only produce forty maps per hour. The enormous length of the Western Front, which ran for 700 km (440 miles) from the coast of Belgium, through France and to the border of Switzerland, meant that this was not a scalable solution even if hundreds of printing stations were set up. Nor could they have kept pace with frequent updates from the pioneering aerial surveyors. A typical lithographic engraver, at speed, could etch a moderately detailed map at the rate of ten minutes per square inch, which means a map on an A4 sheet of paper (much smaller than they would have needed) would have taken sixteen hours.⁴¹

Therefore, to sustain the volume of updates and printing required, the intelligence was initially sent back to the headquarters of Ordnance Survey in Southampton, where it was processed into new maps that could be printed in bulk.⁴² To speed things up further, many of the base maps were pre-printed in black and white and then the new information, such as trench positions, could be more rapidly printed on top using a bright ink (such as red). Some base maps were dispatched to the field in this state for rapid smaller print runs for areas of strategic value.

However, shipping maps across the English Channel between Southampton and the front lines became a risky business, so in 1917 Ordnance Survey set up an overseas branch, staffed by 103 men and 46 women, and based it in an old factory near the Aire Canal at Wardrecques in France. From here they could churn out over 100,000 maps a week without the risk of them falling foul of the German navy.⁴³

Looking at those maps today we are spared the shellshock-inducing sensory onslaught that must have been happening around them as they were being used to navigate what has been described as 'the topography of Armageddon'⁴⁴ and by Wilfred Owen as the 'topography of Golgotha'.⁴⁵

In his award-winning book on First World War literature,⁴⁶ Santanu Das uses the term 'slimescape' to describe the condition of the front line. He opens his book with a quote from a front-line newspaper written in March 1917, which is a sobering contrast to the upbeat commentary that accompanied the *Daily Sketch* map of London: 'At night, crouching in a shell-hole and filling it, the mud watches, like an enormous octopus. The victim arrives. It throws its poisonous slobber out at him, blinds him, closes round him, buries him …'⁴⁷

ARTILLERY.

EMPLACEMENTS visible on photograph ° ° ° °
BATTERY located approximately ⊙
SUSPECTED AREAS ringed ◯

NOTE.

The battery signs give the number of pits and approximate position only. For correct positions see Third Army List of Emplacements

THE LIBRARY OF LOST MAPS

Map 46 It never saw action on the front line, but this First World War map shows the complexity and the density of the German trenches (in red) in a key part of the Western Front. Trench maps were frequently updated and, as you can see at the top, this one was printed on 30 October 1917. A month later the Germans would launch an attack from these trenches that would briefly allow them to take Gouzeaucourt village, before it was won back on the same day by the Irish Guards in a bitter battle; 918 casualties are buried in the British cemetery nearby.[5]

Positions Map 30-10-17 Gouzeaucourt: Geographical Section General Staff (1917) 77.5 × 52.5cm

The maps that portrayed the battlefield were devoid of this 'poisonous slobber' and could never capture the hellish microcosms that the troops were living in. Far behind the lines with the clean maps laid out in front of them, the generals struggled to account for this when they planned fresh assaults. In his analysis of how such maps were used in the First World War,[48] geographer Derek Gregory highlights the two ways that the slimescape experienced in the trenches contradicted the neat and ordered lines of the battle space depicted on the staff officers' maps and plans.

First, 'the paper war was confounded at every turn' as soldiers moved much more slowly across the terrain, sinking deep in the mud as they went. Secondly, Gregory suggests that 'surviving the slimescape required a "re-mapping" ... in which other senses had to be heightened in order to apprehend and navigate the field of battle'.[49] Along large stretches of the Western Front, looking for navigational clues in the landscape based on the maps was no longer an option because the terrain had been bombed beyond all recognition. In such desolation a corpse trapped in barbed wire or a rancid pool in a crater became the grim landmarks by which soldiers would orientate themselves.

Finally, on 11 November 1918, after the armistice was signed in a railway carriage in the Forest of Compiègne in France, the guns fell silent. Initial terms were set out to include the withdrawal of German troops from occupied territories, the surrender of military equipment and the continuation of a naval blockade. But the new borders of Europe were not simply drawn along the front lines that had been recorded on the final day of fighting; there needed to be a wholescale revision of the map. And there was not just the Western Front to consider: major empires had been left crumbling across Europe, their territories extending to Asia and Africa.

These issues became the focus of the Paris Peace Conference, which led to a number of major treaties including the Treaty of

KNOWLEDGE IS POWER

Versailles, signed on 28 June 1919, and the Treaty of Trianon, signed almost a year later on 4 June 1920. It also led to the formation of the League of Nations, which was a predecessor to the United Nations and advocated for closer international cooperation to achieve peace and security.

Of course, maps were needed like never before and it was at this conference that their makers were at the height of their powers and influence.

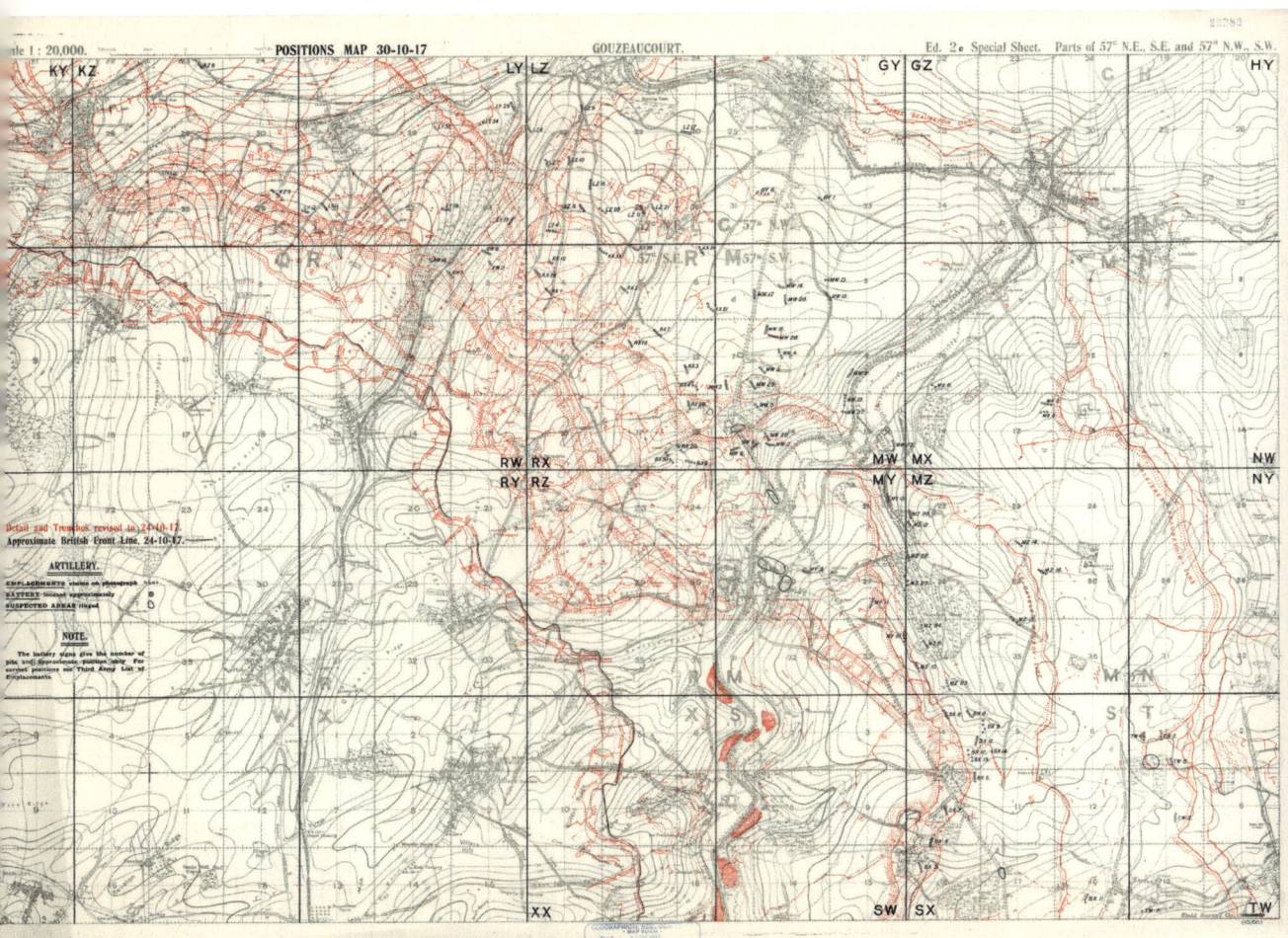

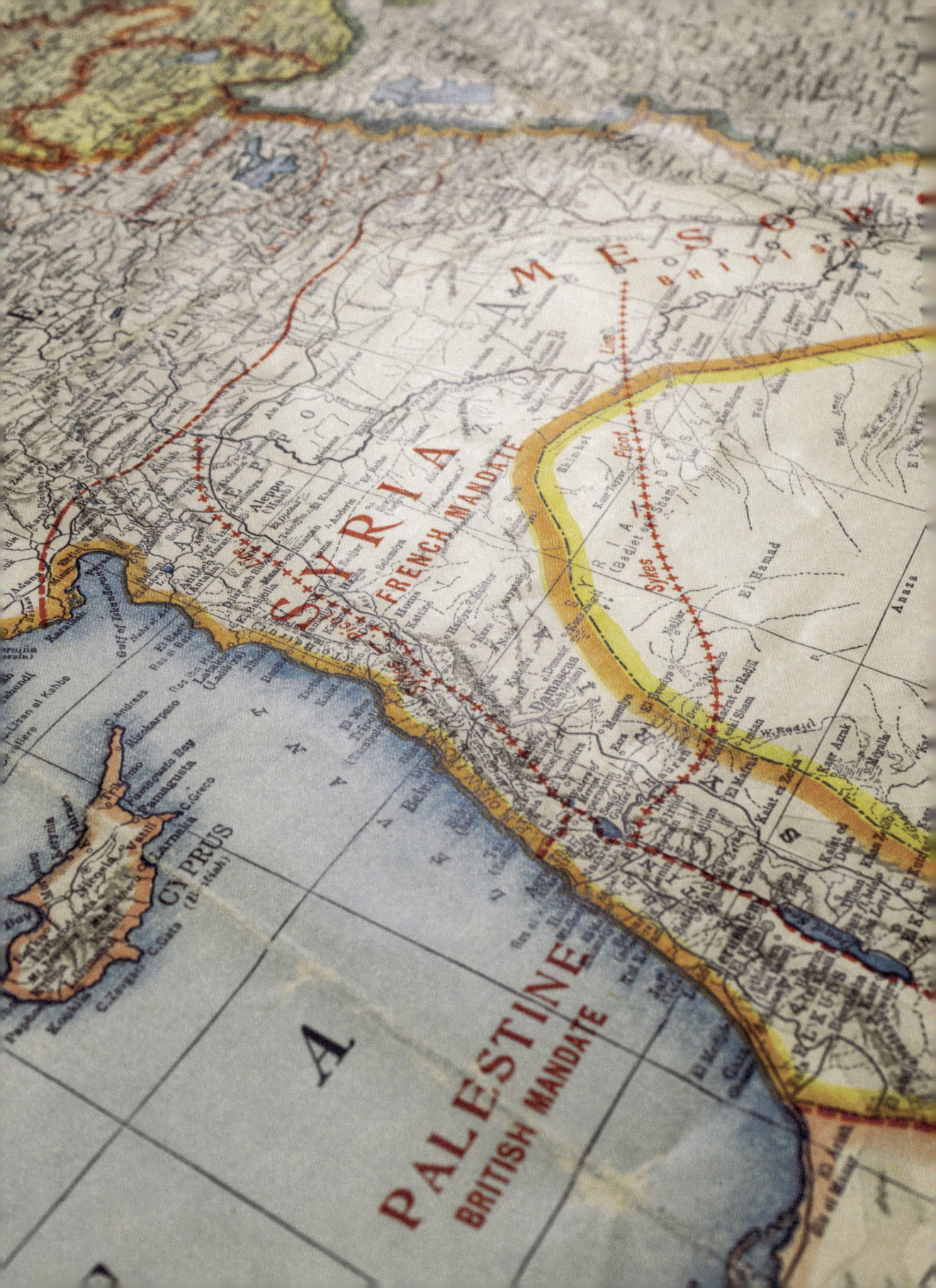

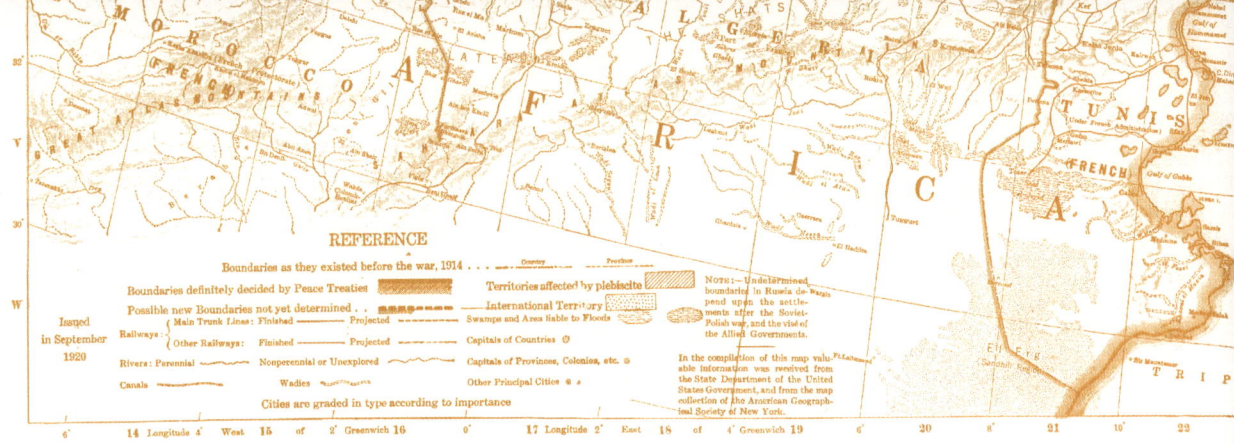

CHAPTER SIX

'Tidying' the Map

'President Wilson takes a map, spreads it on the carpet in an alcove-room, and kneels down. We all squat in a circle round him. It is like hunt the slipper. He explains what has been decided downstairs about the Jugo-Slav frontier.'
Harold Nicolson[1]

The Paris Peace Conference, which began on 18 January 1919, sought to 'unmix' lands and peoples in the defeated empires of Europe by drawing new boundaries that better matched the perceived cultural composition of the continent.[2] But, as Augustus Petermann's 1871 map beautifully illustrated, the European landmass was (and remains) a complex intersection of hard-to-define religious and ethnic groupings, towns and cities linked by a dense network of transport – and economic – interconnections that could not easily be severed by imposing hard borders across them.

Map 49 [Extract] Changing borders in the immediate post-war period demanded new maps, and many media outlets took to printing commemorative editions, using thick red lines and text to overprint the world as it was before the war with the new boundaries and territories that were agreed in its aftermath. The Sykes–Picot Line was created as part of a controversial agreement in 1916 to divide Ottoman lands between the British and French. Sykes specified a line in the sand be drawn on a map he had of the Middle East 'from the "e" in Acre to the last "k" in Kerkuk'.[6]

Liberty Map of New Europe: Literary Digest (1920)
126 × 99cm

THE LIBRARY OF LOST MAPS

Attempting to shift the boundaries on the map was a daunting task that, at the best of times, demanded clear-headedness, but the reality was a foggy sense that old scores needed to be settled and the delusion of a truly 'objective' and 'scientific' peace grounded in the maps and thinking that began in the second half of the nineteenth century. It was, as the first President of Czechoslovakia Tomáš Garrigue Masaryk described, a 'laboratory built over a vast cemetery'.[3]

Undeterred, and far from the carnage, clusters of academics had already begun working within each of the warring states and forming opinions about how the new borders should be drawn. Many were immersed in the geopolitical theories of the likes of Halford Mackinder and Friedrich Ratzel and they assigned themselves the status of world-builders, hovering above a devastated continent and pushing its peoples around like chips in a casino. It was their work that was crafted into the arguments that the statesmen would employ for their territorial claims during the negotiations.

The maps needed for negotiations, however, were not the detailed topographical mapping pioneered on the battlefield, but the ethnographic maps developed in the century before by pioneers like Alexander von Humboldt, Heinrich Berghaus and James Cowles Prichard. They depended on surveys that could not be undertaken by air but through enumerating the peoples on the ground through censuses. And to be done well they required subtlety and nuance: two things there was little time for in the heat of negotiations. Mapmakers had to engage in identity politics, take decisions about who were the insiders and who were the outsiders of a particular nation, and balance the wants and desires of how people would like to be governed. They also had to innovate to produce maps that caught the attention of those negotiating the peace.

Politicians have never liked uncertainty or ambiguity – they need to know unequivocally the limits of their power and how to control their territories. Even today, despite over two centuries of iteration and machination the definition of what constitutes a 'nation' remains amorphous, not least because it is predicated on an even vaguer notion of 'national identity'. In his book *The Image of Europe*, Michael Wintle makes the point that 'it is often the case that it is easier to identify characteristics in an "opposite" group than it is in one's own: the quintessence of Englishness may be hard or impossible to pin down, but the French or Germans (or come to that, the Venezuelans) do not resemble it any way whatsoever'.[4]

So is a nation about the territory or about the people living there? Part of the reason these questions become so controversial and heated is because identity is intensely personal – and can be somewhat fluid. You may, for example, have been raised in a religious family but would now identify as having 'no religion'; your political views may have become less radical over time; you may have moved away from your region of birth decades ago and so feel most at home where you have subsequently spent your life, even though you might be far from your ancestors. But those who seek a clear definition of a nation and its people will want to assign you a clear religion and an exact ancestral location, and even gain a sense of your political outlook.

These were precisely the types of definitions – and decisions – that those drawing the borders at the end of the First World War had to make on behalf of millions of war-weary Europeans. Cartographers had been set the task of creating maps to capture the fundamentals of who was similar enough to be grouped with (or distinguished from) their neighbours in order to dictate who they would be governed by.

From the outbreak of war work was underway to establish the scientific basis to European borders. The maps had to appear to

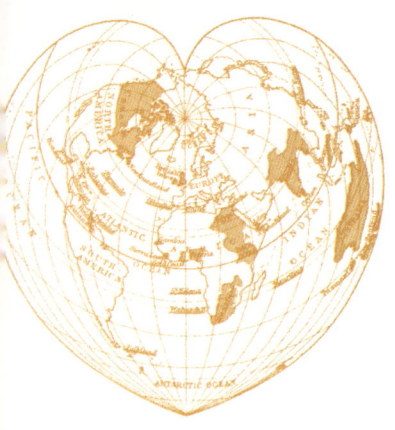

Map 50 The British Empire shown in a heart-shaped map projection. This image of British nationalism adorned the frontispiece of the *Harmsworth Atlas and Gazetteer*, which was published in the early 1900s and claimed to 'provide in a single volume of modest price...a complete encyclopaedia of geographical information.'

The Harmsworth Universal Atlas and Gazetteer: The Amalgamated Press Ltd (1904)
8.5 × 8.5cm

THE LIBRARY OF LOST MAPS

be a truthful reflection of reality, but, although not all of them would have readily admitted it, the scientists did bring with them their own political views, which meant they weren't impartial observers by any stretch. As we shall see, this politicisation is an important moment in map-making and one that would come to haunt geographers during the horrors of Nazi Germany.

Writing for *Geographische Zeitschrift*, a publication he founded, the German geographer Alfred Hettner is unequivocal about the role of his fellow geographers in the face of war. 'It is, after all, quite clear,' he wrote, 'that what is at stake is not merely the interests of Germany and Austro-Hungary, but even their future existence; let love of our countries guide our pens.'[5] Hettner said out loud what many others – on all sides – were thinking: 'Full objectivity in the sense that it stands towards the enemies and its own people with the same attitude, without love and without hatred, is impossible and cannot be strived for.'[6]

Hettner encouraged German geographers to emulate the 'English subjectivity' that permeated books on the British Empire, telling them to 'think and feel nationally'. He and his compatriots had particular concerns about a loss of German territory to a reborn Polish state. This is because at the onset of the First World War the country of Poland did not exist, since it had been dissolved in 1795 and portioned between the empires of Prussia, Austria and Russia. In fact the Polish nationalists themselves had taken Hettner's advice and used it against him and his colleagues. German researchers noted that:

> The sheer number of publications and maps, particularly by Polish authors, concerned with the extent of Polish settlement truly boggles the mind. Nearly all follow the national interest very decidedly, often abandoning the firm ground of objective science. One might achieve

a lot with clever shading methods and skilful colour arrangements ...[7]

One of those authors who had been pushing the boundaries of cartographic objectivity, and in turn those of the Polish state, was a nationalist intellectual named Eugeniusz Romer. Romer had a geography degree, but he specialised in climatology and geomorphology[8] so might seem an unlikely person to intervene on political matters. However, his background proved a source of inspiration in his efforts to map out an independent Poland.

When he began advocating for independence, Romer first had to dismantle the idea of 'Mitteleuropa'. This was another of the geopolitical concepts (in addition to *Lebensraum*) espoused by Ratzel and also enthusiastically adopted by a number of geographers in the first half of the twentieth century. In 1954, decades after the term first emerged, Karl Sinnhuber, an academic in the UCL Department of Geography, tried to define the extent of Mitteleuropa based on the many ways the term had been used. He concluded it was an area that did not have a clear extent, but it was centred roughly over where Czechia is today and might reach as far north as Denmark and as far south as Greece. At its largest, its westerly extent captures the Netherlands and its easterly limits reach Moldova.[9]

It gave Poland the status of broad transitional zone between East and West, for other countries to occupy. While not all Polish geographers agreed that statehood was possible, Romer was clear that any views to the contrary were a justification for German expansion into Polish lands. Rather than a transition zone, he saw Poland as something more self-assured: a bridge state between Western and Eastern Europe.

To make his case, Romer published the *Geograficzno-statystyczny Atlas Polski* (Geographical-Statistical Atlas of Poland)

THE LIBRARY OF LOST MAPS

Map 51 [Extract] Romer's map of Poland, created with his innovative isopleth technique. It's not a map that commands a room, but it marked a big shift in the way that cartographers could adapt their methods to better support the nationalist arguments they were making.

Geograficzno-statystyczny Atlas Polski: G. Freytag & Berndt (1916) 31.8 x 25.9cm

in 1916. It was written in Polish, German and French to ensure – in the same way that the *Atlas de Finlande* did – an international and hopefully receptive audience. Also tapping into the tricks of the development of national atlases, Romer used the name 'Poland' to denote a territory of 800,000 km² that at the time was still politically divided among Austria-Hungary, Germany and Russia. He said, 'Let's call our land by its proper name, because the name is the essence.'[10]

Poland lacked the major landforms that were required to offer a clear bounding of the Polish state. There were few mountain ridges to follow, or rivers to act like a moat to the territory. This fact had been used to endorse the Mitteleuropean and transitional status of the lands.[11] Romer therefore needed to reframe the influence of the environment, and he did so by arguing that

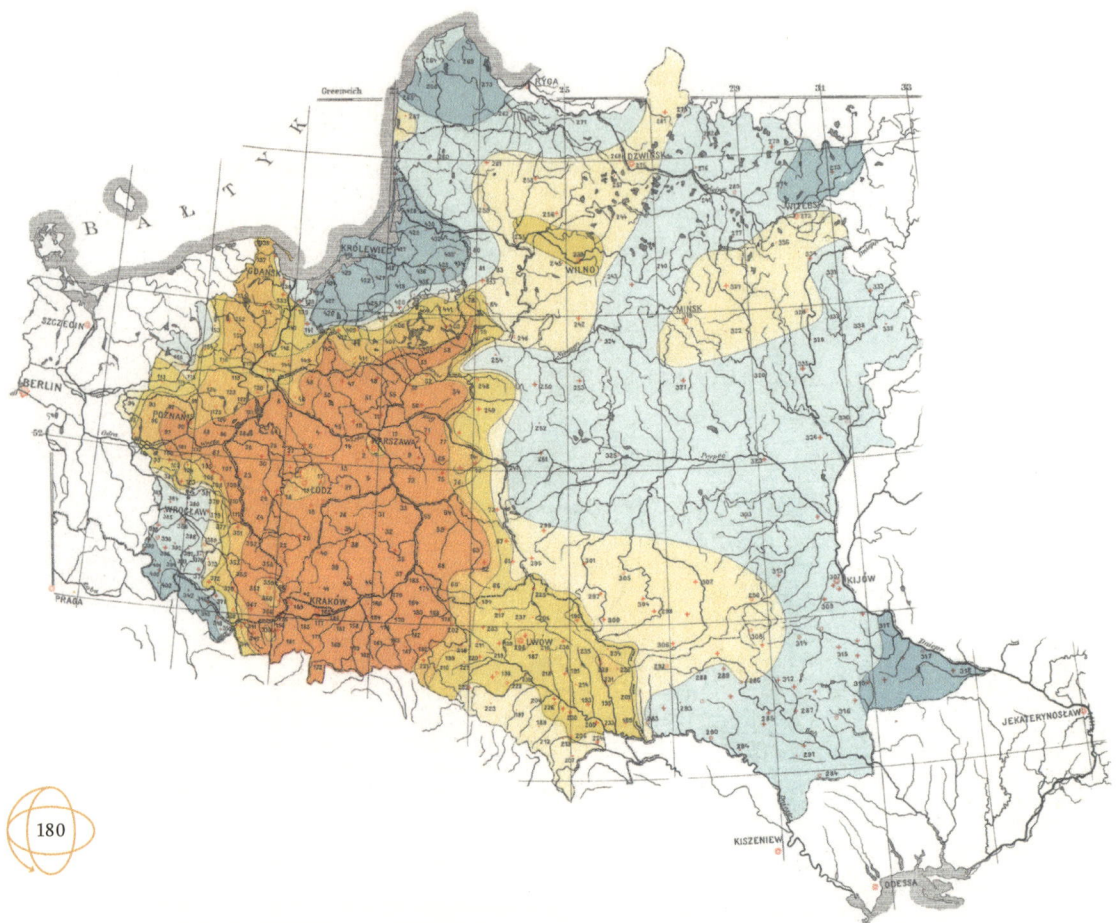

Poland's network of rivers should not be thought of as boundaries, but, rather, as the lifeblood of an organic state.[12]

In addition, debates about environmental characteristics had become important because the ethnographic maps were not indicating a clearly bounded group of people who could be labelled 'Polish', while also matching the territorial ambitions of the soon-to-be nation. To address this, Romer again leaned on his interests in climatology and geomorphology, not least as a student of Humboldt's work, which gave him an idea for a new approach to demarcating the Polish people. He needed to use statistical data from the three states that had emerged from a carved-up Poland for a single map of the Polish population. But to create this harmonious image, Romer could not depend on the official administrative units, for it would have looked more disjointed, and given legitimacy to the idea that there was no meaningful indicator of a Polish state.

Inspired by Humboldt's isolines technique (see Chapter 5), Romer therefore developed a method to erase the political boundaries of Austria-Hungary, Germany and Russia, thus uniting the Polish lands and the people who lived in them. The method itself was explained by Romer as follows:

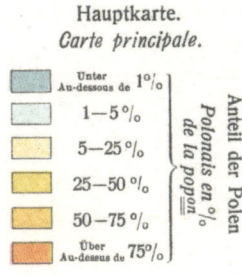

> The statistical value of a district [*powiat*] refers to a point marked by the location of [its] chief town ... We have drawn lines between the two points proportional in length to the value [of the district]. This method is named interpolation, and the resulting lines are lines of equal value ... We named those lines of social conditions isarithms.[13]

Nowadays cartographers call isolines that have been converted into coloured polygons isopleths, not isarithms as Romer did.

But, whatever it's called, the technique is the one Romer used to great effect, creating the impression of a population terrain, that would have been a visual language many of the environmental determinists arguing for Poland's 'transitional' status would have found it hard to disagree with, even though it was depicting population peaks instead of mountain peaks.

The Geographical-Statistical Atlas of Poland played its part in shaping the borders of the reborn Republic of Poland at the Paris Peace Conference.[14] Romer's materials were used by the Allies (including the American, French and British delegations) and by the Polish delegation to support Poland's territorial claims. And Romer himself was present to proactively make new maps when needed to bolster the case.

At the end of the conference, Poland gained its statehood, along the lines of Romer's maps, but the borders were not immediately peaceful. The final shape of the country was determined by uprisings (in Wielkopolska, Silesia and Sejny), armed conflicts (including with Bolshevik Russia, with Ukrainians for Lviv, with Lithuanians for Vilnius) and plebiscites (in Silesia, Warmia and Mazury).[15]

While mainland Europe's intellectuals engaged in the mapping needed to preserve their homelands, the British and Americans were able to remain somewhat aloof and adopted the role of mediators and arbiters, albeit to build a world that suited their geopolitical perspectives. In 1916 (the same year that Romer's isopleth-filled atlas was published), a group of geographers were enjoying an afternoon at the Royal Geographical Society discussing a new kind of map that showed the ethnic groups of Hungary.

The map had been drawn by one Bertie Cotterell Wallis, whose innovations bore a striking resemblance to Romer's isopleth technique. Wallis framed it as purely an experimental idea, saying, 'I believe the method has been suggested in Germany, but I do not know of its application.'[16] Wallis elaborated on his approach as follows:

> The number representing the average density [of each ethnic group] for each district is mapped in the geographical centre of the district and these numbers are treated as 'spot-heights', and used to draw 'population lines' ... On the whole, this map gives the impression that there tends to be a clear line of demarcation between each nationality and its neighbours; with the possible exception of the Germans who have been noted for their ability to disappear into the population by whom they are surrounded.[17]

The first person to offer feedback was Lionel Lyde, founding professor of geography at UCL, who expressed his enthusiasm, before offering his own interpretation and explanation of the patterns the map revealed.

> Mr. Wallis is one of the most ingenious persons in London in this kind of statistical work ... He has two maps there which give information that is most valuable. They suggest the whole basis of the hostility of the Magyars [Hungarians] to the Rumanians. The Magyars and the Rumanians are both essentially lowlanders. Directly you get above the 600 contour-line, west of the Raab, in the Bakony Forest, north of Fünfkirchen and elsewhere, there are no Magyars – they are lowlanders. So are the Rumanians. They only went up into the hills because they wanted refuge, and they have always been trying to come down from them.[18]

James Fairgrieve, a pioneer of geographical education in schools,[19] chimed in with his approval of the use of the line method saying that it came to him 'rather with a shock' that it had not been used in this way before. While a Miss S. Nicolls (women had only been admitted as fellows to the Society three years before, after a twenty-year debate[20]) gave her full backing to the statistical approach: 'If you compare mere descriptions of a region you will find that no two people entirely agree. But if the statistical facts are worked out carefully by the geographer you will get a result that is certainly true and which makes you receive a new impression by a shifting of your point of view.'[21]

Nicolls's presumption that the maps she had seen were 'certainly true' is a reminder of how convincing and reassuring maps can be even if they are considered experimental by their makers.[22] But what is really extraordinary is the way in which cartographic innovations designed in one context – Humboldt's global

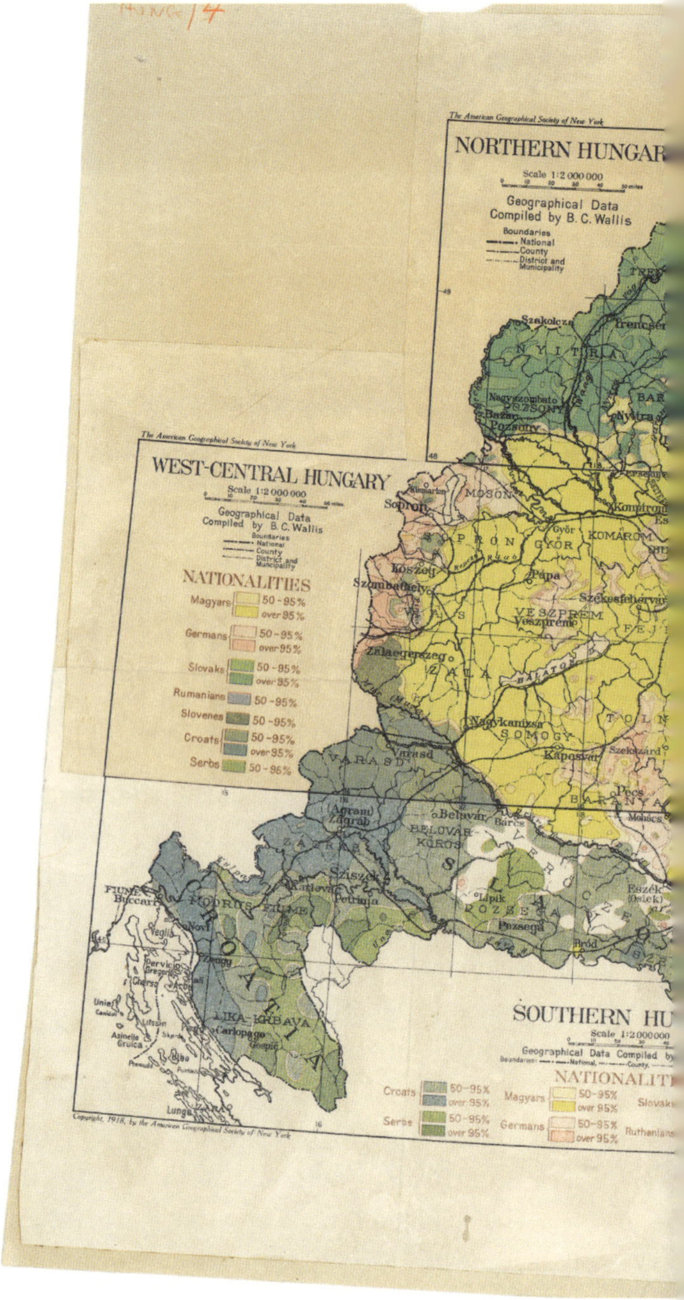

Map 52 B.C. Wallis's ethnographic map of Hungary pieced together from three separate sheets that have been disbound from the American Geographical Society's *Geographical Review* journal and then mounted on thicker paper.

'TIDYING' THE MAP

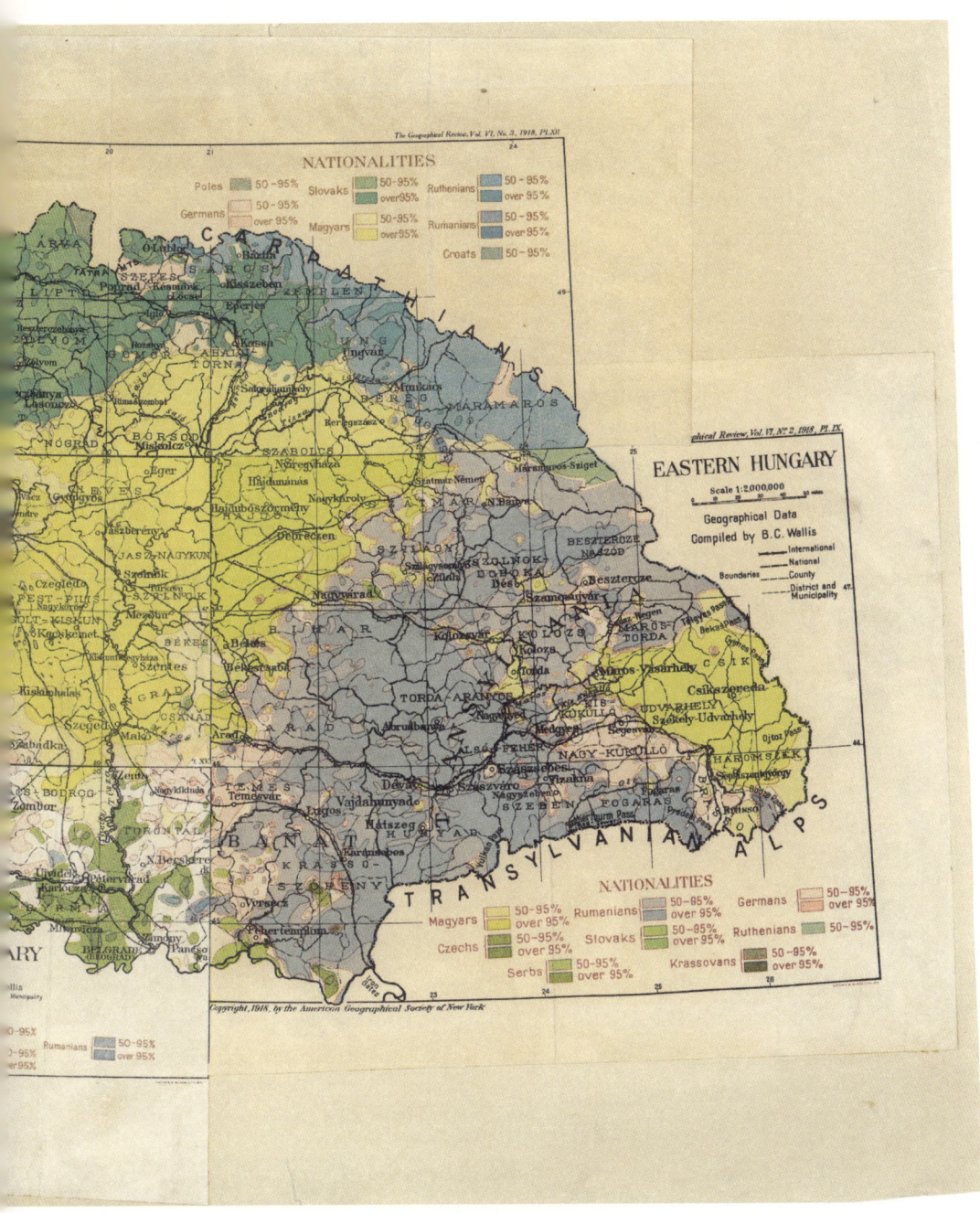

Hungary: Northern Hungary, West-Central Hungary, Eastern Hungary, Southern Hungary: American Geographical Society of New York (1918)
51 × 33cm

THE LIBRARY OF LOST MAPS

temperature mapping – can be adapted and redeployed in a very different one: redefining the shapes of Poland and Hungary.

Buoyed by the positivity in the room, Wallis (who was a schoolteacher by background) went on to publish three separate maps of Hungary (each covering a third of its area) and four articles of commentary in *Geographical Review*,[23] which was the journal of the American Geographical Society (AGS). Readers could then stick the maps together to form a single map of the entire country. This is what someone has done for the copies I found in the Map Library: the three have been neatly glued onto a thicker sheet of paper.

Wallis's approach was emulated in the *British Naval Intelligence Handbook for Hungary* (I was thrilled to find a copy among the atlas collection), which was part of the briefing material prepared for the Paris Peace Conference.[24] But his maps – and commentary – became particularly influential to a mysterious sounding group working on the US's peace plan called 'The Inquiry'.

It was founded in 1917 by President Woodrow Wilson and based at the AGS. In charge of the research was a towering figure in American geography in the first half of the twentieth century: Isaiah Bowman (1878–1950). Bowman[25] was also the editor of the *Geographical Review* and had heard about Wallis's work through the informal networks operating between the various (but surprisingly uncoordinated) groups among the Allies gathering intelligence for anticipated peace negotiations.[26] As a result of his papers and maps in the journal, Wallis was deemed the 'ultimate authority'[27] on the ethnographic composition of Hungary, which was an important consideration for the Treaty of Trianon, that saw Hungary lose two-thirds of its pre-war territory (and the same proportion of its population[28]).

The scope of the Inquiry extended well beyond Hungary and employed dozens of geographers, historians, geologists and economists who produced the 'Black Book', which mapped the most desirable (from a US perspective) territory of every European country.[29] They drew their own maps and collated the work of others, such as Wallis and Romer, with whom Bowman developed a particularly close relationship.[30] Many of these maps were shipped to Paris for the negotiations.

The Inquiry formed its own view on the best approach to borders and used the AGS's publications as a mouthpiece to give legitimacy to its ideas and inform the thinking of others prior to the end of the war. One particularly influential book was written by Leon Dominian, who had joined the staff of the AGS in 1912. Titled *The Frontiers of Language and Nationality in Europe*,[31] it was published for the AGS in 1917 just in time for the USA to enter the First World War. The following year Dominian began working in the State Department on boundary problems, first mediating between Honduras and Guatemala before joining the American Commission to Negotiate Peace at Paris, which became the official name for the Inquiry, in 1919.

This was a fascinating find on the shelves of the Geography Department's reading room, replete with pull-out maps and copious photographs of the peoples and landscapes Dominian was mapping. He sets out in the preface that 'language exerts a strong formative influence on nationality because words express thoughts and ideals'.[32] He then goes on to say that national feeling is 'found in the persistent action of the land, or geography, which like the recurrent motif of an operatic composition prevails from beginning to end of the orchestration and endows it with unity of theme'.[33]

To Dominian it therefore followed that language was also a good approximation of distinct sets of economic and social conditions. Interestingly, this is not something that was met with wholesale agreement at the time. For example, UCL's Lyde had written:

> Although a common language and a common literature, like a common creed, obviously have a real cementing power and value, there is no justification for making language a test of nationality. But the magnificent unity of Belgium in spite of a language line running right across the country from east to west – along the latitude of Waterloo – makes it unnecessary to elaborate the point.[34]

With a range of ideas and no fixed definition of 'race' (or ethnicity), as Jeremy Crampton sets out in *The History of Cartography*, 'the Inquiry chose a more eclectic approach that nevertheless identified clear "zones of civilization"'. It was around these zones that they sought to draw the new borders of Europe to calm what Bowman described as an 'ethnic storm centre'.[35]

On 18 January 1919 the Peace Conference officially began at the Quai d'Orsay in Paris. Delegates from over thirty nations attended, the major powers being the US, Britain, France, Italy and Japan, known as the 'Big Five'. Initially, the main decisions were made by the Council of Ten (two representatives from each country). This later became the Council of Four (Woodrow Wilson of the US, David Lloyd George of the UK, Georges Clemenceau of France and Vittorio Orlando of Italy), as Japan became less involved in European matters.

Accompanying President Wilson was a 'core group of Inquiry men'[36] who fanned out across the territorial commissions charged with drawing up the boundaries of post-war Europe. Bowman also had inputs into Wilson's 'Fourteen Points', which were later cited by the Germans in their armistice agreement and included an emphasis on self-determination as well as the proposal for the League of Nations.[37]

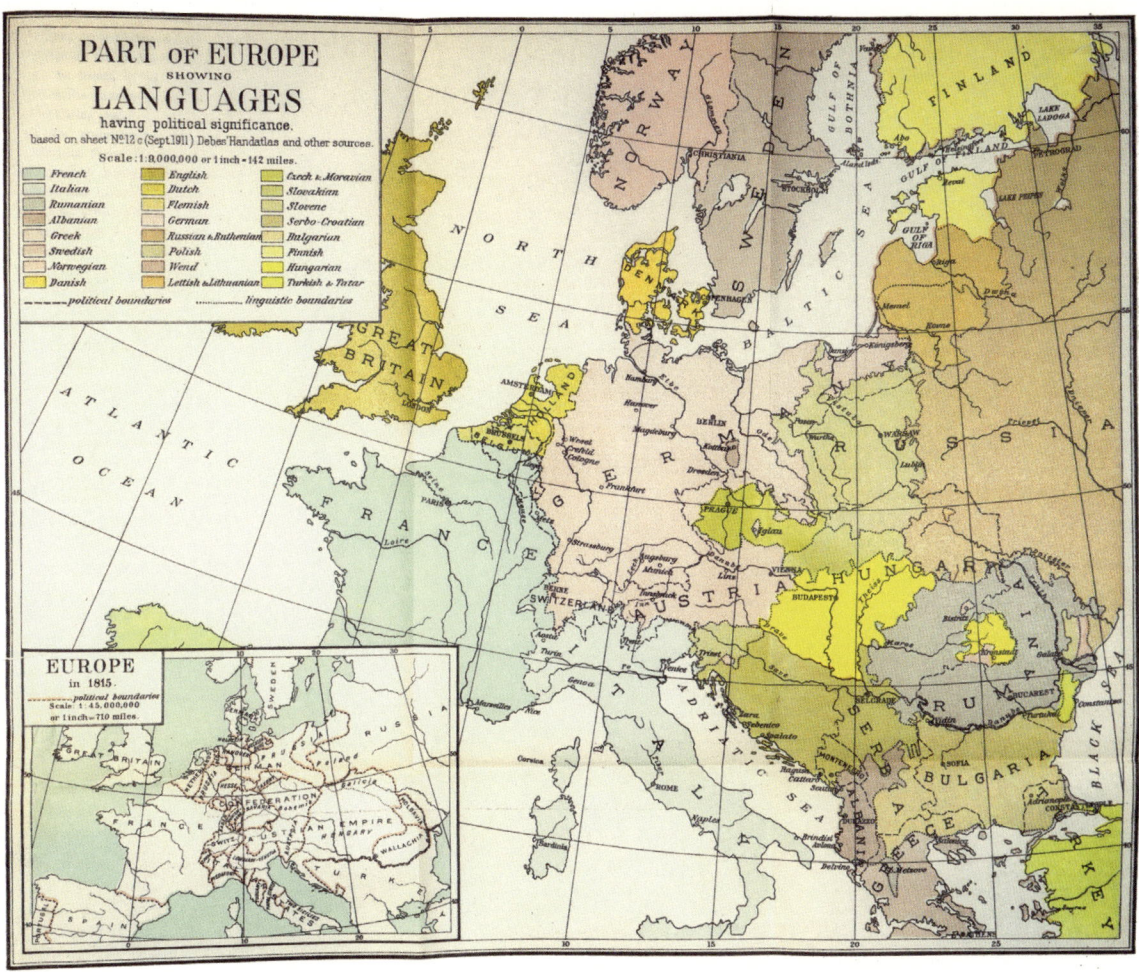

Map 53 Dominian's European language areas of 'political significance' as of September 1911. Perhaps naively, he wrote at the end of his book that: 'The growing coincidence of linguistic and political boundaries must be regarded as a normal development. It is a form of order evolved out of the chaos characterizing the origin of human institutions.' [7]

The Frontiers of Language and Nationality in Europe (Part of Europe Showing Languages Having Political Significance): American Geographical Society/ Henry Holt and Company (1917) 31 × 26.5cm

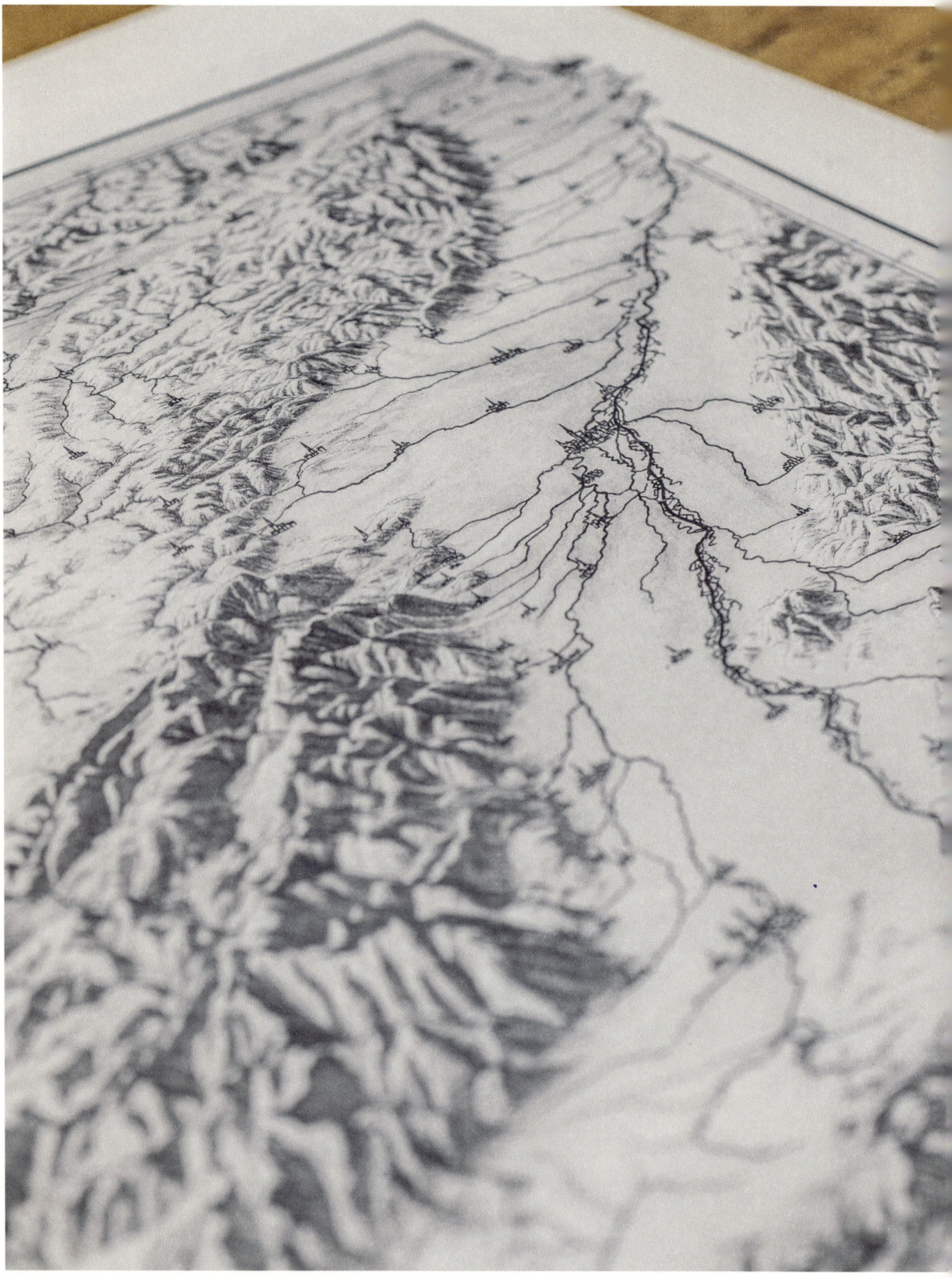

'TIDYING' THE MAP

Cartographic Supremacy

A first-hand account of the US's involvement in the peace negotiations was published a year after the conference in a book intriguingly titled *What Really Happened in Paris*. One of its editors, Charles Seymour, served as the chief of the Austro-Hungarian Division of the American Commission, and in it he sets out the complexities of the work:

> The commissions spent long hours in studying the conflicting claims of the nationalities and in comparing them with the host of statistics which were available. If nothing else interfered the obvious frontier was the line that separated the nationalities, Czechs from Germans, Rumanians from Jugo-Slavs, Jugo-Slavs from Magyars.
>
> But many other factors had to be considered … If a chain of mountains or a river offered a natural frontier, it might seem advisable to depart slightly from the linguistic line … If the linguistic line were crossed and recrossed by a railway or canal, it would be questionable policy not to arrange the political frontier in such a way as to leave the railway or canal entirely within one state or the other, so as to avoid troublesome customs interference with trade.[38]

With such complexity, the Inquiry's experts often held the most sway because it had accumulated the most 'scientific' evidence to inform the positions of the new borders. Throughout its existence,

Map 54 [Extract] The British Naval Intelligence atlas for Alsace-Lorraine was produced in June 1919 (the same month as the signing of the Treaty of Versailles). In common with other intelligence atlases of the period, it opens with a sketch of the region that omits the borders. I took this to be the blank canvas upon which borders were drawn by negotiators. The map here features the city of Strasbourg and the River Rhine.

A Manual of Alsace-Lorraine Atlas: Naval Staff Intelligence Division (1919)
42.5 × 24cm

its staff produced and collected nearly 2,000 separate reports and documents plus at least 1,200 maps.[39]

Seymour and his colleagues must have made an impressive arrival in Paris. The British diplomat and author Harold Nicolson wrote in his diary that on 6 January (twelve days before the start of the conference) he and his colleagues met the three members of the US delegation who were to be their opposite numbers:

> They are all members of Colonel House's 'Enquiry' and university men. There is Professor Clive Day of Yale – middle-aged, pale, slim, arid, decent. There is Professor Charles Seymour also of Yale – young, dark, might be a major in the Sappers. Third is Professor Lybyer[40] – silent, somewhat remote. They show us their maps. There is a vast relief map in sections depicting the Adriatic, very beautiful. They evidently know their subject backwards. Nice people – but we enter into no details. A feeling, however, that our general views are identical.[41]

Nicolson's account also tells us how their maps were being (mis)-interpreted and acted upon by the major figures at the Peace Conference at crucial moments. He writes tellingly of a meeting that took place around a dining table on 13 May between the British and Italian delegations. First, 'They all sit round the map. The appearance of a pie about to be distributed is thus enhanced.' Then David Lloyd George (the British prime minister) showed the Italians the territory he had in mind for them but they ask that given his prominent role as part of the Inquiry for a settlement on Türkiye's western Aegean coast.[42] Nicolson sets out what happened next:

> 'Oh no!' says Ll. G. [Lloyd George], 'you can't have that

– it's full of Greeks!' He goes on to point out that there are further Greeks at Makri, and a whole wedge of them along the coast towards Alexandretta. 'Oh, no,' I whisper to him, 'there are not many Greeks there.' 'But yes,' he answers, 'don't you see it's coloured green?' I then realise that he mistakes my map for an ethnological map, and thinks the green means Greeks instead of valleys, and the brown means Turks instead of mountains. Ll. G. takes this correction with great good humour. He is as quick as a kingfisher. Meanwhile Orlando and Sonnino chatter to themselves in Italian.⁴³

Concerningly, this gathering was taking place to help defuse one of the biggest crises of the conference, which erupted on 24 April when the Italian delegation stormed out in protest at the terms they were being offered.⁴⁴ They had only returned to discussions a few days before this meeting around Nicolson's table.

It's a fascinating example, too, of the way that an ethnographic map could radically change the perspective of the statesmen involved's assessment of the trade-offs required in negotiations. As we have seen with Eugeniusz Romer's and Bertie Cotterell Wallis's efforts, a new kind of map could give a revelatory impression of a problem and radically change a position. From Isaiah Bowman's reflections, also written in *What Really Happened in Paris*, it is clear he saw himself as the ultimate arbiter of what was truthful and what was not and was clearly irritated by eye-catching maps that contradicted his opinions on a matter:

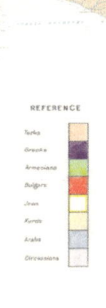

Map 55 [Extract] This ethnographic map has had watercolour applied to a standard military sheet that depicts Anatolia, a region that is now part of Türkiye. It may have been produced at speed to help inform discussions at the Paris Peace Conference in 1919–20 (it's a region not far from the one Lloyd George was discussing with the Italian delegation). The dots across the map came to mind when I read this harsh description of a teenage soldier by Nicolson: 'A young French subaltern of some nineteen years of age comes to our table. He is unshaven, greasy, and covered with spots like an ethnographical map.' ⁸

Anatolia Ethnographical Map: War Office (1919)
73 × 60.5cm

THE LIBRARY OF LOST MAPS

Each one of the Central European nationalities had its own bagful of statistical and cartographical tricks ... A new instrument was discovered: the map language. A map was as good as a brilliant poster, and just being a map made it respectable, authentic. A perverted map was a life-belt to many a foundering argument.[45]

When he wrote these words, Bowman was probably thinking of a map of Hungary produced by Count Pál Teleki for the Hungarian delegation in 1918. Teleki's map was shared at the request of Lloyd George[46] and grabbed attention because it depicted the Magyars – the group most closely associated with Hungary – in a blood-red colour and the remaining ethnic categories in more muted tones. Unfortunately, the UCL Map Library does not have a copy of this map, but I was lucky enough to see Bowman's copy in the archives of the American Geographical Society. And no scan can do justice to the rich red colour the Hungarians achieved with their printing. It's no wonder that the map became known as the 'Carte Rouge'.

In the austere surroundings of the Peace Conference, Teleki's map would have certainly stood out against the other maps of the time. His use of bright colours might be considered trickery – or brilliance, depending on which side you were on. Teleki also deployed white space in regions that

- Magyars
- Germans
- Slovaks
- Roumanians
- Servians
- Croates
- Ruthenians
- Bunyevaces (Catholic Servians)
- Vends (Slovens)
- Bulgarians
- Poles
- Others
- Uninhabited

other maps had filled with ethnic groups different from the Magyars, and thus gave further prominence to the red areas set against a plain background. In addition, its scattered red shapes at a glance unite the western concentration of Magyars with the smaller cluster to the east, which had been allocated to Romania. It was this that most irked Bowman, who had already determined drastic changes to the Hungarian border based on his own analysis (supported by Wallis's map).

Map 56 [Extract] Count Teleki's 'Carte Rouge' map of Hungary. This copy belongs to the American Geographical Society.

Ethnographic Map of Hungary, Based on the Density of Population According to the Census of 1910 (1919)
79 × 57cm

> This idea has occurred to me since examining a map by Count Teleki which gives altogether a wrong impression of the distribution of Magyars in Transylvania that we ought to keep a fairly good collection of propaganda maps … I should like to write a little paper on the various types of lies and liars that I met in this form of cartography.[47]

Bowman was a man who did not like to be contradicted and would have been envious of the attention the map attracted, especially from those outside the US delegation. Lloyd George's request that it be shared, for example, was a telling endorsement of the approach, as it was clear he was struggling to read the maps and get his head around the complexities of the negotiations, apparently once asking his delegation 'Who are the Slovaks?'[48]

For Hungary the negotiated borders signed into the Treaty of Trianon were a fait accompli that no map or impassioned speech was going to shift, and it had to surrender 64 per cent of its pre-war territory and 72 per cent of its population to neighbouring countries. But, unlike any of Bowman's preferred maps, the Carte Rouge has stood the test of time and assumed an almost mythical status, with twelve editions inspiring countless other ethnographic maps.[49] Teleki even went on to become the prime minister of Hungary from July 1920 to April 1921 and from February 1939 to

April 1941. For some today it remains an important statement of Hungary's ethnic identity and territorial integrity, despite it evoking a moment in time over a century ago.[50]

In his diaries, Nicolson admitted his disillusionment with the role that the maps were playing in the key decisions and the impacts they might have on the ground. Only one month into negotiations, on 24 February, he wrote:

> A disheartening job. How fallible one feels here! A map – a pencil – tracing paper. Yet my courage fails at the thought of the people whom our errant lines enclose or exclude, the happiness of several thousands of people. How impossible to combine speed with examination![51]

As Lloyd George's poor map-reading abilities demonstrated, there was a lack of geographical expertise among the British delegation. With no grand commission from Britain,[52] the influence of geographers was therefore more indirect – by supplying maps to the likes of the Inquiry – and the sharing of ideas, as was the case with Wallis's maps of Hungary.[53]

But of course, that is not to diminish in any way the importance placed on geography and maps throughout the negotiations more generally. Prior to the end of the war, the French had formed a group that included Paul Vidal de La Blache (before his death in 1918) and his son-in-law Emmanuel de Martonne, who had worked with some of the most famous French and German geographers of their day. Poland had the talents of Romer, and Hungary, of course, had Count Teleki, while the US's experts from the Inquiry were frequently seen at President Wilson's side supporting what seemed to be a real enthusiasm for maps.[54] Seymour later recalled of his time in Paris that meetings were often held in the front room of the president's house, where:

one might have seen President Wilson himself on all fours, kneeling on a gigantic map spread upon the floor and tracing with his finger a proposed boundary.[55]

Bowman also recounted the same image, but added 'there were ten million dead peering through the windows'.[56]

This marks a climax in the history of maps and map-making. Less than a century before, many of the kinds of maps that graced the floor of Wilson's front room were little more than ideas in George Bellas Greenough's notebooks or experiments from Alexander von Humboldt and Heinrich Berghaus. Of course, maps of terrain, landscape and land ownership were certainly well developed by the middle of the nineteenth century. But the idea that they could be scientific, data-driven and expressive of more abstract questions, such as 'what makes a nation?', was still unfathomable. A technological revolution had also taken place not just in the way lands were surveyed but also in the way that maps could be printed quickly and cheaply to meet the demands of warring nations and a more geographically literate public.

Maps could now visualise a new world and, in the hands of the powerful, give the certainty to take that vision and implement it. But to what end? As Geoffrey Martin sets out in *The History of Cartography*, prior to the Treaty of Versailles, the countries of Central Europe were 'delineated with 8,000 miles of boundary. After Versailles, there were 10,000 miles of boundary, and of this some 3,000 miles were newly created, thus lengthening the "zones of friction"'[57] and, as we shall see, it was not long before mapmakers began aggravating rather than soothing these.

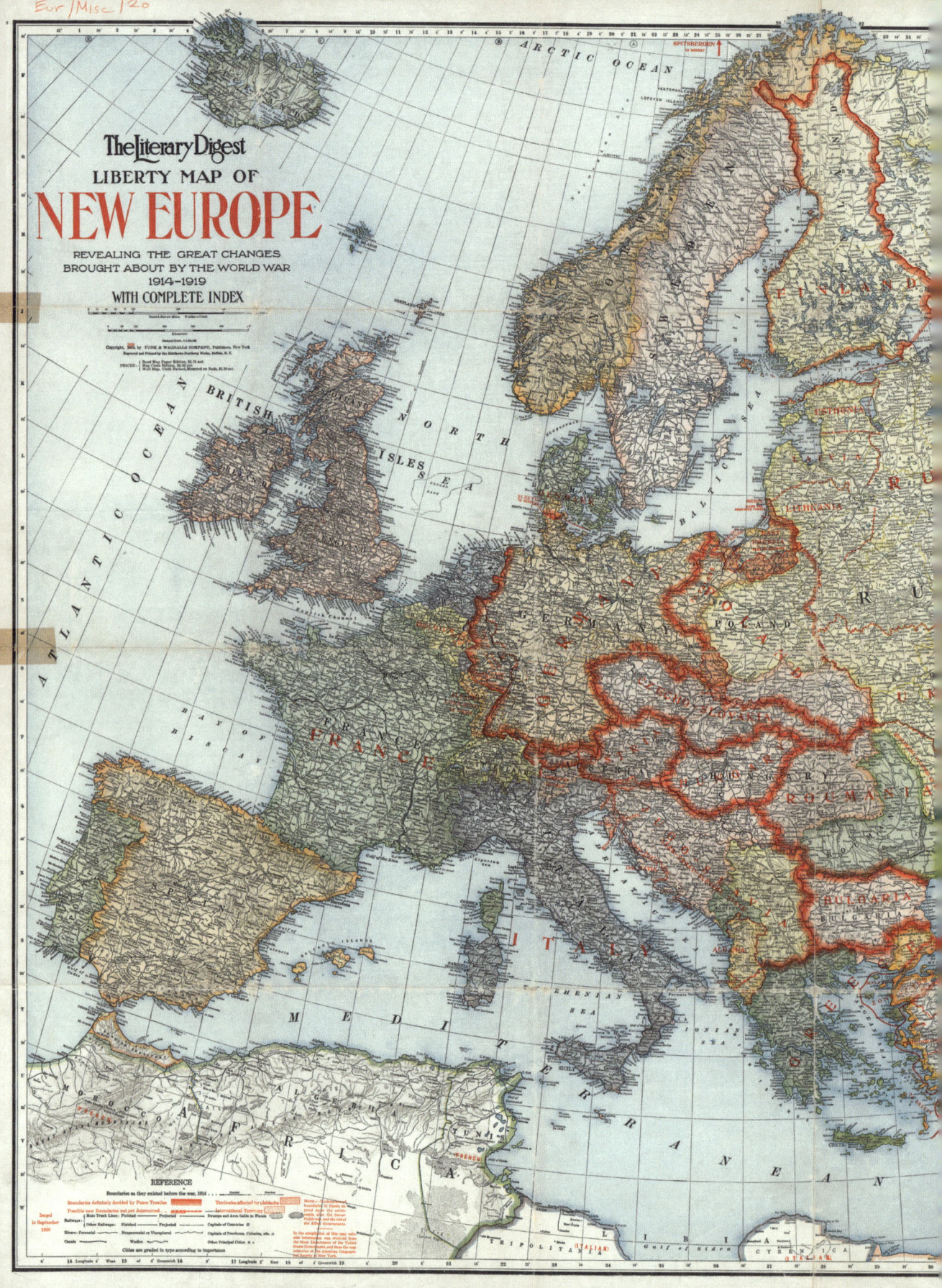

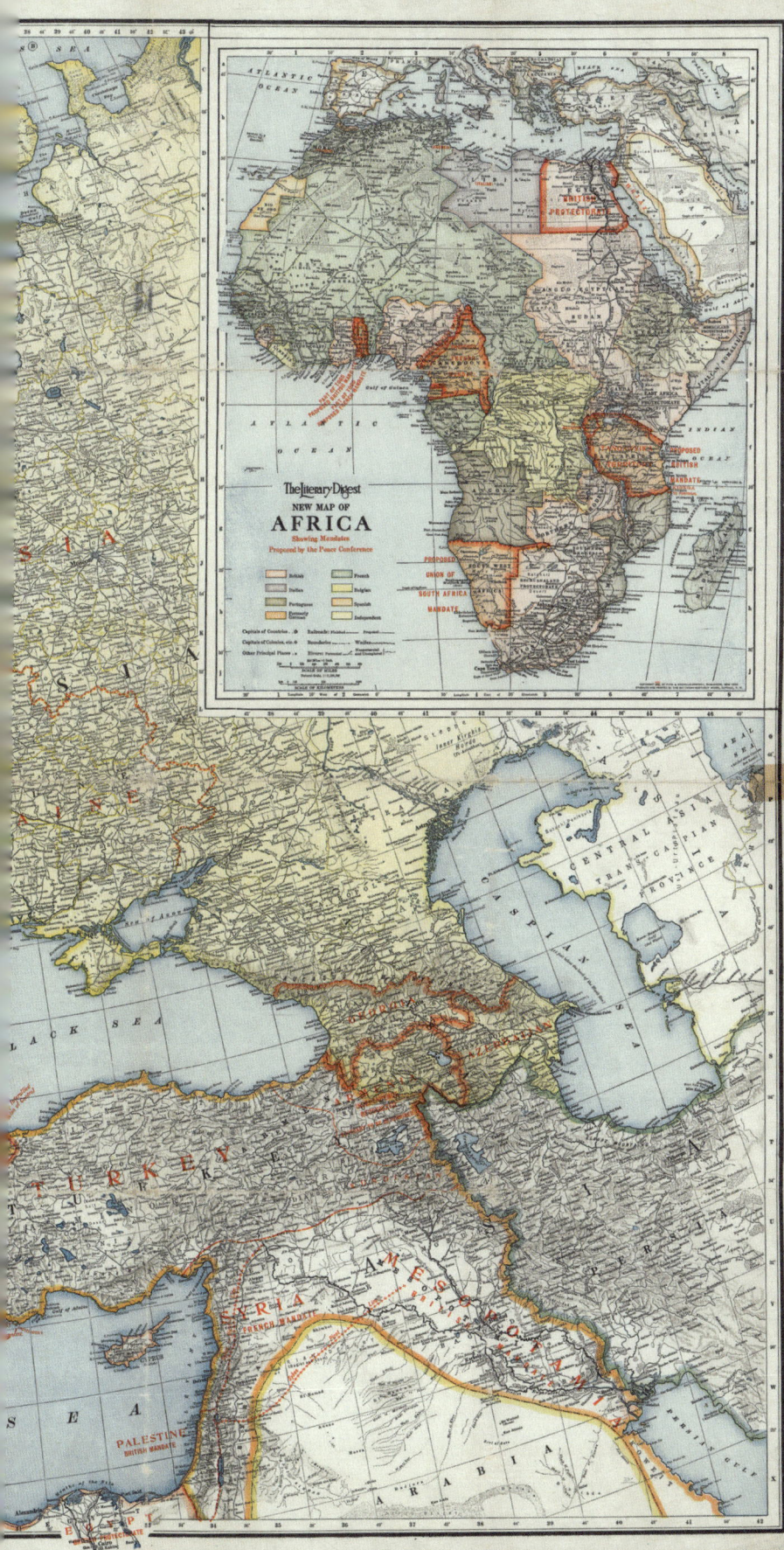

Map 49 The climax of the Paris Peace Conference was a proliferation of new borders and the creation of a new world order, with the League of Nations to police it (although the US never formally joined). While I've focused on Europe, the new borders extended to the Middle East and Africa too. As this 'Liberty Map' from the *Literary Digest* shows, German territories were portioned out to further grow the empires of the victorious Allies.

Liberty Map of New Europe: Literary Digest (1920)
126 × 99cm

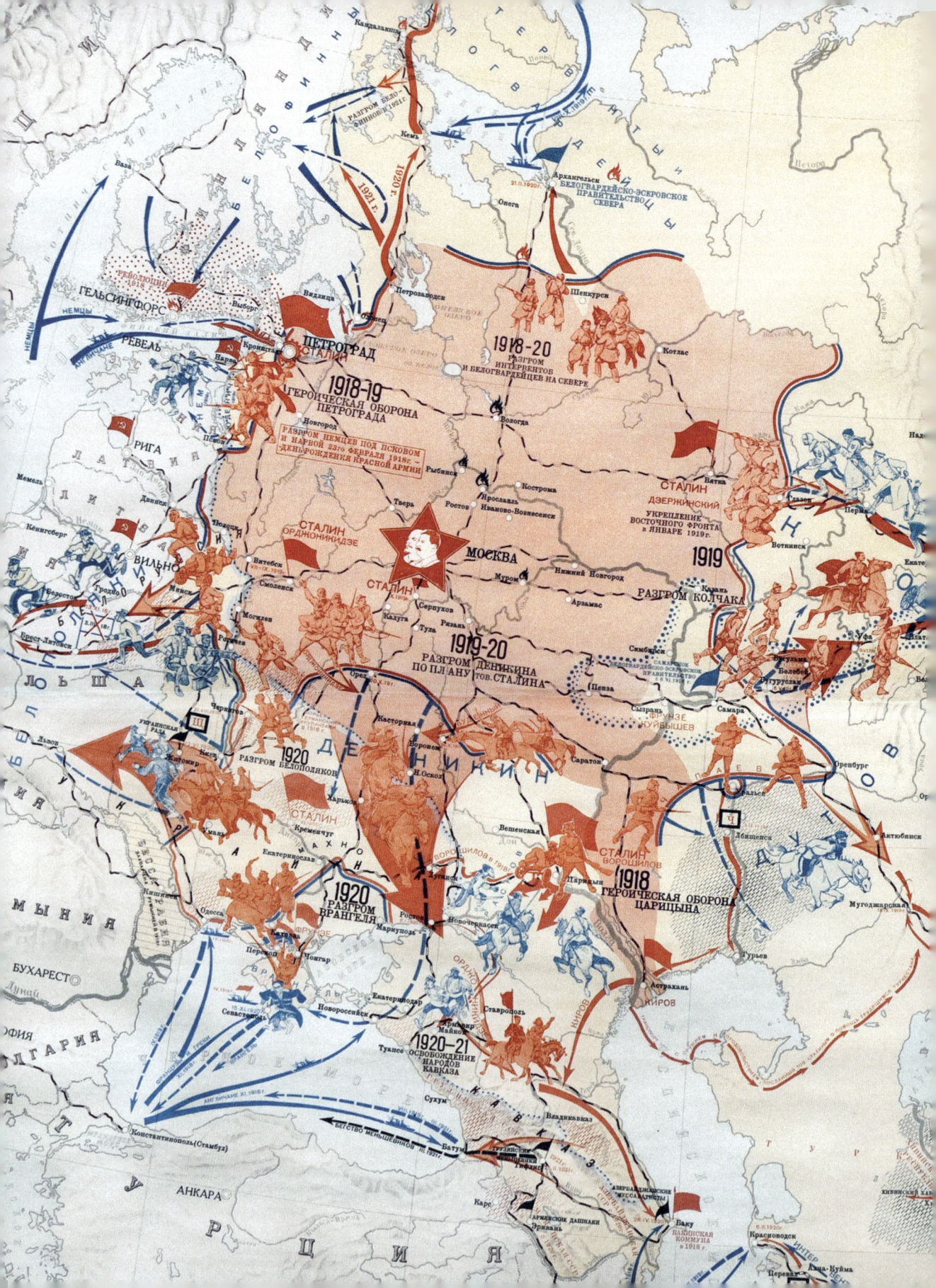

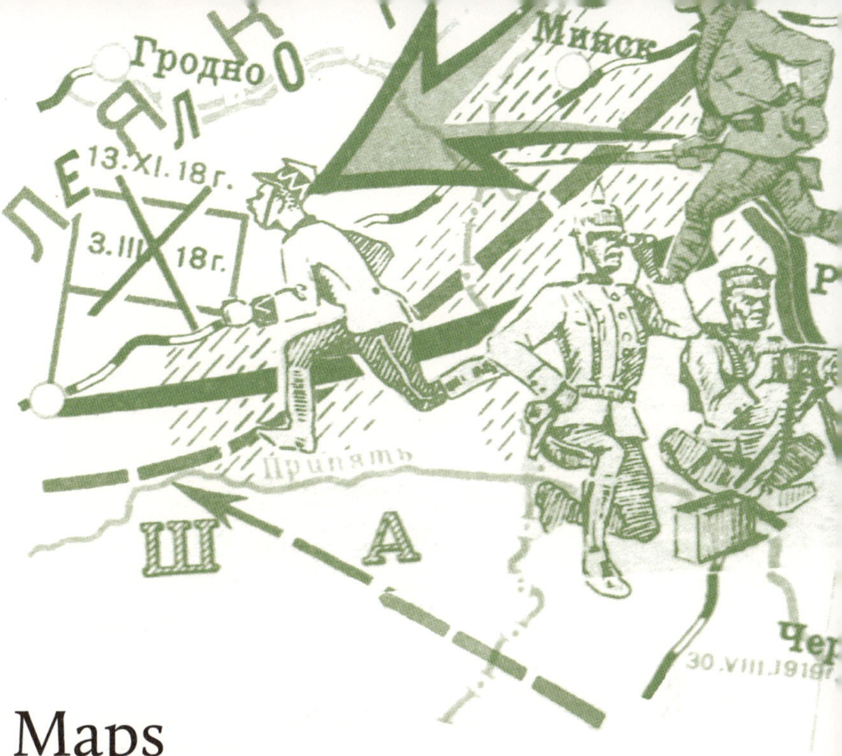

CHAPTER SEVEN
Manipulative Maps

'The first lie of a map – also the first lie of fiction – is that it is the truth. And a great deal of a map's, or story's, or poem's authority results from its ability to *convince* us of its authority.'
Peter Turchi[1]

I had assumed that all the teaching with maps took place in the Map Library itself, but, of course, maps would have provided colourful backdrops to lecturers as they imparted their knowledge in classrooms across the university. In their heyday the most popular maps would have been on permanent display in the lecture halls and when their time came to be front and centre of a class they were manoeuvred into place by a series of ropes and pulleys like scene changes between acts of a play. For more occasional use, or in rooms without dedicated pulleys, the maps would have been attached to a metal hanger with a very heavy base and clamp system. I found this in one of the cupboards and painfully established that it was just as effective at clamping fingers as it was maps!

Map 57 [Extract] A spectacular wall chart entitled 'Civil War in the USSR in 1918–1922' that tells the stories of key moments in the history of the Soviet Union. The illustrations were based on a famous 1940 poster 'Stalin on the Fronts of the Civil War' and the base map was taken from the *Great Soviet Atlas of the World*, which I'll introduce to you shortly.

ГРАЖДАНСКАЯ ВОЙНА В СССР В 1918–1922 ГОДАХ:; *Главное управление геодезии и картографии при Совете Министров СССР* (1949)
192 × 106cm

In her role as map librarian, Anne Oxenham would have hauled this stand (and the map to be hung from it) to a classroom, where she'd have to reach up and put the top of the map in the clamp and then extend the shaft, by means of another lethal catch, in order that the map could hang freely. I also happened upon some wooden canes that I presumed were used by professors to point at the cartographic backdrop Anne had prepared for them. The physicality of it all is completely alien to me: the most I exert myself prior to a lecture is plugging a cable into my laptop.

I was astonished to discover that a vast collection of these wall maps was still present in the Map Library. Anne told me that they were hidden in plain sight in one of the stuffy classrooms on the floor above, and, sure enough, down one side of the room there was a line of built-in cupboards, each bulging with rolled wall maps. Many had been produced overseas,[2] so luggage labels had been affixed to them with a neatly written English translation of their subject matter. Alas, rubber bands, which had disintegrated with age, were used for the attachment so most of the labels had fallen off and were among the accumulated debris on the floor. It was another puzzle to solve.

The cupboards themselves have special rails running at ceiling height and from these many of the maps are hung, now furled and resembling leafy brown chrysalises. Unhooking a map and opening it was like watching bright butterflies emerging from their metamorphosis. With each map spreading its wings, the room was filled with colour, conjuring everything from the natural vegetation of Africa to the industrial outputs of Poland. In all there were 220 of these brightly coloured maps, with the largest being two metres across, and at least as tall.

As I furled and refurled these giants, it dawned on me the wall charts were the most widely seen maps of the collection, since in one hour of teaching they would have had more eyes on them than

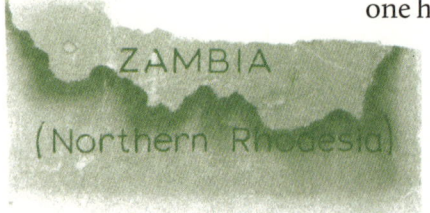

MANIPULATIVE MAPS

in the lifetime of most of the paper maps in the drawers. With each point of a professor's cane, the maps would be enlivened as they were explored and explained. I knew it was in these moments that students became knowledgeable about the world they inhabited and formed their opinions about it. Teachers know this power – and responsibility – as views shaped in a classroom can last a lifetime. For this reason, what we use to educate the young has always been a battleground for political ideals (and remains so today).

It is why after the Paris Peace Conference countries on all sides invested in maps for education and why the Map Library stores many such maps from the interwar period. For the victors it was a chance to imbue students with a sense of their nation's achievements. But for the defeated it was a way to communicate a sense of injustice and nationalism. For example, in May 1921 the convention of German geographers passed a resolution that proclaimed: 'It is a national necessity and duty that the link to Germandom of the areas which were torn from the German Empire in the Treaty of Versailles ... remains clearly visible in atlases, and advocates that only those works for which this is the case, be used for instruction in all school grades.'[3] This was a way to build up grievances over many years: by the time Hitler came to power in 1933, pupils for twelve years had been taught which lands were 'really' theirs.

This would have been the intention of 'Professor Dr Arthur Haberlandt', whose name appears beneath the title of one of the most impressive wall charts in the collection. Published in Austria in 1927/1928 it shows the ethnographic make-up of Europe. Setting aside its subject matter for a moment, this map is a masterclass in map design.

Experienced cartographers will tell you that it is hard to make a thematic map that remains clear and easy to read with more than a handful of colours to distinguish between the different

(Overleaf)

Map 58 Haberlandt's ethnographic map of Europe is a triumph of design through its use of colour. At the top of the map is written 'Approved for general use in intermediate educational institutions and secondary schools by decree of the Federal Ministry of Education dated March 26, 1928'.

Karte der Völker Europas nach Sprache und Volksdichte: Freytag & Berndt (1927) 162 × 132cm

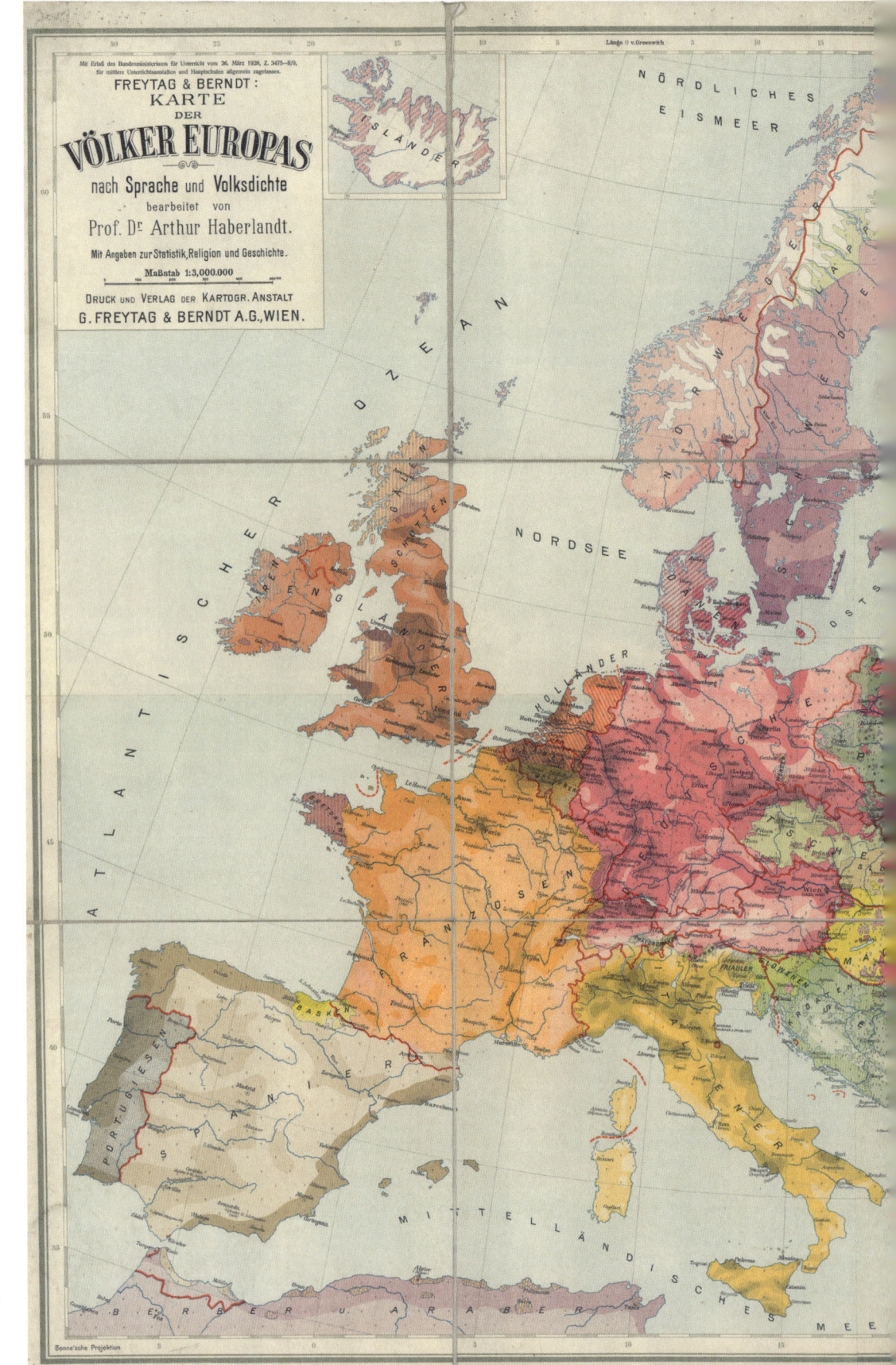

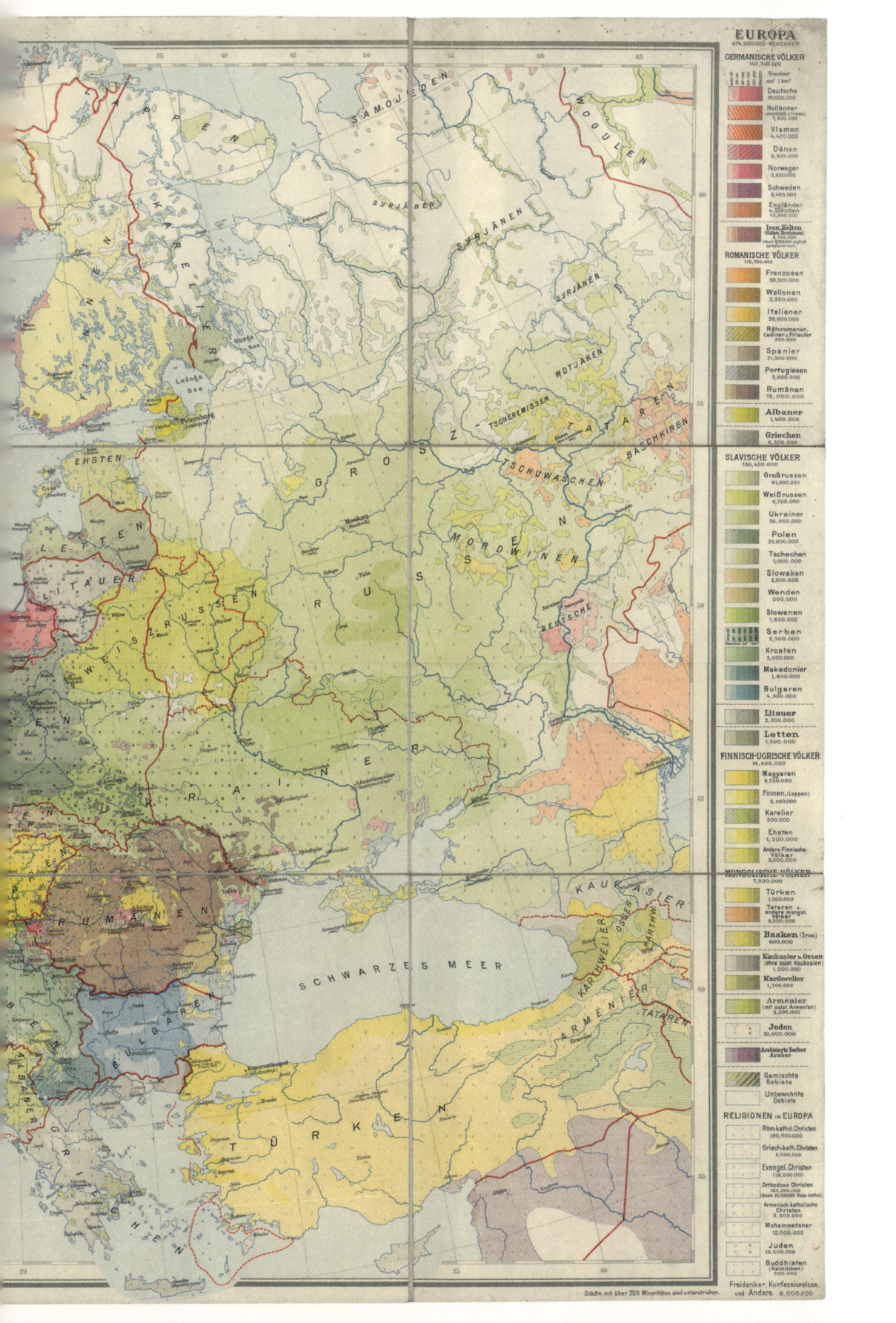

categories within the data.[4] If too many are used, or they are too similar, then the human eye struggles to tell one colour from another and therefore the map stops being effective.

In this kaleidoscopic wall map there are forty-four different colour classes, one for each ethnicity, plus one set of symbols to represent 'Juden' (Jewish people). But each colour also has six different intensities to indicate the population density (number of people per square kilometre), so this means there are 264 unique categories on this map using colour alone. Haberlandt goes even further with the inclusion of eight different symbols for religion (Jewish is included twice) so, in theory at least, there are 1,848 (264 × 7) different combinations of values on this map! For example, London is 'über 200 Engländer u. Schotten' and 'Evangel. Christen' (over 200 English and Scottish, and Protestant) per square kilometre; Madrid is 20–60 'Spanier' and 'Röm.-kathol. Christen' (20–60 Spanish and Roman Catholic Christian) per square kilometre. As a cartographer, I cannot tell you how much effort it would have taken to work all this out (manually) to produce a map that looks fantastic from a distance, while having immense detail close up.

The density approach is not unlike those used by the Humboldt-inspired efforts of Bertie Cotterell Wallis's maps of Hungary and Eugeniusz Romer's of Poland. It gives this map a sense of smooth undulations of colour within the larger ethnic groups and then the more occasional abrupt transitions into minority groups. Top of the list on the right-hand side and depicted in bold pinks are the 'Deutsche' group who dominate the map in Central Europe. They extend into France (Alsace-Lorraine), blur the borders of Switzerland and Austria and reach over Poland into 'East Prussia' adjacent to Lithuania. Look carefully and you can make out the much more muted yellows of the 'Magyaren' group that Pál Teleki had shown in blood red on his 'Carte Rouge'.

Haberlandt was an Austrian ethnologist and geographer who studied anthropology and prehistory at the University of Vienna. He became a lecturer and was then made Director of the Museum für Volkskunde (Museum of Folk Life and Folk Art) in 1924. He spent the First World War sourcing data on ethnic groups that might then have been used to negotiate borders for the ailing Austro-Hungarian Empire.[5] He proudly travelled with a small group of 'scientists with guns' in the unexplored mountains and valleys of south-eastern Europe.[6] After the war, Haberlandt was among a group of Austrian geographers who had subsequently shifted their attentions to the idea of 'Anschluss' or connection to Germany. They were united with many of their German counterparts by the idea that maps could – and should – be the basis to the argument for a 'Greater Germany'. When Hitler eventually annexed Austria, Haberlandt was supportive and allowed the Nazis to store their looted artefacts in his Museum für Volkskunde.[7]

This map was designed for the classroom so it would have fallen to teachers to explain why, thanks to the Paris Peace Conference, such a clear block of the German pinks had been dissected by borders: a point emphasised by the way the map omits country labels, instead including the names of ethnographic groupings, further diminishing the sense of the post-war national identity. This is especially the case for the contested 'Mitteleuropa', where the boundary of Poland surrounds the intersection of the 'Pollen', 'Weiszrussen' and 'Ukrainer' groupings.

Edicts such as the use of the old borders in schools and maps like Arthur Haberlandt's were early examples of the use of maps as part of a 'massive map campaign' pursued by '*völkisch* (German nationalist) activists advocating the supremacy of Germans'.[8] As we shall see, it was a campaign which reached its climax towards the end of the 1930s with terrible consequences.

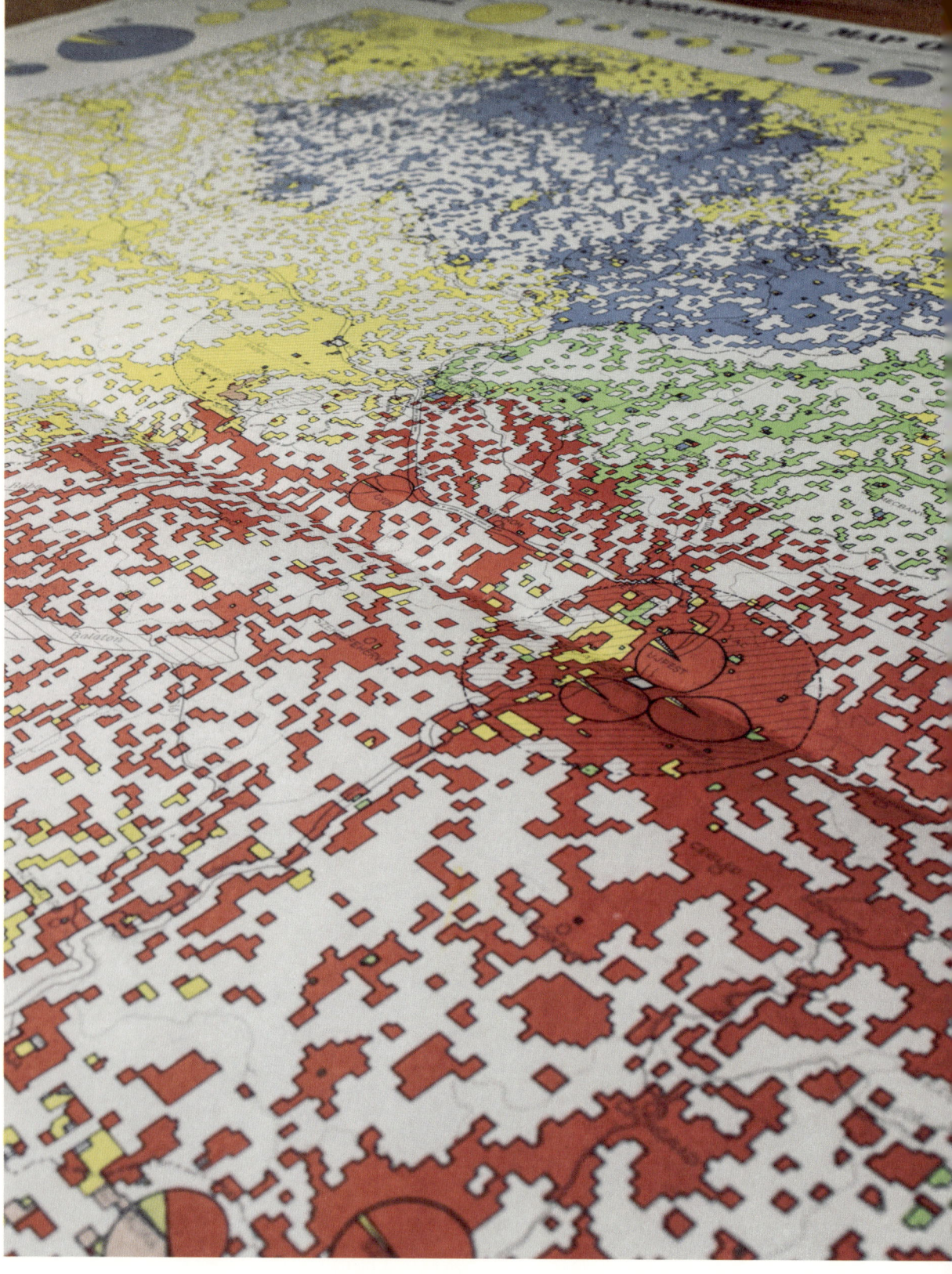

MANIPULATIVE MAPS

Charting the Way to Fascism

The charts produced in the second half of the nineteenth century and in the years preceding the Paris Peace Conference are proof that maps were always political but they either hid behind a veneer of scientific respectability (even if their creators were not at all impartial) or were blatantly satirical or propagandistic.

The German representatives in Paris had felt caught out by the gloss and spin of some of their adversaries' maps brought to argue for territorial claims, not least because, unlike those across the table from them, they were willing to be honest about the fact that their maps were not detailed enough.[9] They were particularly annoyed by the Polish efforts from the likes of Eugeniusz Romer, which they saw as veering from the overly simplistic to deliberate falsifications.[10]

Spurred on by this, it was a German professor named Karl Haushofer who changed the game when it came to using maps to argue a (German) point of view. While the likes of Haberlandt were covert in their mixing of science and politics in the maps they created, Haushofer wanted to overtly combine the connotative power of maps with that scientific respectability and weaponise them for the state. He called this 'Suggestive Cartography'[11] and argued that maps could represent *aspirations* as well as facts. He leaned on the artistic nature of maps, talking up their subjective interpretation and the creative decisions that can go into them so as to maximise their educational and political influence. To Haushofer, suggestive maps were a key

Map 59 [Extract] This map is a descendant of Teleki's 'Carte Rouge' and was published in 1942. Its bright colours show the ethnographic groups of the region but omit a dark history: the destruction of the Jewish communities to the west of Hungary, in Austria and Germany, which by this time had become too small to feature on the map, despite once being populous enough to do so.[9]

Ethnographical Map of Central Europe – East & West Sheet: War Office (1942, copy)
140× 90cm

ingredient to restoring Germany to greatness following its humiliation in Paris.

This sense of defeat was especially raw since Haushofer had been an officer in the German army, and in 1908 he had travelled to Japan to study what he saw as its successful military culture. He returned to Germany on the eve of the First World War and soon became a senior commander with postings to the Western and Romanian fronts. By 1916 he was a colonel and ended the war as a major responsible for a command post in the Alsace region. Following the armistice, it fell to Haushofer to lead 45,000 defeated German soldiers eastwards across the Rhine back into Germany.[12]

By this time Haushofer, who was approaching his fiftieth birthday, had realised that the world of hierarchy and structures that he was so fond of was collapsing. He was also weary and bruised from battle, so he decided to resign from the army.[13] His first two attempts to leave were refused and it was only after a damning medical report – with a list of ailments including eyes damaged by shrapnel and concussion from heavy artillery fire – that his third letter of resignation was accepted.[14]

Following his military service, Haushofer took an unpaid position[15] as an academic at the University of Munich, where he turned his attentions to geopolitics. His enthusiasm for maps came from the unlikeliest of places: Robert Louis Stevenson's *Treasure Island*. Or to be precise, a commentary on Stevenson's use of maps in his books that was buried deep in the January 1895 edition of the Royal Geographical Society's *Monthly Record*:[16]

Mr. R. Louis Stevenson on Maps. - The latent poetry and profound suggestiveness of a map has never been more gracefully exemplified than by the late Mr. Stevenson ... His description of the origin of his own first novel

MANIPULATIVE MAPS

is as follows: 'On one of these occasions I made the map of an island; it was elaborately and (I thought) beautifully coloured. The shape of it took my fancy beyond expression; it contained harbours that pleased me like sonnets, and, with the unconsciousness of the predestined, I ticketed my performance Treasure Island. I am told there are people who do not care for maps, and find it hard to believe …'[17]

It is not clear when Haushofer read this, but it is credited in his article 'The Suggestive Map', published twenty-seven years later, that sets out his ideas for more effective mapping for geopolitical ends. Using the same critique of German mapping efforts that Alfred Hettner had made in 1914, Haushofer highlighted the contrast to the 'Anglo-Saxon' approach to map-making, which converted maps into important propaganda tools that, for example, celebrated the greatness of the British Empire.

His argument was simple: the German focus on detail and accuracy over broader, suggestive imagery was detrimental to influencing political narratives and connecting with a wider public. The cartographic arms race, which started with the Society for the Diffusion of Useful Knowledge bemoaning the lack of appetite for maps in Britain compared to Germany (see Chapter 4), had swung the other way and now it was the German geographers who felt their country was falling behind.

'Let the Anglo-Saxon Stevenson tell us,' wrote Haushofer in a direct quote from the Royal Geographical Society, 'It may be found that geography, used as an instrument of intellectual training, will produce results unobtainable by other means.' He signed off his article with a reference to a speech given by British politician Robert Lowe on the passing of the Second Reform Act in 1867: '"Let us educate our masters!" says the Briton!'[18]

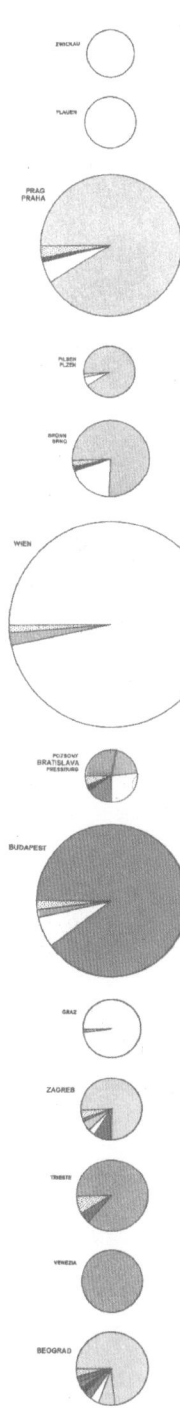

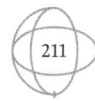

These could easily have become the forgotten utterances of an embittered professor, but a year after Haushofer published them he was visiting Landsberg Fortress Prison in Bavaria to tutor two men who would change the course of history: Adolf Hitler and Rudolph Hess.

Haushofer had been introduced to Hess at a dinner on 4 April 1919 and from this point on the two men became very close. They would take daily walks together to discuss the issues of the day, with Haushofer becoming something of a father figure to Hess. His influence became clear through Hess's letters, which moved from the personal to more geopolitical in their subject matter.[19] Hess's intense anti-Semitism did not seem to concern Haushofer, even though his wife Martha was half Jewish.

In 1923, the newly formed Nazi Party, led by Hitler, attempted to overthrow the German government in what became known as the 'beer hall putsch'. Hitler and his accomplice Hess (who briefly hid in Haushofer's home[20]) were found guilty of treason and sent to Landsberg. They served only eight months of a five-year term, but this short spell behind bars was long enough for Hitler to dictate his infamous book *Mein Kampf* to Hess. It is during this time that the two men were visited in prison by Haushofer, who tutored them in a range of subjects, including geopolitics. Hitler would later recall this time in prison as his 'university education at state expense'.[21] In his book *The Demon of Geopolitics* Holger Herwig reflects on this period:

> Adolf Hitler gained immensely from the 'down time' at Landsberg Fortress Prison between April and December 1924 ... Never before had he had a teacher of Haushofer's stature (army general and university professor); never before had he given such a person his undivided attention for hours every Wednesday for half a year; never before

had he been assigned readings on a highly specific topic. And never again would he take the time to repeat such learning.[22]

Looking back on his time in Landsberg, Hitler said: 'Without my imprisonment, *Mein Kampf* would never have been written; and, if I may say so, during this time, after constant rethinking, many things that had earlier been stated simply from intuition for the first time attained full clarity.'[23]

One of the things that had been clarified was the idea that *Lebensraum*, popularised by Friedrich Ratzel and enthusiastically adopted by many German academics including Haushofer, should be sought not just through expansion as part of foreign policy, but also through the destruction of the Jewish population as part of domestic policy. This became particularly acute in Hitler's anti-Semitic assessment of the 'Jewish Bolshevism' that had overrun Russia, providing additional impetus to the sense that expansion was needed towards Moscow and across the states in between.[24]

Throughout the 1920s and 1930s, Haushofer promoted the development of aspects of national socialist ideology as well as his suggestive cartography in his publication *Zeitschrift für Geopolitik*. For maximum effect, Haushofer and his acolytes argued that geopolitical maps had to be simple, demonstrate a clear message and be cheap to produce in large quantities. Their design made regular use of bold, generalised arrows and drew their persuasive power by suggesting, for example, the encirclement of Germany and conveying its vulnerability to attack from neighbouring states.[25]

When the Nazis came into power in 1933, map and atlas production was supercharged as they demanded more detailed and effective maps, not just to promote their horrific policies and ideas but also to inform them.[26] Geographers – both in Germany and in

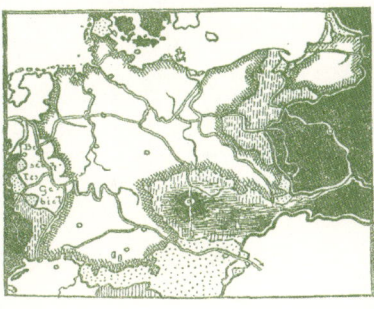

Map 60 The map Haushofer used to illustrate his article espousing the need for more 'suggestive' cartography.

'Die suggestive Karte'? (Die Einschränkung u. Bindung Des Deutschen Lebensraums seit 1918): Die Grenzboten (1922)
13.5 × 10cm

THE LIBRARY OF LOST MAPS

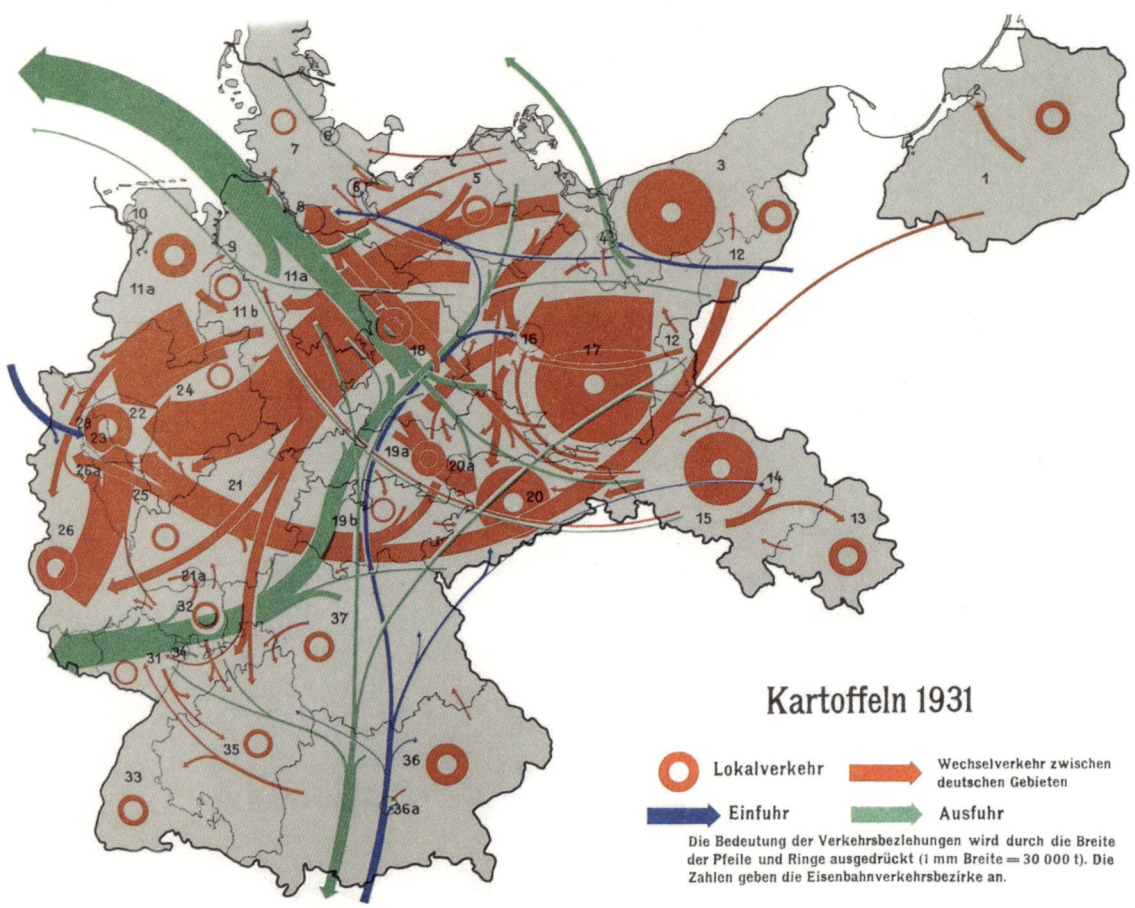

a sympathetic Austria after the 'Anschluss' – saw an opportunity to increase their influence and garner resources to embark on ambitious programmes of regional research. Their results were shared through impressive-looking atlases that would have been hugely expense to produce, but that gave a sense of credibility to the Nazis' abhorrent endeavours.

The Map Library has a surprising number of these atlases, their bland titles and austere jackets acting as camouflage for the brightly coloured and consequential maps they contain. Their route into the Map Library was almost certainly via the military disposals that took place at the end of the war. A couple have had their pages carefully removed from the original binding and mounted on board with handwritten translations, probably by British or American military staff, while a copy of the 1934 *Atlas Niedersachsen* (Atlas of Lower Saxony) has scorched edges as if extracted from the embers of a burned-out building. Even though several of the atlases were started shortly before the Nazis came to power, their authors were given funding to finish them, and the value of this work to the aims of the Third Reich is made clear in their opening paragraphs.

The 1934 *Deutscher Landwirtschaftatlas* (Agricultural Atlas of Germany) begins with the words: 'The government of the new Reich has one of its greatest and most urgent missions in rescuing the German peasant. After a long time, German agriculture is once again at the forefront of economic policy.' It features maps that Haushofer would have approved of. For example, the (perhaps slightly uninspiring) topic of the regional integration of potato markets becomes a dynamic tangle of bold arrows showing internal flows of potatoes (red), imports (blue) and exports (green).²⁷

The charred copy of the *Atlas Niedersachsen* describes the need for comprehensive data about the nature of the land and its population, as well as its settlement, economic, educational and cultural conditions and their interrelationships. With such data it was intended to 'provide wider circles with a deeper knowledge of Lower Saxony and, in the spirit of our Leader, deepen and strengthen the connection to the whole nation through love for the local homeland'.

Huge teams were assembled to create regional atlases of border areas that were considered 'German' but that were disputed with neighbouring countries. Another atlas I found

Map 61 One of four maps from the *Deutscher Landwirtschaftatlas* (1931) showing the regional integration of agricultural markets, in this case for potatoes, in 1931. Arrows are coloured according to internal flows (red), imports (blue) and exports (green).

Deutscher Landwirtschaftatlas (Regionale Verflechtung landwirtschaftlicher Märkte): Statistisches Reichsamt (1931) 19.5 × 15cm

was the *Burgenlandatlas* (Atlas of Burgenland, 1940) that offers a study of every aspect of life in the region following its transition from Hungarian to Austrian control after the First World War.[28] Previously part of the Hungarian side of the Austro-Hungarian Empire, Burgenland was annexed to Austria on 13 October 1921, despite armed resistance. A referendum in Ödenburg, the town destined to be the regional capital, resulted in that town reverting to Hungarian control, thus adding to a sense of precarity on the outer reaches of the pan-German population. On 1 January 1922, Burgenland became an independent federal state within Austria and in 1938, after the Nazi takeover of Austria, was among the first regions seized and its status dissolved. By the year's end, Burgenland's Jewish population had had their property expropriated or been expelled.[29]

The copy of the *Burgenlandatlas* in the Map Library bears three different stamps. The first is 'Geographisches Institut der Universität Wien' (Geographic Institute of the University of Vienna) written around the coat of arms of Austria, while the second credits the same institute but at its centre bears the infamous Nazi insignia. The third stamp is the words 'Ausgescheidene Doublette', indicating it was cancelled from the library it was in as a duplicate. When I looked closely I noticed that this stamp appears on the top of the second Nazi one and could have been added at the same time, so I presumed that there must have been a particularly diligent librarian who sought to ensure the 'correct' stamps were present, even as they were arranging its removal, perhaps to another library under Nazi control.

The Geographisches Institut at the University of Vienna hosted the Südostdeutsche Forschungsgemeinschaft, SODFG (South-east German Research Association), which was founded in the autumn of 1931 by a group of Viennese scientists. It was one of six research associations (mostly based in Germany) that were concerned with the study of German ethnic and cultural territory. Petra Svatek writes in the *British Journal for the History of Science* that the SODFG:

> focussed on the neighbouring states of Germany, the successor states of the Austro-Hungarian monarchy and the overseas territories populated by German emigrants … Until the end of the Second World War, they collected basic territory and culture-related data to justify or falsify territorial claims (which were needed for the ethnic segregation of the individual population groups), planned mass relocations, and acted as consultants to SS and Wehrmacht intelligence services or other political institutions.[30]

The SODFG also became a clearing house for looted maps. They were stored at the Abbey of St Lambrecht in Steiermark, Austria, which also housed an outstation of the Mauthausen and Ravensbrück concentration camps, with at least four of the prisoners tasked to support the SODFG.[31]

The SODFG stamp appears on several historic maps of Eastern Europe in the Map Library as well as on the *Burgenlandatlas* and the *Atlas Niedersachsen*, which would indicate a high probability that they came through this system and then were seized at the end of the war. So my initial assessment that a 'diligent librarian' was behind the stamps may have been horribly wide of the mark: the SODFG stamps could be evidence that these maps were processed via a system of forced labour under the most atrocious conditions.

The preface to the *Burgenlandatlas* leaves the reader in no doubt about the views of its editor, an Austrian geographer named Hugo Hassinger: 'Just as the political Anschluss of Burgenland to the German Empire was realized in the spirit of National

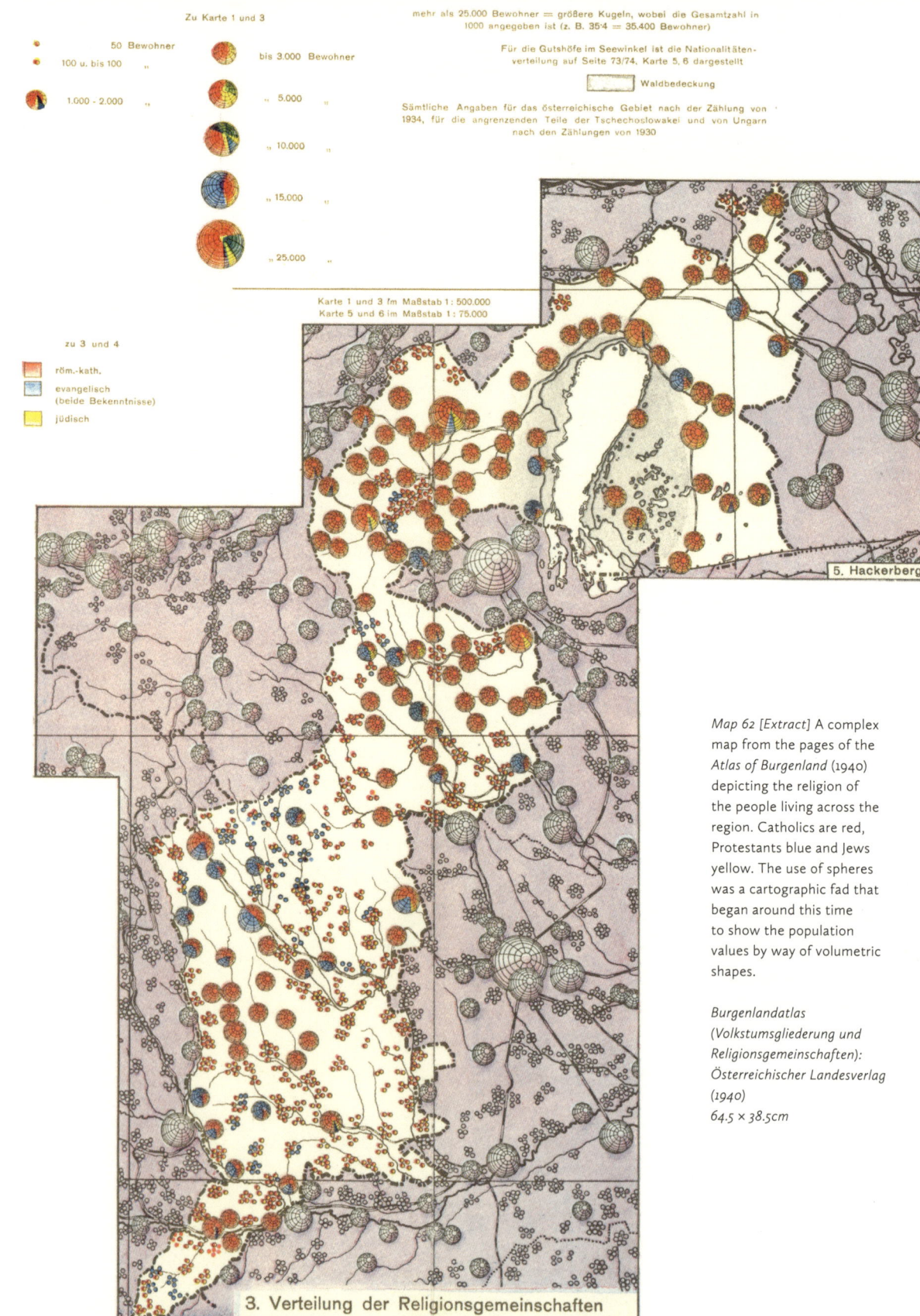

Map 62 [Extract] A complex map from the pages of the *Atlas of Burgenland* (1940) depicting the religion of the people living across the region. Catholics are red, Protestants blue and Jews yellow. The use of spheres was a cartographic fad that began around this time to show the population values by way of volumetric shapes.

Burgenlandatlas (Volkstumsgliederung und Religionsgemeinschaften): Österreichischer Landesverlag (1940)
64.5 × 38.5cm

Socialism, so too might this atlas establish a connection to the science of Greater Germany.'[32]

Hassinger was a senior member of the SODFG and he and his colleagues were able to build a team of fifty people to work on the atlas, which shows in the volume and complexity of the maps and graphics it contained. The result, as Svatek observes, was its 'significant contribution to the National Socialist concept of living space (Lebensraumpolitik) even before 1938' and its use as a critical tool for schools, academic institutions and public authorities.[33]

As a mapmaker I find the contrast between the cartographic skill of these maps and their consequences very hard to come to terms with. The maps are brightly coloured and, in the safety of the Map Library, it's easy to forget their historical context. For example, the *Burgenlandatlas* map of religion gives the region the look of a billiards table, with the balls scattered across it. Each sphere comprises coloured segments to show the number of Roman Catholics (red), Protestants (blue) and Jews (yellow) enumerated from a census taken in 1934. They are framed by a light purple colour that fills Burgenland's neighbouring regions.

Somehow, these are not the colours I'd expected to see on a map that would have been in genocidal hands, tasked with eradicating any hint of the third of those religions. But it was not all bright colours: the Roma were depicted in black and were described in the text of the atlas as a serious burden that had 'over-multiplied' and should disappear from Burgenland.[34]

The researchers left no stone unturned when it came to understanding the lives of the Burgenlanders. I found another map tucked in the corner of one of the pages of the atlas that showed how far spouses had travelled to be with their partner in the small town of Lutzmannsburg. Each couple was assigned a colour based on their ethnographic grouping, with 'Germans' being the brightest red (and most prevalent) to enforce the message that the town was well connected to other German areas. Non-Germans were coloured in blue circles, and Jews are marked with an X.

When Hitler embarked on his reign of destruction and terror, the intellectual elites of Britain and America (and their allies) – so comfortable with the previous notions of ethnographic mapping and their self-declared scientific approaches to statehood – were shocked at the way he was pursuing their ideas. Many academics sought to distance themselves from these methods, in light of how they were being applied by the Nazis.

There was also a curiosity about what could be learned, both in terms of improving their own communication approaches but also to combat the Nazis, about the evolving thinking in geopolitics, as well as Hitler's apparent success in using it to garner support. So German atlases were reviewed in other parts of the world. For example, in 1934 the American geographer Richard Hartshorne reviewed the *Saar Atlas* which mapped the Saar border region between France and Germany that Hitler was keen to annex. After praising the scientific impartiality of the work, he naively remarks that in some cases there is a 'certain exaggeration, to show a closer connection with Germany than with France'.[35]

Within a decade of his review Hartshorne, who spent time researching in Germany in 1931–2, would have co-authored a book called *German Strategy of World Conquest*[36] which, without a hint of self-awareness, expresses surprise that: 'Germany intends to dominate the world and has a plan for doing so. Americans, far from Europe, and preoccupied with their own affairs, have been unaware of this intention ... The swift march of conquest has stunned or dazzled the onlookers ... They do not realise that the method of piecemeal conquest is a well-established procedure in German history.'[37]

Such was Haushofer's reputation that myths swirled around a grand 'Institut für Geopolitik' at the University of Munich,

MANIPULATIVE MAPS

which was said to generate the concepts that were core to Hitler's ideological thinking.[38] The truth was that no single entity existed but, as is evidenced on the pages of some of the Map Library's atlases, there were well-resourced research groups, like the SODFG, that contributed to a 'Brains Trust'[39] that informed the machinery of the Nazi state and told it what it wanted to hear.

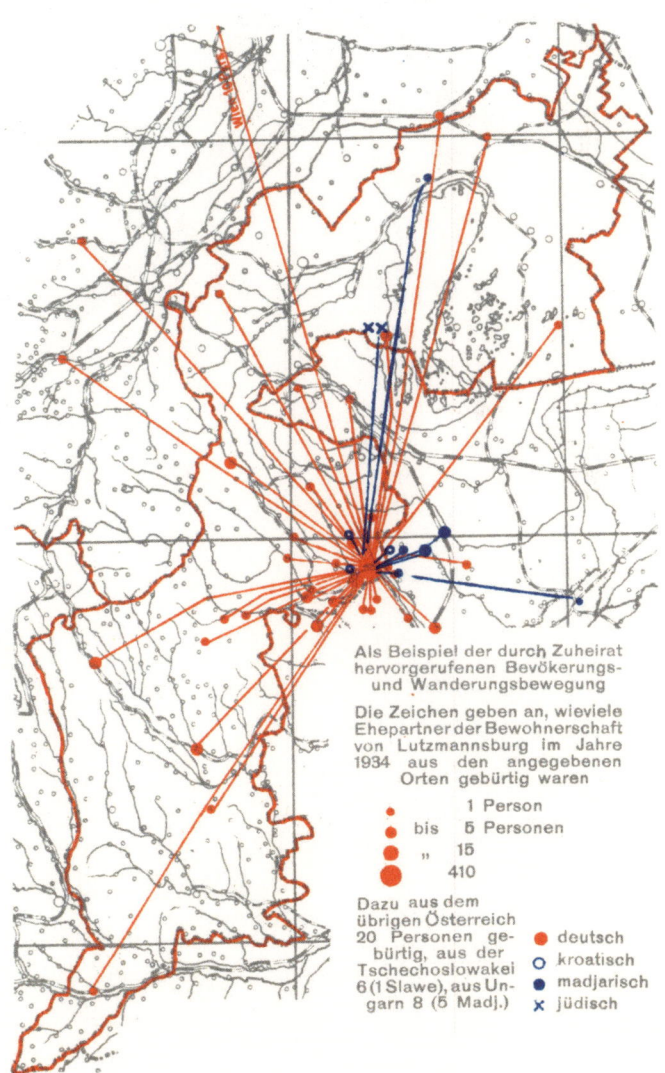

Map 63 [Extract] The connections between areas show where spouses had travelled from to live with their partner in Lutzmannsburg, a village on the Austria-Hungary border. Such maps were part of a data-driven exercise in delineating those areas that should be part of a 'Greater Germany' and determining the rights of the 'Germans' living there.

Burgenlandatlas (Heiratseinzugsbereich von Lutzmannsburg): Österreichischer Landesverlag (1940)
9 × 14.5cm

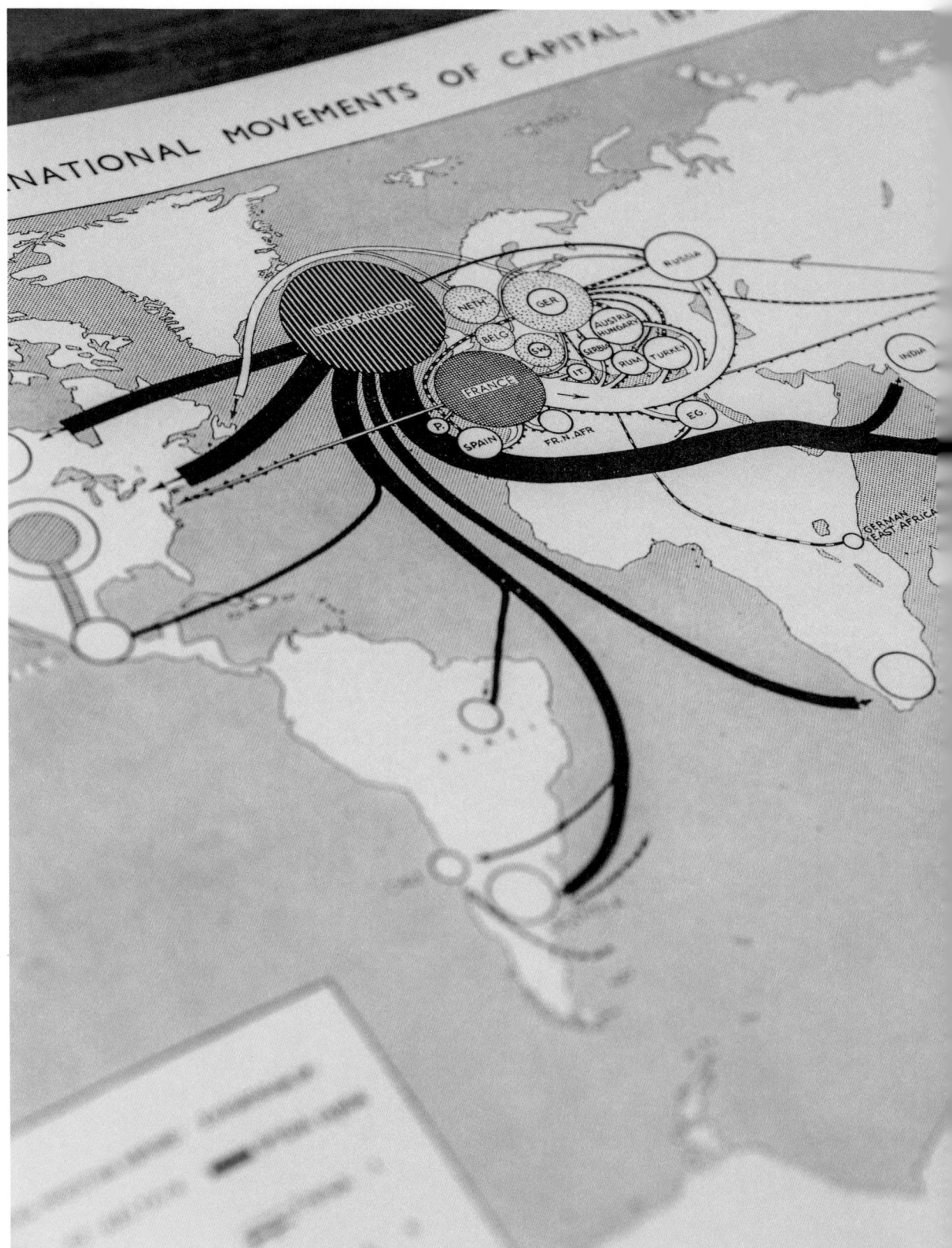

MANIPULATIVE MAPS

Beyond Germany

While Karl Haushofer and his associates were pioneering their use of persuasive cartography, especially the deployment of maps distilled to deliver a clear message to readers, others working outside Germany were using similar techniques for radically different politics. Their maps and atlases rub shoulders with the German creations in the Map Library and add an alternative perspective to the geopolitical ideas (and ideals) that swirled around the creation of maps in the build-up to the Second World War.

    For example, I wasn't sure what to make of a very tatty atlas on the shelves of the USSR section of the atlas cabinets. It was crudely bound in cloth with a spine held together with duct tape. The maps inside had been printed on very thin, shiny paper and in such poor-quality black ink that they were almost illegible, not least because they clearly needed to be in full colour to be understood. The text was Russian, but I spotted some English translations neatly written in pencil in the margins, and there was a green stamp that said 'duplicate copy number 2' on the title page. Intriguingly, a small square had been cut out, perhaps to remove another marking.

    My initial reaction was simply to dispose of this atlas because it looked to me as if it was something that an academic in the department had cheaply photocopied. Fortunately, one of the students I'd recruited to help with cataloguing the Map Library's collection spoke Russian and she was quick to enter the

Map 64 [Extract] The *Atlas of To-Day and To-Morrow*, which contains some of the first maps to show the flow of money between countries.

An Atlas of To-day and To-morrow (International Movements of Capital 1914–1930): Victor Gollancz (1938)
23 × 16cm

THE LIBRARY OF LOST MAPS

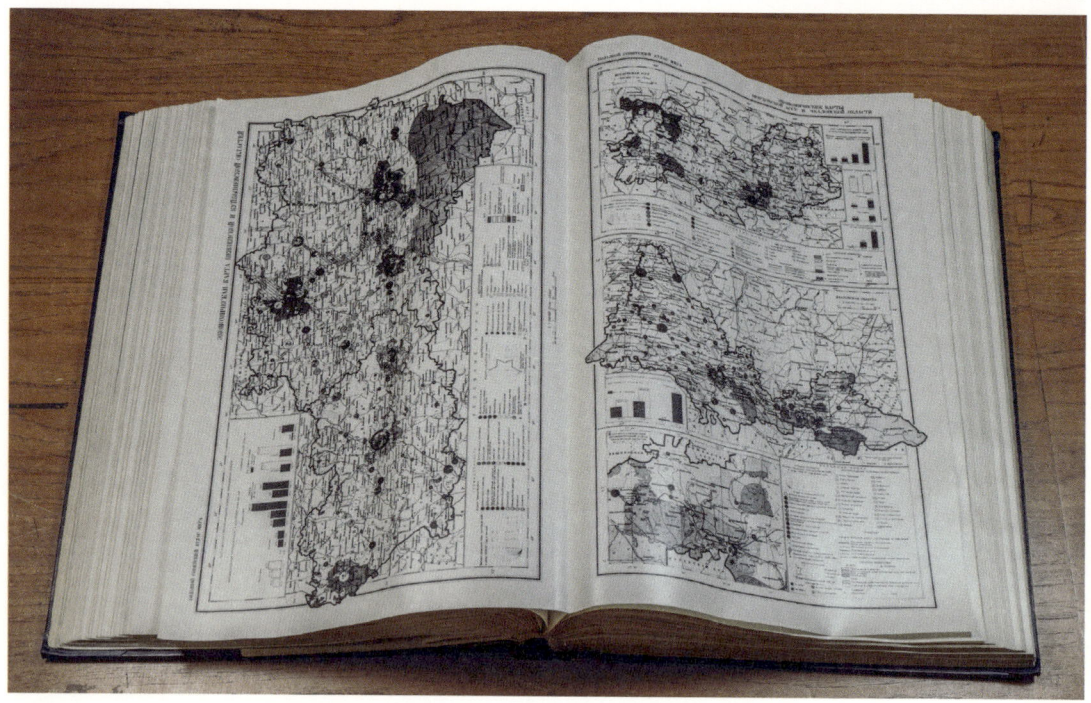

Map 65 A photocopied version of the *Bolshoi Sovietskiy Atlas Mira* (Great Soviet Atlas of the World) volume II (published 1939), which was taken by the intelligence services from one of the full-colour versions that made it out of Russia before Stalin censored it. This atlas is practically illegible but, at a time when detailed maps were scarce, anything was better than nothing.

БОЛЬШОЙ СОВЕТСКИЙ АТЛАС МИРА: ГЛАВНЫМ УПРАВЛЕНИЕМ ГЕОДЕЗИИ И КАРТОГРАФИИ при СНК СССР (1939)
59 × 49cm [Full Double Page]

atlas's details into our spreadsheet as Большой Советский Атлас Мира (Great Soviet Atlas of the World volume II) with the year 1939, and it got reshelved alongside an original copy of volume I of the atlas, which we had also found. The students also did some background research on both volumes and mentioned, almost in passing, 'the second volume is considered interesting as it was at one point temporarily made secret and removed from all libraries'.

Relieved I hadn't thrown it away, I had to find out more and discovered that *The Bolshoi Sovietskiy Atlas Mira* (Great Soviet Atlas of the World) was a significant cartographic project that was initiated in 1933 with volumes I and II published in 1937 and 1939 respectively. The ambition of the project was astounding, and its production involved 175 cartographers and cost around 5 million rubles (at the time).[40] Volume I took a global perspective but also had maps of the Soviet Union at various scales, reflecting economic, physical and historical themes influenced by Marxist ideology.

Volume II focused on the regions of the USSR and featured the usual maps of topography alongside very detailed thematic maps of data portraying characteristics like economic outputs. Its publication at the start of the Second World War caused Stalin to halt the distribution of the atlas almost immediately for fear its rich information would aid the enemy, but at least two copies had made it into the hands of US Intelligence (Germany had secured copies, too, that they had carefully disbound and translated[41]). The detail of volume II of the atlas was unsurpassed and it became an important source of information for military planning, with photocopies made for this purpose.[42]

Somehow, the Map Library had come into possession of one of these still extremely rare copies, with a more interesting history than an original version. The story is a reminder of a time when there was not universal access to geographic information, long before countries could send up satellites to snoop on their enemies. But it also reminds us of the importance of atlases, which might seem quaint and partial by today's standards, as weapons of the state.

Just like Nazi Germany, the Soviet Union was investing huge sums to gather information and express it as maps to help further its aims, which, alas, amounted to further death and destruction. For example, the population maps of the Great Soviet Atlas included a category for the 'bourgeoisie', which included 'land owners, large and small bourgeois, tradesman and kulaks (peasants with over eight acres of land who employed nonfamily labour)'.[43] The kulaks were a social category that Stalin had been seeking to eliminate through his notorious purges and around the time of the atlas's publication he had issued an order to arrest 270,000 kulaks, of whom 73,000 were executed.[44] These targets were 'fulfilled and over-fulfilled' by provincial officials who bolstered their own position by requesting ever higher quotas to impress the

THE LIBRARY OF LOST MAPS

Map 66 + 67 Two pages from Frank Horrabin's 1933 *Plebs Atlas*. Text was kept to a minimum and maps were extremely simple to convey a single message.

The Plebs Atlas (Pages 16 & 17): N.C.L.C Publishing Society Ltd. (1933)
12.5 × 12cm (left);
15 × 10.5 (right)

central authorities.[45] This was just one phase of 'de-kulakisation', a policy that lasted a decade[46] and that would almost certainly have required detailed maps to enact.

Away from the horrors of state-sponsored murder, maps were also being created with much broader audiences in mind, with a view to helping their readers keep pace with the nerve-racking geopolitical developments of the era. James Francis 'Frank' Horrabin was a British socialist associated with the left wing of the Labour Party, becoming the Member of Parliament for the town of Peterborough between 1929 and 1931 (after a brief stint as a member of the Communist Party).[47] Alongside his political activities, he became involved in the Plebs' League, which was a Marxist educational initiative that inspired him to create a series of atlases under the Plebs banner. He was also a successful illustrator, running several comic strips as well as illustrating books for the likes of H. G. Wells.

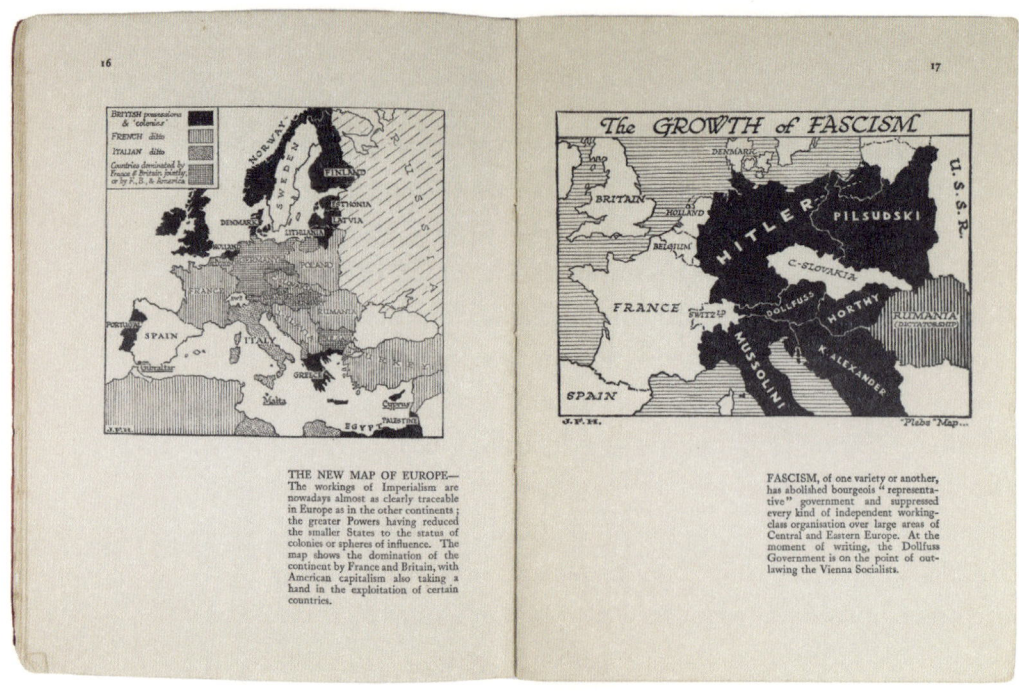

226

Perhaps Horrabin's biggest, yet most overlooked, achievement in the context of map-making was that he was one of the first faces to be seen on British television, and the first to use maps for current affairs with his 'Television News Map' series, which began in 1937.[48] During the war years, the BBC's regular broadcasts were switched off, but on the day of their triumphant return on 7 June 1946, Horrabin's *Newsmaps: 1939–1946* was in the listing at 10 p.m.,[49] and he proved a natural for this new medium. The *Daily Telegraph* even commented at the time that 'among those who have acquired the completely confident television manner are bearded Philip Harben the cook and J. F. Horrabin, the map man. Watching them, one is never ill at ease.'[50]

Horrabin was the master of recycling his maps across multiple publications and formats. They appeared in the *Plebs* journal and then were collated into atlases or shown on television. The simple style of their production would have facilitated this since the maps were intended to be shared as cheaply as possible. In contrast to the grand tomes of the rest of the atlas collection, Horrabin's publications are little more than pamphlets, stapled or stitched together with jackets made of paper that's only slightly thicker than the pages they protect.

As he sets out in his 1933 edition of the *Plebs Atlas*:

> I am a firm believer in the theory that a map should be designed to make some one point clear – and other points be left to other maps. Not only elementary students but older readers are befogged by the wealth of detail, all of it emphasised equally, in an ordinary map. For the same reason I have made many of the maps diagrammatic, and used arrows and different sorts of shading to make their meaning clear … I hope students will use this book when reading the foreign news day by day or week by week,

and will use the margins on each page for making their own supplementary notes. Every map illustrating current history needs to be kept up to date in this way, and the keen student will, by the use of red or blue pencil, make his own additions to this atlas.[51]

The ethos behind these maps – to make them accessible to as many people as possible – was the same as the one pioneered by the Society of the Diffusion of Useful Knowledge a century earlier (see Chapter 4). However, they marked a different style of geographical education. With no claims to being detailed and definitive, Horrabin was encouraging interactivity and edits from his readers, which made his maps living documents for their individual owners to adapt for themselves.

Horrabin also shares some of Haushofer's approach of leaving out details he deemed non-essential and emphasising a particular dimension of the message. And rather than taking months and years to complete they could be turned round in a matter of hours for print (or indeed television).

There was another atlas I found in the collection that touched on many of the same issues as Horrabin's did and in a similar style. Its author was Alexander Radó, and it was published in 1938. It evoked the present and future with its grand title, in block capitals, 'THE ATLAS OF TO-DAY AND TO-MORROW'. Radó's preface strikes a similar tone to Horrabin's:

> The purpose of this Atlas is to provide, so to speak, a snapshot photograph of our rapidly changing world ... [it is] a cartographic record of all the bewildering mass contemporary political and economic problems [which] are yet of the most utmost importance of our own generation, for our present and our future.[52]

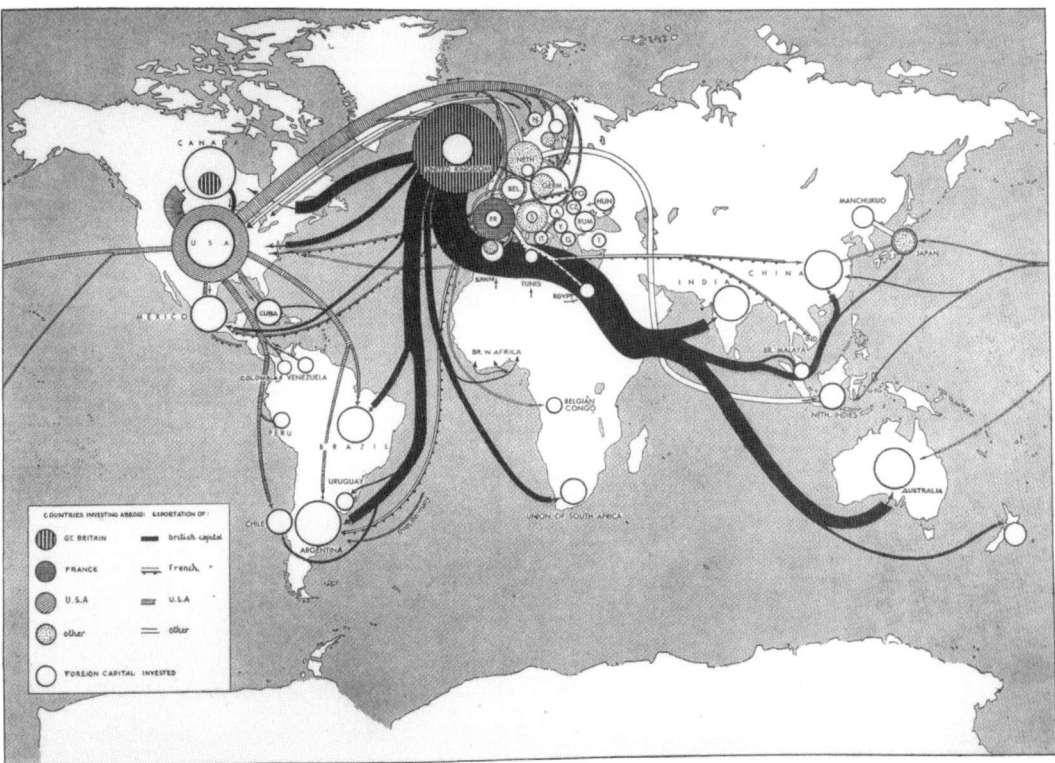

No. 5. INTERNATIONAL MOVEMENTS OF CAPITAL, 1914-1930

It's an easy atlas to miss on the shelves of the Map Library, smaller than a sheet of A4 paper and with a simple cloth binding that's kind of a dusky blue colour. The spine bears the publisher's name of Victor Gollancz, whose authors included George Orwell and Horrabin. The maps, in black and white, were for the time radical in their subject matter. In some of the first maps ever to show financial flows,[53] there are great streams of money radiating out from the UK to its empire and a hint at the coming new world order with the United States depicted as a major global investor during the interwar period, not least in South America. The maps also show the burgeoning new fronts of geopolitical tensions that began in the latter half of the nineteenth century with pages

Map 64 This map shows the dominance of the UK (thanks to its empire) between 1914 and 1930. But it, and the accompanying text, hints at the growing importance of the USA, which had not long transitioned from a debtor to a creditor state according Radó and Rajchman's atlas.

An Atlas of To-day and To-morrow (International Movements of Capital, 1914–1930): Victor Gollancz (1938)
23 × 16cm

dedicated to showing the 'struggle' for control of communications and a particular focus on a new form of transport: transcontinental airways.

Towards the end of the atlas is a map titled 'The Form of Government in Europe' and its accompanying text reads: 'The political and social tension which hangs over Europe is reflected in the different forms of government of the nations. Fully fledged parliamentary democracy is confined almost entirely to northwestern Europe while Czechoslovakia [is] the sole democracy of central Europe.' The final map charts 'The Struggle of Ideologies' and the text alongside ends with an observation that came to pass the following year: 'The world threatens, more and more, to be divided into two ideological blocks – the Fascist and the Socialist and Democratic.'[54]

The book was positively reviewed. The *Financial Times* commented that while the 'use of the diagrammatic form is not altogether new … [it is] carried a degree farther than hitherto in this work',[55] and the *Illustrated London News* featured it as one of their 'Books of the Day', noting the 'elucidating signs and keys'.[56] It seemed to be a successful format, as this innovative atlas sold well – 5,300 copies in 1939 alone – generating royalties in excess of £400 (roughly £22,000 today).[57]

Radó's politics had formed at a young age and would define his entire life. After joining the Hungarian Communist Party at nineteen, he fled to Austria in 1919 because of his political involvement in the short-lived Hungarian communist revolution. He then studied geography in Vienna and worked for the Soviet telegraph agency ROSTA before moving to Germany to continue his studies. He relocated to Moscow in 1924, where he edited the first civilian map of the Soviet Union and wrote a pro-Soviet guidebook. His interest in – and knowledge of – transcontinental airways for the *Atlas of To-Day and To-morrow* may well have

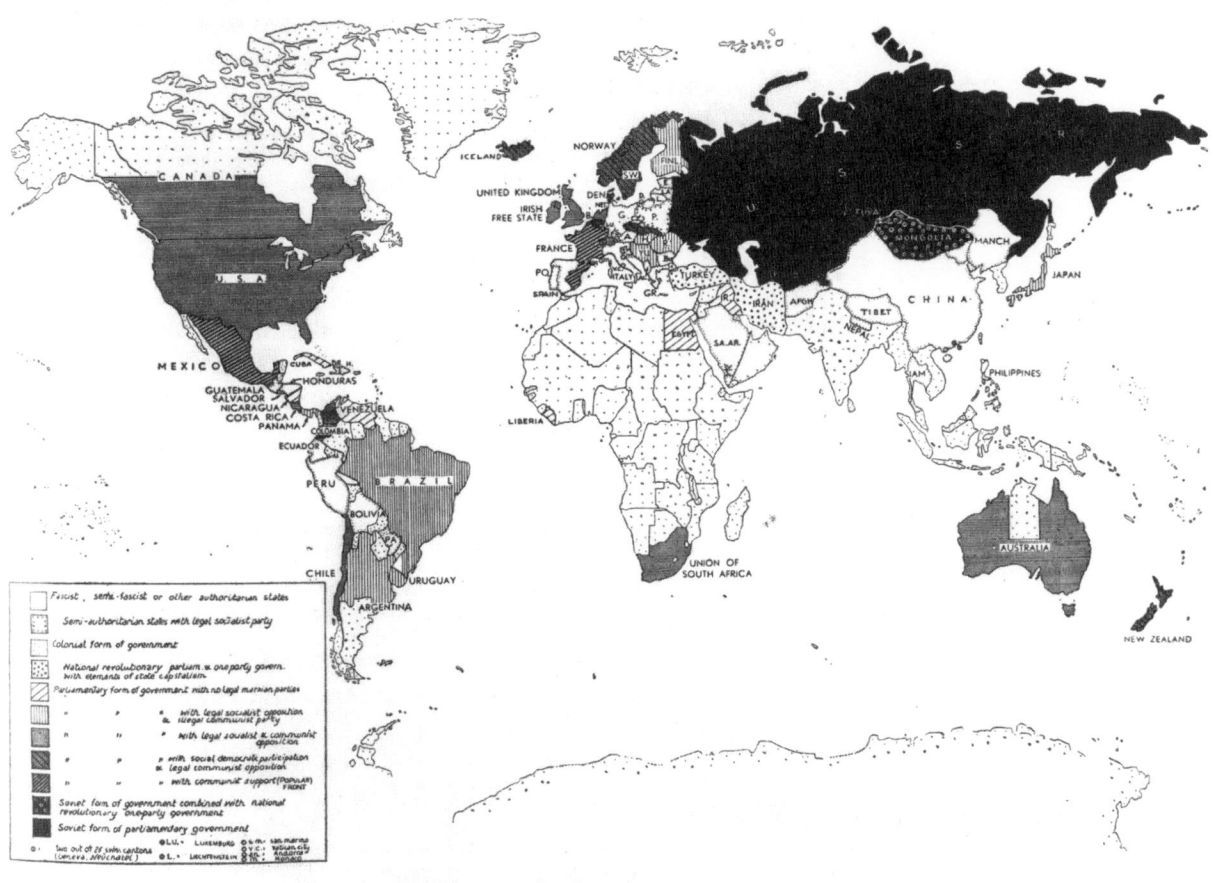

stemmed from his work in the early 1920s preparing pioneering flight route maps for companies like Lufthansa, which were credited to 'A. Radó, geographer'.[58] In 1930 he published his first atlas, the *Atlas für Politik, Wirtschaft, Arbeiterbewegung* (*Atlas of Politics, Economy, and the Labor Movement*), which created waves due to its strong anti-capitalist stance and bold cartography.

The difference in style and tone of the *Atlas of To-Day and To-Morrow* was not what first caught my eye, however. It stands out because it is one of only two atlases[59] in the Map Library's collection of 769 that had a woman's name on the front, that of Marthe Rajchman.[60] The acknowledgement pages of many other

Map 68 Another innovative map from *An Atlas of To-day and To-morrow* that captures the differing political ideologies that had emerged in the interwar period and that would be so consequential to the remainder of the twentieth century.

An Atlas of To-day and To-morrow (The Struggle of Ideologies): Victor Gollancz (1938)
23 × 16cm

THE LIBRARY OF LOST MAPS

Map 69 One of Rajchman's maps adorns the front cover of *Europe and the Czechs*. Her cartography in this book helped to set the scene for readers who were keen to learn about the likely consequences of the policy of appeasement towards Hitler.

Europe and the Czechs: Penguin Books Limited (1938) 11 × 7cm

atlases expressed gratitude to Miss X or Mrs Y for their 'preparation' of the maps, but women were invisible on the covers. Rajchman, by contrast, became extremely high-profile and extraordinarily prolific as a cartographer in her own right, albeit for a fleetingly brief period.[61]

She was trained in the prestigious School of Cartography of the Paris Sorbonne University and had met Radó while there.[62] Her training gives her maps a more formal style than Radó's, and design aficionados may spot a resemblance to Otto Neurath,[63] with restrained use of icons in places, but also the extensive use of arrows and symbology espoused by Haushofer (with whose work Radó was familiar and Rajchman must surely have been, too).[64]

Her distinctive style cropped up in several Penguin Specials, a series of books published from 1937. The most famous of these was *Europe and the Czechs* by the journalist Shiela Grant Duff, who was the sole regular newspaper correspondent in Prague in 1938, when the book was published. It was written in only a few weeks in response to Hitler's threats to move into Czechoslovakia. The intention was to sound a warning about the dangers of the policy of appeasement, which had allowed the Nazis to annex the Sudetenland in the hope that this would sate their territorial ambitions.

On the day when British prime minister Neville Chamberlain returned from meeting Hitler in Munich and famously declared 'peace for our time', free copies of *Europe and the Czechs* were given to all parliamentarians. The book itself was not positively reviewed when it was first published,[65] but it became a huge bestseller as the public became more critical of Chamberlain's approach to appeasing Hitler.[66]

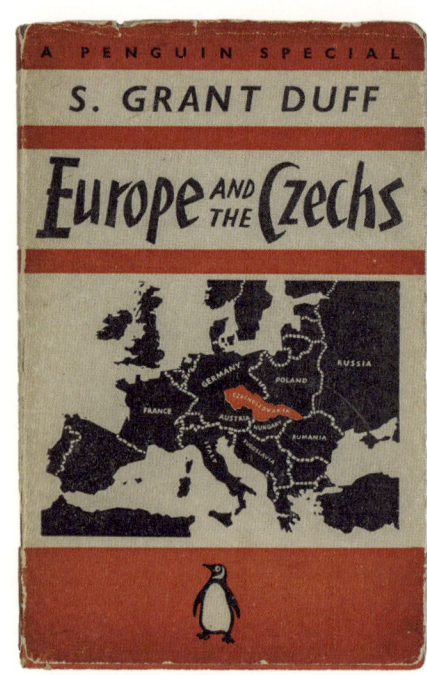

In many ways the book is a distillation of the geopolitical debates that had taken place in the decades before, with its blurb going as far back as the second half of the nineteenth century and referring to Otto von Bismarck (see Chapter 5): 'The book specially emphasises the strategic importance to Hitler of what Bismarck called "a fortress erected by God in the centre of Europe," since Hitler like Bismarck considers that he who is master of that part of the world is master of Europe.'[67]

Rajchman contributed all nine maps to the book and its popularity raised her profile sufficiently to get her name on the front of future publications.[68] Between 1938 and 1944 she published around 350 maps,[69] an extraordinarily productive period in which she helped to create a genre of maps for 'world strategy', to equip the public with the information they needed to become pundits to discuss the next move of their nation in the game of geopolitics. Rajchman's 1942 book *Global War: An Atlas of World Strategy* was a classic of the genre and is dedicated to 'The uncounted other amateur strategists of the United Nations'.

For good or ill, it was the likes of Haushofer, Horrabin, Radó and Rajchman who showed how maps could appeal to the masses as a window onto current events. They were united in their desire for maps to be engaging and participatory and, crucially, political. Haushofer stood apart in the way he promoted the idea that maps could portray an aspiration rather than a fact, a subtle distinction that the Allies were slower to appreciate, and which must have contributed to his mythical status as the intellectual mastermind behind Hitler.

At the onset of the Second World War, the public became armchair generals to follow the action with simpler, clearer maps, and the real generals were being equipped with maps and atlases of an unprecedented quality and effectiveness.

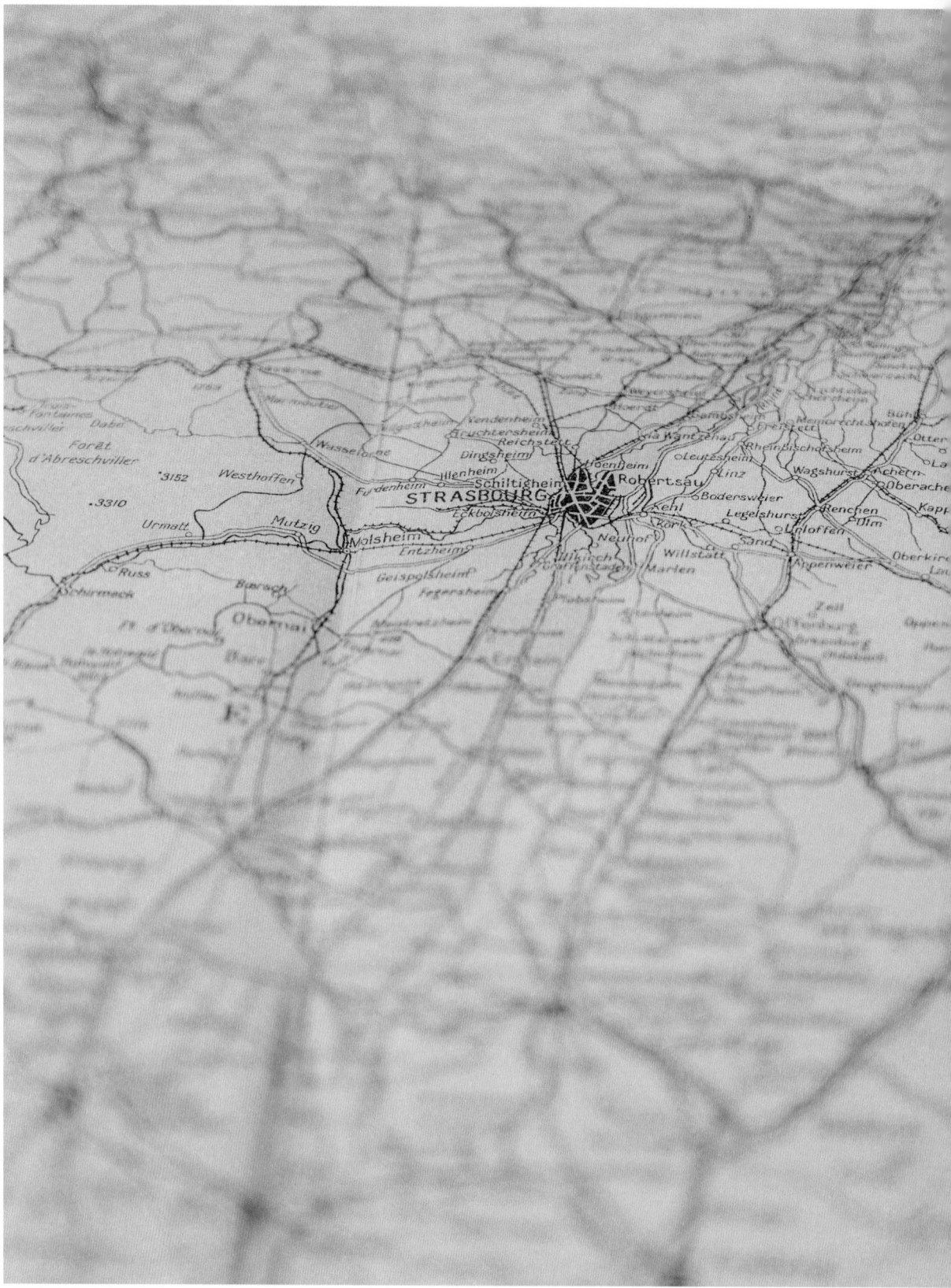

MANIPULATIVE MAPS

Maps at War. Again.

It is no surprise that, at the outbreak of the Second World War, geographers and those who could make maps were in high demand. With a greater need than ever to communicate with the public, we see the changed wartime lives of those whose maps now reside in the Map Library.

In a drawer full of miscellaneous maps of Asia, my eye was drawn to a map of the Pacific War in February 1942. Its style looked familiar and then I saw the name at the bottom: J. F. Horrabin. With his TV show taken off air throughout the war, he began working to communicate developments on the battlefield for the *Serial Map Service* (SMS). This was a subscription-based publication that ran from 1939 to 1948 and offered regular maps that would be sent out with free global postage to subscribers in 'neutral countries'[70] as far away as Australia.[71] Alongside the maps, detailed commentary would show the progress of the war and the major geopolitical issues at play. Its first maps published in September 1939 were indistinguishable from the pages of a generic atlas, but then between 1941 and 1945 there was a shift towards maps that were much more eye-catching thanks to the employment of Horrabin and a few other designers and illustrators.

Meanwhile, many of the decision makers in the military and the government would have remembered the value of maps at the Paris Peace Conference twenty years before and were now conscripting academics into military service, often in intelligence roles, to help make the maps needed.

Map 70 [Extract] On first sight, I assumed that this 'Stanford's Map of the Western Front' was from the First World War, but in fact it was printed in 1940. The front as mapped here is the French border, which was meant to be defended by the Maginot Line, built to deter a German invasion directly into France. However, on 10 May 1940 Hitler's troops skirted round it to invade France via the Low Countries and the Ardennes Forest. Within two weeks they had made it as far as Dunkirk.

Map of the Western Front: Edward Stanford Ltd (1940) 44 × 49.5cm

SERIAL MAP SERVICE

FEBRUARY, 1942

THE PACIFIC WAR 124-125

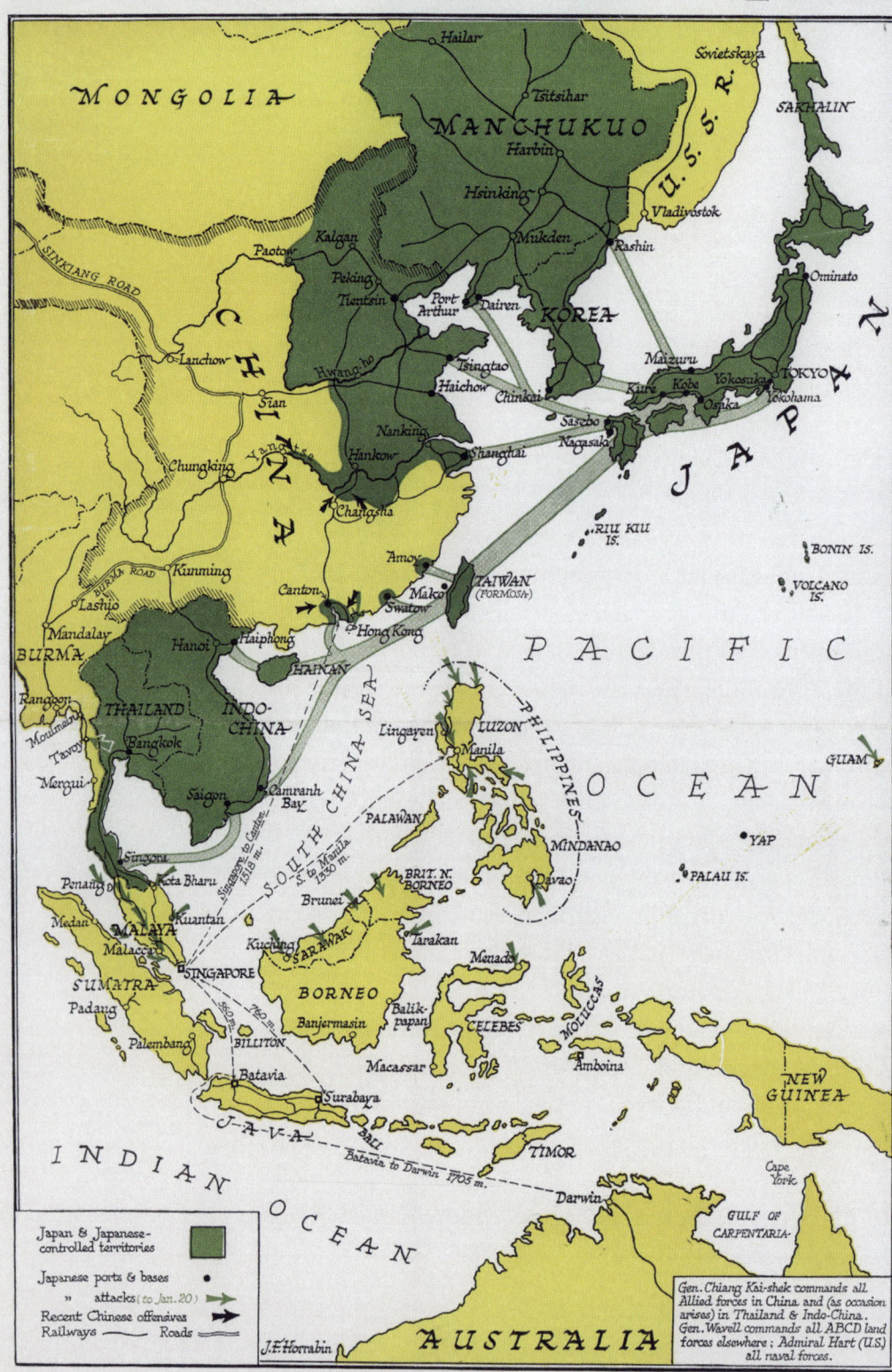

MANIPULATIVE MAPS

One professor in the UCL Department of Geography at the time was Robert Dickinson, whose unpublished autobiography is stored in the departmental archives. It reveals his bewilderment at his new life after 'receiving a mysterious call' to join 'a super-secret inter-service intelligence group, set up at the order of Winston Churchill when he came to office after the Norwegian fiasco'.[72]

The 'Norwegian fiasco' he refers to was the Allied failure to defend Norway from Nazi invasion. One of the points of weakness was how inferior their intelligence was compared to the Nazis'. A post-mortem of the campaign published by the British government after the war said: 'Our leaders and their troops were again and again handicapped by their ignorance of climatic and geographical peculiarities, by the lack of detailed knowledge of harbours, landing grounds, and storage facilities, and even by ignorance of the general qualities and prejudices of the Norwegian people.'[73]

This description seems to be a marked contrast to the German efforts. The Map Library received fifteen German intelligence handbooks as part of its post-war acquisitions, covering France, southern Britain, the Netherlands and European Russia, among other areas, and they are extremely impressive in terms of their detail. The manual for Ireland contains a map of its coastline with the geology of the beaches, which one presumes is in preparation for any amphibious landings. There are postcard-like pictures of everyday life and a dizzying array of facts and data about the regions they feature. The guide to Belgium, produced in 1939, had comprehensive street plans for the country's cities and towns as well as national maps that were copies from Michelin road maps.

Dickinson was based in Oxford for the war and describes his day-to-day work as responding to requests from London without knowing, for example, 'who wanted, or why, a terrain appreciation of Sardinia ... [it was] months before one realized their military

Map 71 One of Frank Horrabin's maps for the Serial Maps Service (1942). The magazine was especially popular during the war when its subscribers hoped to get a global perspective on events. Maps such as this were accompanied by detailed articles from the experts of the day. The SMS continued until 1948 when it abruptly ceased publication after suffering a large reduction in subscribers following the end of the war.

Serial Map Service (The Pacific War) (1942)
21.5 × 33cm

THE LIBRARY OF LOST MAPS

significance'.[74] But Dickinson was a great expert on German towns and cities, so he was also tasked with making maps for the RAF's bombing of Germany, a fact he mentions in passing while recollecting some of the 'hilarious moments' of his military career:

> I recall the time when after spending several months on the preparation of a 'model' map (that I subsequently learned formed a basis of our bombing policy) I journeyed by packed train from Oxford to London, thence to a meeting of some 12 chiefs of Bomber Command in Whitehall. I didn't have the cardboard map roll. I'd left it lying in the luggage rack in the train compartment! An officer called the Lost Luggage Office at Paddington. It was there, delivered by a railway porter and was dispatched immediately by special messenger. These two or three minutes were the most terrible experience of my life, for (such was not my thoughts at the moment – when I had none) I would at least be court marshalled [sic] or dismissed from the service.[75]

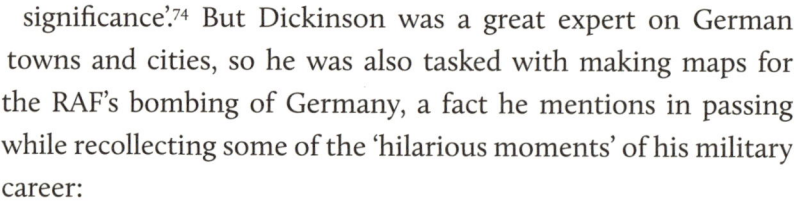

Map 72 [Extract] A fold-out map of the coast of Ireland contained within a German military guidebook. It shows the different coastline features such as cliffs, sand and mudflats, and areas of dunes. It also identifies coastal fortifications 'as of early July 1940'. This extraordinarily detailed map offered the basis for plans to invade the island of Ireland, should they have been needed.

Militärgeographische Angaben über Irland Süd- und Ostküste (Von Mizen Head bis Malin Head) (Beschaffenheit und Gliederung der Ostküst Irlands im Freistaatgebiet von Carnsore Point biz zum Carlingford Lough): Generalstab des Heeres Abteilung für Kriegskarten und Vermessungswesen (IV. Mil.-Geo.) (c.1940)
25.5 × 45.5cm

MANIPULATIVE MAPS

It must have been hard for him to adjust from the study of German cities as an intellectual pursuit[76] to the task of destroying them. He must, too, have been deeply disappointed in the gravitation towards National Socialism of those German academics he had worked with and admired before the war. He describes one of Hugo Hassinger's essays as 'masterly'[77] and I wonder if he had made a special effort to select for the Map Library those items from the SODFG, like Hassinger's *Burgenlandatlas*, that were being distributed at the end of the war. At the very least he would have been one of the best placed to appreciate their significance.

When the war came to an end, there was a period of soul-searching as many who had been involved, both in its build-up, and then in its dreadful conclusions, looked back on what they had done. Some geographers like Dickinson did so with a sense of pride, others sought to rewrite (or forget) their histories to avoid the consequences of their actions, while still others, as we shall see, simply could not continue living with their guilt.

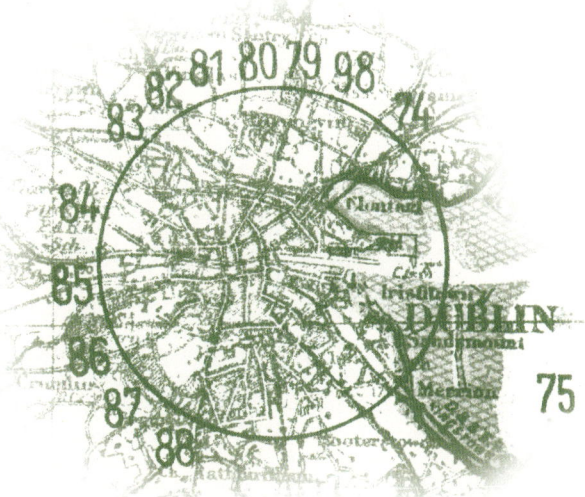

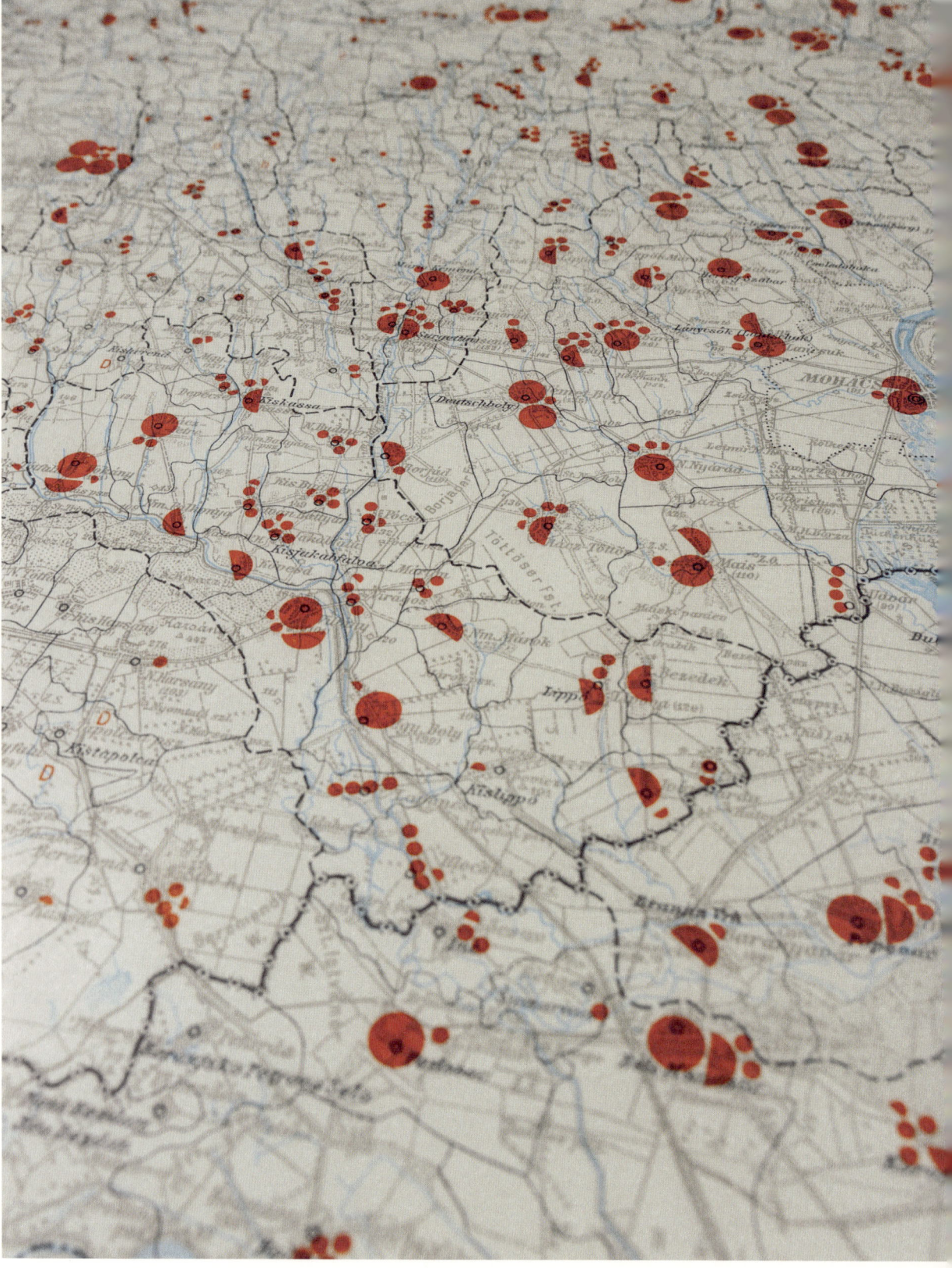

MANIPULATIVE MAPS

Data-Driven Destruction

'Today, maps give us tales of broken men. They allow us into distant worlds, if only for a glimpse.'
Steven Seegel[78]

The Second World War casts a long shadow over the Map Library. This, of course, is partly because so many of the maps were sourced from this period of prolific map-making, but it goes deeper than that. Geography itself was shaken to its core by the fact that many of the ideas that had led to the horrors had not just permeated the discipline but evolved from it (as they had from other sciences). In the light of this, many geographers wanted to forget and move on. But in my exploration of the Map Library I keep finding small reminders of not just the discipline's complicity, but also the remarkable influence that many geographers had at all levels of the military and government.

I happened upon a little-watched interview on YouTube[79] (it had fewer than 100 views when I came across it) with an American geographer named Peter Nash. Nash (who died in 2011) is being quizzed by Anne Buttimer at the University of Geneva in 1984. The video was uploaded to Buttimer's YouTube account in 2016, a year before her death. Over the course of their conversation, Nash details what an extraordinary time the last days of the Second World War were for him, when he worked as an interrogator on behalf of the US's Office for Strategic Services (OSS).

Map 73 [Extract] A sheet from a series of detailed ethnographic maps prepared by the SS in 1941. This map shows the number of people deemed to be of German ethnicity near the Croatian–Hungarian border around the town labelled Fünfkirchen (just off this crop and known as Pécs today).

Volkstumskarte von Ungarn und Jugoslawien, Stand 1930/31 Sonderausgabe Deutschtum: W. Krallert/ Publikationsstelle Wien (1941) 40.5 × 57cm

241

I was asked by my commanding officer to interview Karl Haushofer. And so I spent about ten days with him and his wife ... and wrote up a complete account on my findings, including that there was really no institute for Geopolitik. It was a top-secret document, which was later taken by Father Walsh and published mostly under his name in *Life* magazine. But I couldn't do anything about that.

But that will always be something that I will remember very much. I learned a lot about Haushofer, and even today in my classes on the history of geographic thought, when I speak about [Friedrich] Ratzel I always tell them about what Friedrich told Karl, and then what Karl [Haushofer] told Rudolph [Hess] and then Rudolph told Adolf [Hitler] and how ... some thoughts of a geographer really influenced the history of the world.

The Father Walsh article Nash referred to was a gripping write-up of Haushofer's interrogation in his Bavarian home that was published in *Life* magazine by the 'Geographer-Priest', the Very Reverend Edmund A. Walsh.[80] Despite Nash's assessment that there was no single influential 'Institute for Geopolitics' serving as the engine for Hitler's warped ideology, this was not the position that prosecutors took in the Nuremberg Trials.

In September 1945, the Office of the US Chief of Council published their indictment against Haushofer. They were unequivocal that: 'Haushofer was Hitler's intellectual godfather. It was Haushofer rather than Hess, who wrote *Mein Kampf* and who furnished the backbone for the Nazi bible, for what we call the criminal plan ... Really, Hitler was largely only a symbol and a rabble-rousing mouthpiece. The intellectual content of which he was the symbol was the doctrine of Haushofer.'[81]

Nash's interrogation must have taken place only a few months before Haushofer and his wife Martha ended their own lives on 10 March 1946, Karl aged seventy-six and Martha sixty-nine. They clearly saw no way out and were emotionally unable to face the consequences of their actions in being so complicit with the Nazi regime.[82]

The Haushofers' shock at how events had caught up with them would have been compounded when one of their sons, Albrecht, met a grim end in the final days of the war. Albrecht was also a geographer but, unlike his father, he became involved in a resistance effort. Things came to a head on 20 July 1944 with an assassination attempt against Hitler, which failed after the briefcase containing the explosives was moved behind a hefty table leg just before detonation, shielding Hitler from the worst of the blast. Albrecht went into hiding but was found and arrested at a farm in Bavaria on 7 December 1944 and sent to Moabit Prison in Berlin.

Within a few months, the Russian army had reached the suburbs of the city, so at midnight on 22/23 April 1945 Albrecht and fifteen other inmates were escorted to the prison gates where they were met by a squad of SS-Sonderkommando. They were told they were to be transferred to another facility for their own safety, but instead the SS took them to a patch of open land nearby and 'shot them through the nape of the neck'.[83] Somehow one of the prisoners managed to survive[84] and was able to alert Albrecht's brother, who found his body almost three weeks later on 12 May. In Albrecht's right hand, which was hidden under his coat, were five folded sheets of paper upon which he had written eighty sonnets.[85]

Tellingly, Albrecht ends one of the sonnets about his father with:

> He did not see the rising breath of evil.
> He let the daemon soar into the world.[86]

While Nazi ideology was not shaped exclusively by Haushofer's thinking, as some of the maps and atlases that now repose in the Map Library prove, there were many research outfits that, in their own significant ways, fed the daemon of the Nazi regime. Once the Nuremberg Trials got underway, Haushofer could not live with himself, but there were many others who were complicit in the mapping and data-gathering activities that the Nazis depended on, with no qualms about embracing a future in which they did not need to be fully accountable for their actions.

Cartographers have tremendous power to conjure lasting impressions of the world as they see it, but the consequences of their work often rest in the hands of the decision-makers using their creations. For this reason, mapmakers simply had to pivot their stated *raison d'être* away from informing the machinery of the Nazi state to supporting the West's efforts in countering the post-war threat from the East. Those who made the case that their intellects and skills were best kept on the western side of the descending Iron Curtain, and too big a risk if they were to move over to the East, could count on leniency in the years after the war.[87] Offering mapmakers this way out confirmed the ongoing status of maps as an integral part of warfare. They were considered assets in just the same way as high-profile German and Austrian scientists, engineers and technicians, many of whom were recruited by the US at the end of the war under 'Project Paperclip'.[88]

For example, there is a large pile of ethnographic maps in the Map Library (which I now can't bear the sight of) that were created under an SS officer named Wilfried Krallert, who was one of Hugo Hassinger's students and a member of the SODFG. They cover parts of eastern Europe and the Balkans and map ethnicities (including Jews) in excruciating detail, down to the smallest village. Krallert's job was to obtain as much ethnographic social, cultural and economic information as he could and feed that back

MANIPULATIVE MAPS

to Berlin. Over the course of his SS career, he oversaw the looting of Jewish bookshops and the raiding of census offices, and at the end of the war he was responsible for hiding the maps the SODFG had accumulated through these activities.[89]

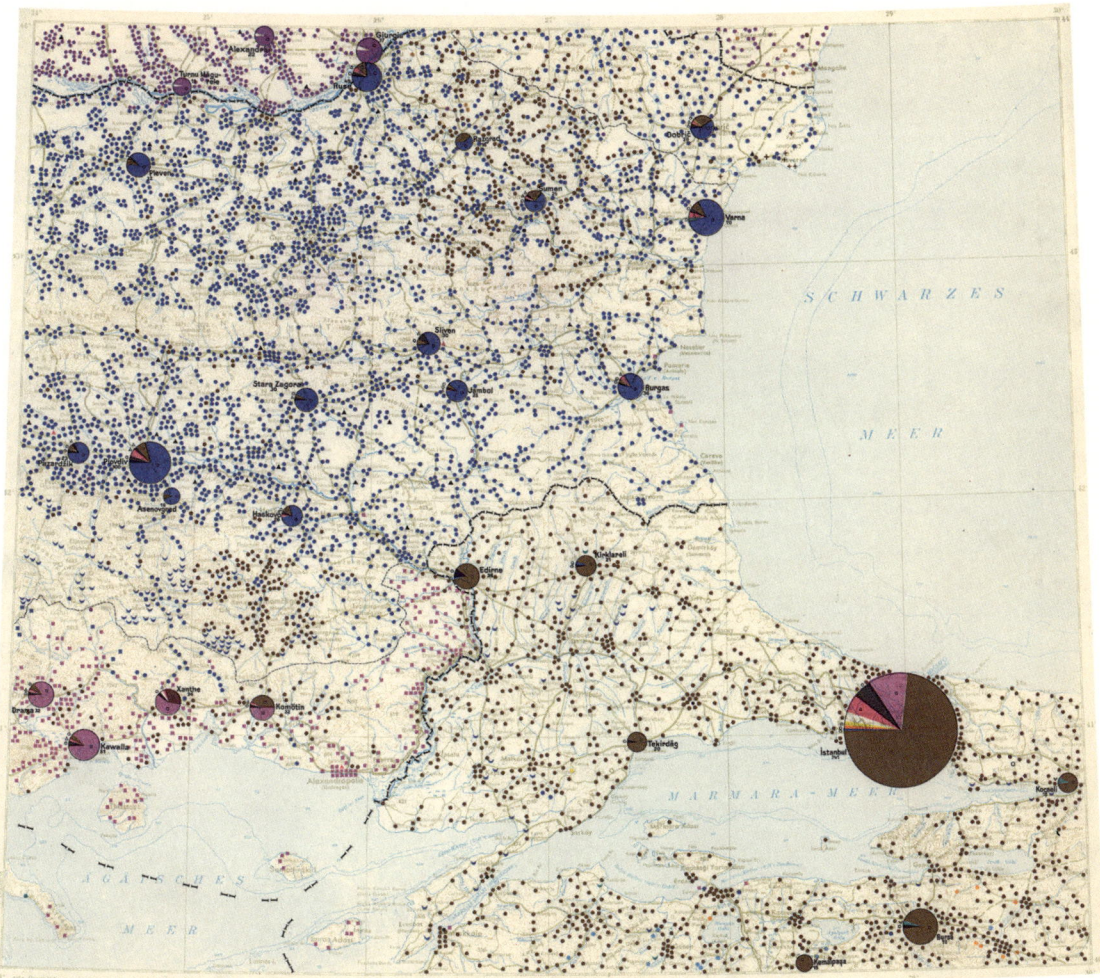

Map 74 Wilfried Krallert oversaw the creation of detailed ethnographic maps across a broad region that included Yugoslavia, Romania, Bulgaria and, as can be seen here, part of Türkiye. I've seen these maps appear in several map libraries, but few mention their connection to the Nazis.[10] I think this is, in part, because the efforts to move on in the post-war period reframed such maps as a useful resource for population research, rather than a direct link to one of the most sinister eras in history.

Istanbul: W. Krallert/ Publikationsstelle Wien (1942)
52.5 × 46cm

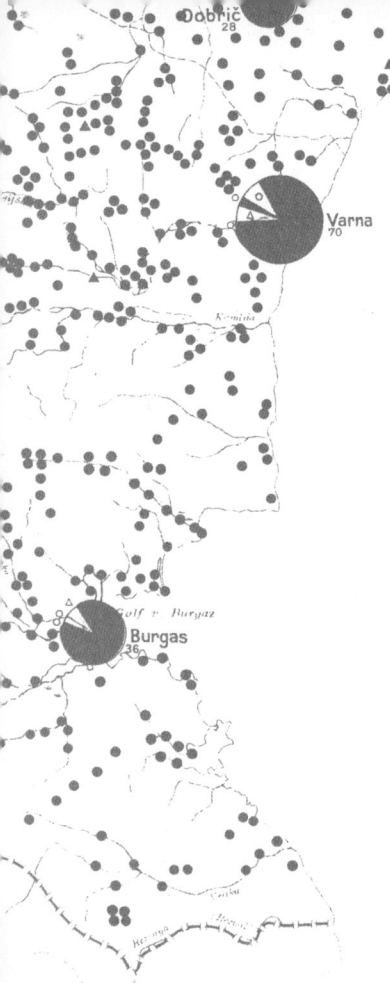

THE LIBRARY OF LOST MAPS

As the Allies advanced, Krallert was captured by the British and was interrogated for three years. Despite being a convicted Nazi, he managed to avoid prosecution by being recruited into the British Intelligence Service where he was subsequently fired for poor behaviour. The British summed him up as 'a Nazi of the worst type ... An intriguer who uses his position to introduce his friends, and his friends to increase his position. An expensive and unproductive luxury for any organisation.'[90]

Krallert was nonetheless able to work with other European agencies, where he got by on his expertise in Soviet mapping (at that time in high demand) and the populations of Eastern Europe, but was noted for his general untrustworthiness.[91] Despite his capture, he must have kept a stash of the ethnographic maps he had created as a salaried member of the SS, since he sold them on to other organisations (including the American Geographical Society (AGS)[92]) in the 1950s. He continued to mix in some German academic circles but was monitored by the US and British Intelligence agencies into the 1960s.

In another example, several high-ranking Nazi scientists were involved in the creation of a multi-volume atlas of disease titled *Seuchen-Atlas* (Atlas of Epidemic Disease) published by Justus Perthes. I discovered two of the volumes (released in 1942 and 1943) on the Map Library shelves, which were incomplete with their pages disbound and stored in an unassuming folder: another rare find as the atlas was classified during the war and few copies were known to have made it out of Germany. Across both volumes there were maps to show the distribution and spread of ailments such as plague, malaria and leprosy that were accompanied with detailed commentaries and tables of data.

The name on the front of the atlas was Heinz Zeiss, infamous for his involvement in the Holocaust,[93] and the publication was just one part of a broader scheme of work the Nazis were pursuing around 'geomedicine'. Tellingly, while I was in the AGS archive in Milwaukee, I also unearthed a letter from the medical historian Henry Sigerist, written in 1943, in which he says, 'I strongly dislike the word "geomedicine", not only because it is a mixture of Greek and Latin that sounds badly to my ear, but also because it is a nazi product. The founders of nazism used the term a great deal in the 1920s, particularly Heinz Zeiss, who later turned out to be one of the worst nazis.'[94]

Zeiss's collaborators included another medic, Ernst Rodenwaldt, with whom he jointly published a textbook in 1942.[95] There was also the statistician Friedrich Burgdörfer who created the statistical definition of what it meant to be a 'Jew' in order that the people who fell into this category could be located and counted in the Nazis' notorious censuses. Burgdörfer also used exaggerated versions of the data to make anti-Semitic arguments about the size of the Jewish population, advocating its decline while encouraging the growth of the German birthrate.[96]

Zeiss was imprisoned in the Soviet Union at the end of the war and died in 1949, but Rodenwaldt and Burgdörfer were able to offer their services as experts in population and disease mapping to the US Navy. The Navy were impressed and even encouraged American researchers to collaborate with them: Henry J. Alvis, Commander of Medical Corps, US Navy, wrote to the AGS in 1948 about developing efforts to create a global disease atlas and advocating for the German scientists to play an integral part:

> The group ... is still intact and a number of the original workers are available, notably Prof. Rodenwaldt of the University of Heidelberg. Would your organization be interested in this sort of thing? Do you think that they might care to enter into sponsorship for such a project on the basis of assisting scientists in Europe, specifically

Zeiß / Seuchen-Atlas

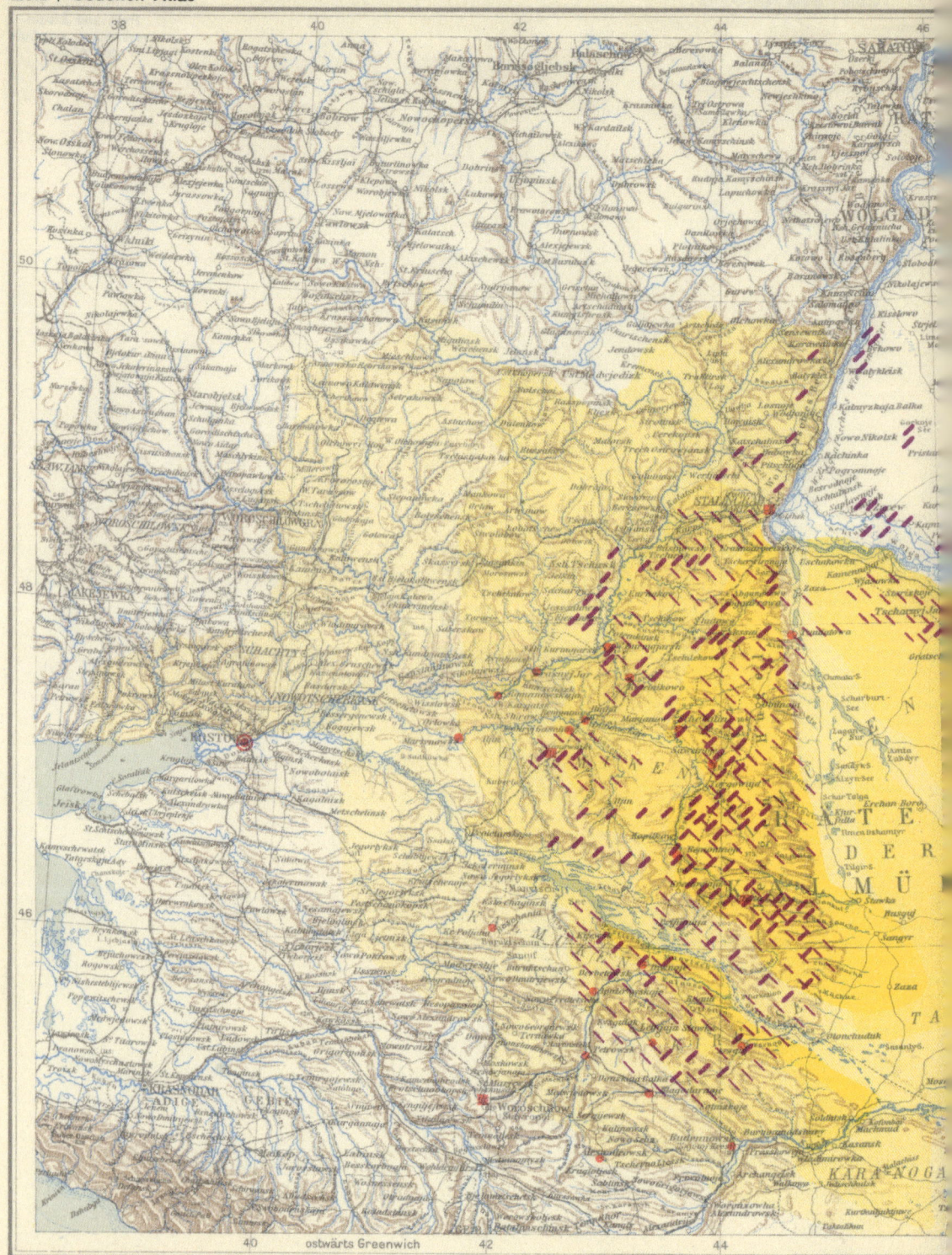

Grundlage: Stielers Hand-Atlas

1 : 3 000 000

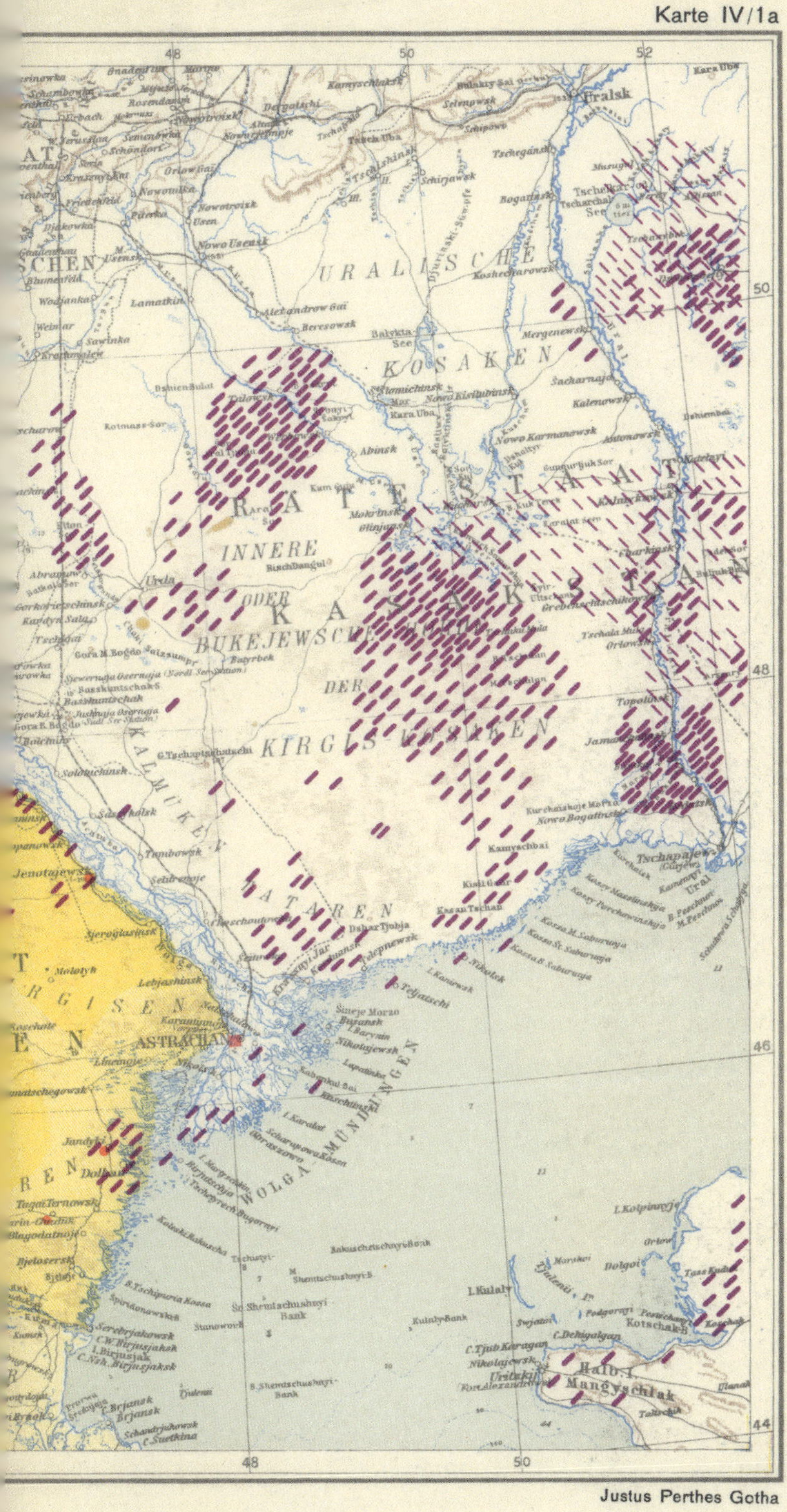

Karte IV/1a

Pestvorkommen und Zieselverbreitung
westlich der unteren Wolga
(K. G. Grell)

Beobachtete Pestfälle
(1877–1933)

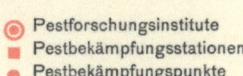
beim Menschen
bei Nagern (Epizootie)

⊙ Pestforschungsinstitute
■ Pestbekämpfungsstationen
● Pestbekämpfungspunkte

Dichte der Zieselhöhlen
je Hektar

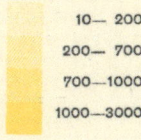

10— 200
200— 700
700—1000
1000—3000

Map 75 Heinz Zeiss wrote that the purpose of the *Seuchen-Atlas* (Atlas of Epidemic Disease) was 'to protect the German Army from damage by hidden enemies', commenting that it was 'a happy beginning of a new line of research in German and European hygiene' – a concept that was used to justify many unspeakable acts. The atlas was kept secret during the war but a few copies were discovered and shared by the Allies. It was given a favourable write-up by the *Geographical Review* in 1947.[11]

*Seuchen-Atlas
(Pestvorkommen und Zieselverbreitung)*: Justus Perthes (1942-1943, incomplete)
36.5 × 27.5cm

Justus Perthes Gotha

● Pestforschungsinstitute
■ Pestbekämpfungsstationen
● Pestbekämpfungspunkte

in Germany? Some of the group may be successful in migrating to the States at a later date and are interested in carrying on this project over here. Would they be of any interest to you?[97]

This request appeared to be ignored, but the US Navy still funded the German scientists' project to the tune of at least $15,000.[98] The first of three volumes was published in 1952 and was directed by Rodenwaldt (then a full professor in the Medical Faculty at Heidelberg) who reincarnated the *Seuchen-Atlas* as the *Welt-Seuchen-Atlas* (World Atlas of Epidemic Diseases).[99] Two further volumes were published in 1956 and 1961.

While many of the worst ideas had fermented under the Nazis, they had not all originated with them. Geography and cartography, alongside many other disciplines, had not just witnessed but fuelled the ideologies that evolved into atrocities. Let's consider just one example. You will recall that in Chapter 6 I mentioned Leon Dominian, who published a map-filled book for the AGS that argued that the languages within areas should be used as a basis to the 'scientific' drawing of borders. The introduction to Dominian's book was written by a man named Madison Grant, who, despite now being notorious for his racist ideology, was at the time embraced by the AGS, particularly by Isaiah Bowman (AGS president and senior member of the US delegation sent to the Paris Peace Conference).

Grant wrote his own book called *The Passing of the Great Race* in 1916 and the AGS published a series of eye-catching maps to accompany it.[100] Grant summarised his beliefs in the AGS's *Geographical Review* that the decline of the 'Master Race' alongside the growth of more 'backward races' was 'undoubtedly one of the causes of the present [first] world war'.[101] In the subsequent decades, Grant went on to gain plaudits from Hitler, being quoted in his speeches and having one of his books referred to by Hitler as his 'bible'.[102] A reviewer in the *New York Times Book Review* section said of Grant's writing: 'Substitute Aryan for Nordic and a good deal of Mr. Grant's argument would lend itself without much difficulty to the support of some recent pronouncements and proceedings in Germany ...'[103] Perhaps even more extraordinarily, lawyers defending the Nazis referenced Grant's work to make the case that Nazi eugenics programmes were in part inspired by ideas from the US,[104] and that it was therefore hypocritical to prosecute the regime that enacted them.[105]

In the words of Albrecht Haushofer, geography had let the daemon soar, leading geographers to reflect on how it had been allowed to happen. When they looked back over the previous century to the development and application of geopolitical ideas and the creation of the maps that were used to promote them, the explanation was plain to see. It would be decades before 'geopolitics' was taught again in geography departments and now it is largely taught in opposition to the more 'classical' ideas, even though, as Putin's full-scale invasion of Ukraine has shown, there are those in power who still subscribe to them with the same deadly consequences.[106]

gen, den stürmisch vordrängenden
Vereinigten Staaten von Amerika,
chaft an sich gerissen haben, ein
nun vollständig unter den Einfluß
ie Feste der kapitalistischen Welt-

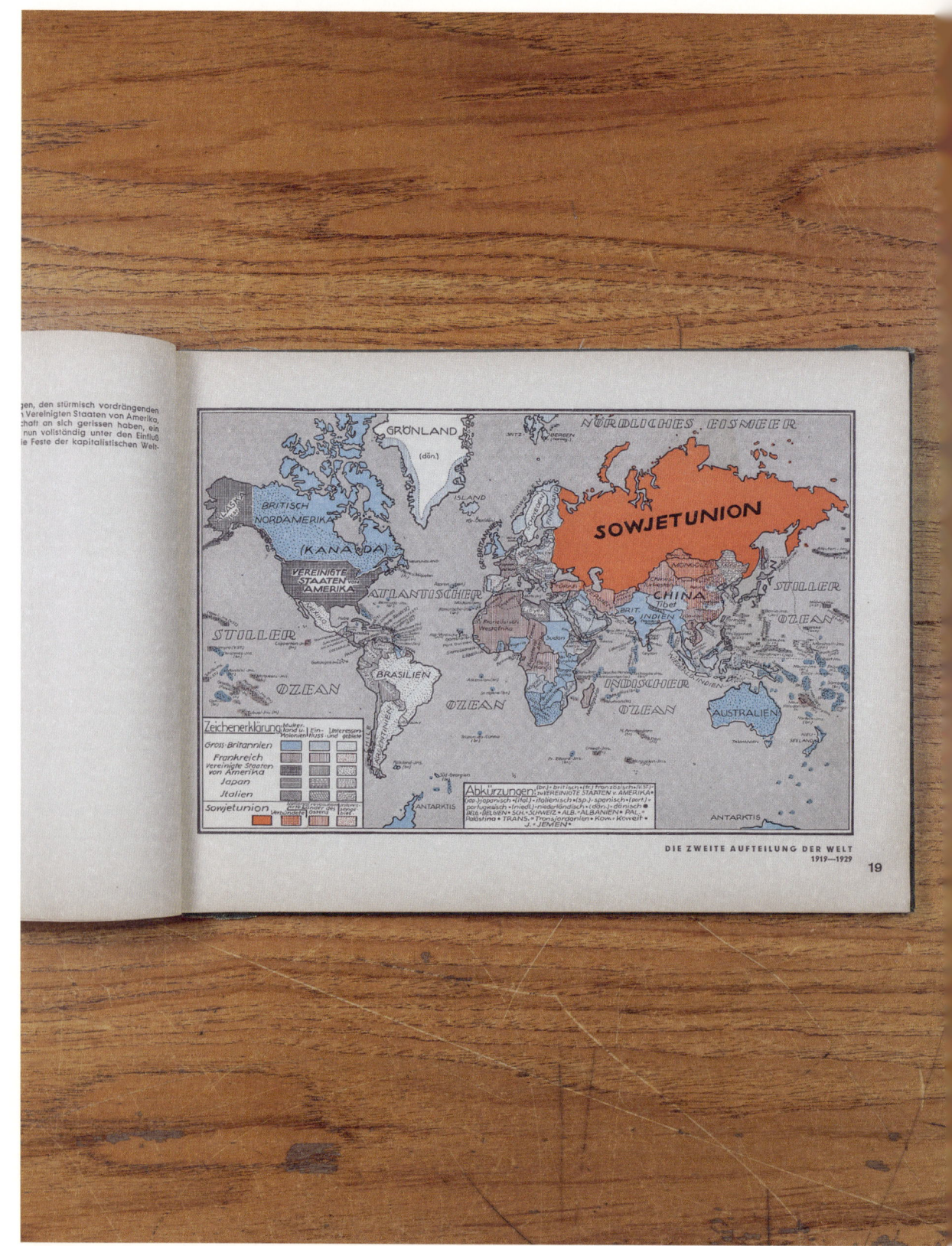

DIE ZWEITE AUFTEILUNG DER WELT
1919—1929

MANIPULATIVE MAPS

Radó

We will never know the full extent of the role that most individual mapmakers, and their maps, played in the Second World War, but there is one man in particular whose story is too good not to tell. When I talked about the *Atlas of To-Day and To-Morrow*, I left out some very important details about its author, Alexander 'Sándor' Radó. You'll recall he was a pioneer of communist geopolitical mapping, but what I didn't mention was that Radó was one of the most successful Soviet spies of the era. He used his maps as cover for his espionage activities, which became legendary in the intelligence world. In many ways he was not just a great survivor of the tumultuous first half of the twentieth century: he provides an important bridgehead into what life was like from the 1950s onwards.

Unsurprisingly there is no definitive account of his espionage activities and there are many unanswered questions, but in his version of events he was recruited to spy for the Soviet Union in 1935, when he was in Moscow advising on *The Bolshoi Sovietskiy Atlas Mira*, the same one Stalin withdrew that I mentioned earlier in this chapter.[107] However, this may not be entirely true as Radó set up a map publishing company, Inpress, in Paris two years earlier in 1933, which was a very well-funded front for the Soviets. Given the tight margins for journalism and map-making, Radó could afford a somewhat implausible middle-class life in the Paris suburbs, and the firm even had sufficient funds to employ sixteen journalists, who themselves were mostly spies. It was also in 1930s Paris that

Map 76 'The Second Division of the World 1919–1929' as shown in Alexander Radó's socialist atlas. The main colours of the map represent the major powers (Great Britain is blue, USSR red, France brown, US black cross-hatched) and additional hatches are applied to show their relative influence on other regions as 'motherland and colonies', 'areas of influence' or 'areas of interest'. Those countries allied to the USSR have been given a slightly different categorisation to encompass 'Soviet Union treaty countries', 'national revolutionary countries of the east' and 'Soviet Union interest areas'.

Atlas für Politik Wirtschaft Arbeiterbewegung (Die Zweite Aufsteilung Der Welt 1919-1929): Verlag Für Literatur und Politik (1930)
29.5 × 19cm

THE LIBRARY OF LOST MAPS

Radó started to build his credentials, becoming a fellow of the Royal Geographical Society in London in 1936 thanks to support from the president and secretary of the Paris Geographical Society. In the November of that year, Radó moved his family to Geneva, a city that would have had better access to Nazi intelligence, which was the principal focus of Moscow at the time.[108]

After moving to Geneva, Radó set up Geopress, which operated under the same auspices as Inpress in Paris. This was also the time when he worked with Marthe Rajchman on the *Atlas of To-Day and To-Morrow*. From the outbreak of war, Geneva became a hotspot for espionage, and so Radó was kept busy running a group of communist agents that continued to expand through to 1941 and whose coded reports were dispatched to Moscow via intermediary locations as Geopress communications. Radó became an important node in the network of a major Soviet intelligence operation that British intelligence had badged the 'Red Orchestra'. At its peak, the organisation included 48 agents in Germany, 35 in France, 17 in Belgium and 6 in the Netherlands.[109]

Radó went by the names of Dora (an anagram of his actual name) and Albert, although the circumstances in which one was used rather than the other were never fully established.[110] The Red Orchestra's transmissions were discovered and shut down in Germany and occupied Europe at the end of 1942, and many of the agents were captured and imprisoned. Radó and his spies had some protection thanks to Swiss neutrality, but this proved an insufficient defence in the autumn of 1943 when the Swiss police, acting on behalf of the authorities in Berlin, arrested most of the unit. With the help of local French resistance fighters, Radó and his wife fled over the border to France and worked with their rescuers until Paris was liberated the following summer, where they moved to a newly rented apartment in the city.

In the final months of the war, Moscow had concerns about

MANIPULATIVE MAPS

double agents within Radó's network, so they picked him up with six other former agents in Paris in January 1945 to return them to the Soviet Union to make a more detailed collective report.¹¹¹ To avoid the battles still raging in what were to be the last weeks of the war in Europe, the flight to Moscow went via Cairo. It's not entirely clear what spooked Radó, but one theory is that another of the spies on the flight who resented Radó's authority threatened to undermine him to their bosses by alleging corruption and financial mismanagement within the ring. The spy thought to have issued the threat was Alexander Foote, who went on to write a book titled *Handbook for Spies* and has his own intriguing backstory as a possible double agent.¹¹² When the plane landed in Egypt, a terrified Radó managed to escape his Russian handlers and, for reasons that are also unclear, fled to seek help at the British Embassy.

On Friday 12 January he was taken to the British General Headquarters Middle East,¹¹³ where he offered a fake name, 'Lane', and told a variety of inconsistent stories to his interrogators about his wartime activities. Radó was clearly in a state of distress, as within a few hours of his arrival he attempted to end his life by cutting his throat. He was discovered just in time and taken for treatment. However, that evening while he was recovering in hospital, Radó grabbed a pistol from a guard and made a second attempt on his life, which was again thwarted as his minders were able to restrain him.¹¹⁴

The British reasoned that such extreme acts of self-harm were not the behaviour of someone as inconsequential as 'Lane' had suggested, and that he must have been a much more significant individual. He was therefore

Map 77 A tiny Geopress map that Radó sent out during the war as part of his cover for being a spy. This one was received on 31 May 1939 by the Library of Congress in Washington.

Extension of the Italian "Mother Country": Geopress (1939)
5.5 × 8cm

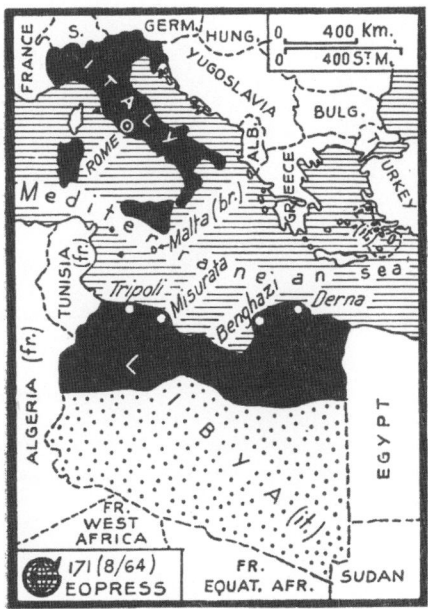

Extension of the Italian "Mother Country"

subjected to a period of interrogation. This eventually extracted more realistic and detailed stories about his actions during the Second World War.

There was not much coordination between the different branches of British intelligence, and there was also the small matter that the notorious double agent Kim Philby was the senior officer responsible for counterintelligence against the Soviet Union and oversaw the case. To their later regret, Radó was released by British intelligence and handed over to the Egyptian authorities, who in turn handed him to three Soviet officials, who on 30 July 1945 put him on a small military plane back to Moscow.

Much of what we know of Radó can be extracted from the declassified files[115] kept about him by British and American counterintelligence operations, which particularly benefited from considerable amounts of information from Radó during his interrogations in Egypt.

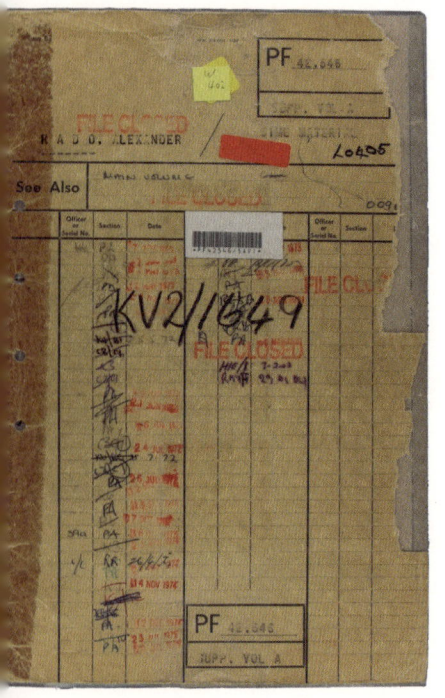

It turns out that Radó's British intelligence file started to fill up when, in the autumn of 1940, he sent two maps to a Miss Myers of 23 St James's Place, London SW1. The first map showed the war in Africa and the second was titled 'The New Naval Bases of the USA in the Atlantic'. Sending maps without further explanation seemed to be something Radó did, perhaps to advertise his services, but perhaps also to generate a paper trail to demonstrate the legitimacy of his activities.[116] On this occasion, the letters were intercepted because Ms Myers was the secretary to Dorothy de Rothschild, the sister of James Armand de Rothschild who at the time was a Member of Parliament and residing at the same address. An officer cautions in their report, 'It'll be noticed that there is a Member of Parliament living at this address. In view of this and the possibility of arousing suspicion, it was considered inexpedient to make inquiries at the house.'[117]

MANIPULATIVE MAPS

The intelligence services weren't sure what to make of the maps, with the initial report noting that they 'appear to have a pro Axis bias as one shows rather unnecessarily occupied France and two stresses British Africa occupied by Italy'.[118] A follow-up note describes the maps as 'primitive', 'obviously worthless and of little interest to anybody' and 'not particularly inspiring' as a form of propaganda. This assessment led the agents to wonder if the maps were in fact a cipher and embedded in them was the 'key to some code to employ in the transmission of future messages'.[119]

In the end, 'discreet inquiries' were made and Ms Myers was judged to be an innocent recipient of the unsolicited maps and therefore not considered a threat. Radó, too, was deemed fairly innocuous, his publication record and company registration offering reassurance. At that moment in time, the main concern was Radó's Communist leanings, but the letter was let through. Not much else appears until his arrival in Cairo in 1945.

What's fascinating about the Radó intelligence reports, and any dealings he has with the British state or elsewhere, is the extent to which he talks up his status as a geographer/cartographer, and indeed this is something that is frequently commented upon in his file. One British officer notes that Radó was 'a member of the Royal Geographic [sic] Society of London' and this membership enabled him to enter England 'without any difficulty whatsoever'. Another, in notes made after the war in 1947, strikes a much more complimentary tone about his map-making abilities than the reports of the maps sent to Ms Myers by describing Radó as 'a former Soviet agent who was a famous cartographer'.[120]

On 11 April 1945, in what must have been one of his darkest moments midway through his detention by the British, he wrote to a friend[121] pleading for information about his family but also requesting help with a move to England or the USA. Radó makes a case that he would be welcomed as he was invited in 1941 to

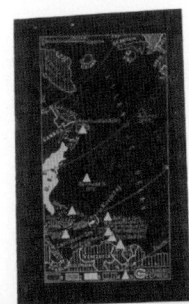

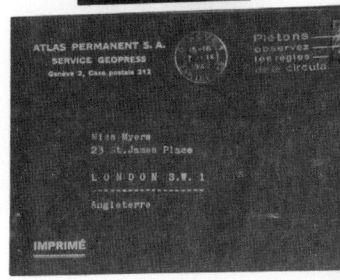

The cover of one of the folders of intelligence collected on Radó by British spies (left). It includes a photocopy of the intercepted map he sent to Ms Myers (above).

give a talk in New York on political geography, but he could not accept this invitation due to the difficulties of wartime travel. He then goes on to say that perhaps Victor Gollancz, 'who published my 1938 Atlas of To-Day and To-Morrow', could help provide a reference, or even Ellen Wilkinson MP or J. F. Horrabin MP (he had not been an MP for fourteen years at this point), who 'both wrote very flattering articles about this atlas', could offer to help him return to his scientific work. His final request is for them to seek out Rajchman and her father Ludwig (who Radó lists as the former director of the Health Department of the League of Nations) as he felt sure they would help him if they 'are informed about my bad luck and my whereabouts'.[122]

Radó's maps were much more than ways of communicating the issues of the day with broad audiences: they were in fact providing cover for his spying activities and then became his only lifeline as he tried to seek the support of the people who had helped him to make them. But it was no use. Radó was deported back to Russia, seemingly against his will, and disappeared for the best part of a decade. Many had assumed he had been shot, his family thought he had been in a plane crash (they sought clarification from the British state but were told nothing[123]), while others inferred he was in a labour camp.

Remarkably, after the release of many prisoners following Stalin's death, he reappeared in 1954 and rapidly became a very well-known cartographer in his native Hungary and an influential member of the scientific establishment, which permitted him frequent travel on the geography/cartography conference circuit (while still being trailed by the CIA[124]). There were rumours of his past, which reached fever pitch in 1960 when a Hungarian delegation arrived to a media frenzy at the International Geographical Conference (IGC) in Stockholm. Such events aren't known for their newsworthiness, but this was triggered by the Swedish press getting wind of who Radó was and printing headlines announcing the arrival of a 'superspy'.[125]

To take back control of the narrative, Radó penned his own state-sponsored 'tell-all' biography called *Dóra jelenti* (Codename Dora[126]) and published it in 1971. No mention was made of his detention in Egypt, perhaps because 10 per cent of the book had been censored by Soviet and Hungarian authorities: the full extent of his attempts to evade Soviet capture only became publicly known when the interrogation reports were declassified. The book became a publishing success, running to four editions in Hungarian and with translations in twenty-three languages. Radó was portrayed as a socialist hero and the book was even made into a film in 1978[127] (although, unlike the book, this was not a box office hit).[128] As the *New York Times* reported at the time: 'The Soviet Union's decision to publish Professor Radó's account appears to be part of a campaign to publicize the exploits of former Soviet agents and add prestige to the Kremlin's intelligence system.'[129]

Despite his fame, Radó remained an entrepreneurial cartographer, setting up an agency that collated information about changes needed to maps (nudged borders, new roads, etc.) and then selling the updates to subscribers. The CIA, which continued to show an interest in him to the end of his life, suspected this was another front for espionage activities on behalf of Hungary. It described it as a

> very clever collection device. For a $10 subscription (payable in hard currency) readers obtained maps designed for ease in reproduction ... A normal issue contained perhaps two or three maps of communist countries and about two dozen from the West. Radó, on the other hand, obtained an extensive and systematically organized file of Western data (which the East could use to keep its maps accurate

and up-to-date), a plausible basis for close contact with Western sources (and a potential for their recruitment), and considerable prestige as a world clearing house for geographic information.[130]

As they wrote in a recently declassified CIA obituary, a US Army attaché (their name is redacted) recalled meeting Radó at the French ambassador's residence in Budapest in 1980, just a year before his death. The attaché was being sent to another posting and obviously sensed this could be the last time they might meet, so they asked Radó again about his 'post-war adventures': 'Professor Radó, will we ever know the complete story of the Dora network? What happened to you in the Soviet Union after 1945? What about your ten years in the Lubjanka prison? Was that not a strange reward for someone who did so much for the Soviet war effort?'

In response Radó apparently 'pursed his lips, folded his hands across his generous paunch' and gave a terse reply: 'I wrote a book; you have seen it – *Dóra jelenti*. Others have also written books all of which tell of many things. It is enough for everyone to know that I worked in a common cause with all those who were fighting against Hitler – all of the accounts agree on this, if nothing else. I hope I will be remembered mostly for my contributions to geographical science.'[131]

In the Map Library, at least, this is how he is remembered, his maps (which include a good number of wall charts used for teaching) and atlases left on the shelves remaining tight-lipped about his secret life. But as the author of the CIA obituary concludes he was also an '... individual caught up in the history of his time, involved in an epic struggle between great powers – as a spy ... In Western traditions, there are Valhallas for the fallen warrior; cavalrymen have their Fiddler's Green and sailors, Davy Jones's

MANIPULATIVE MAPS

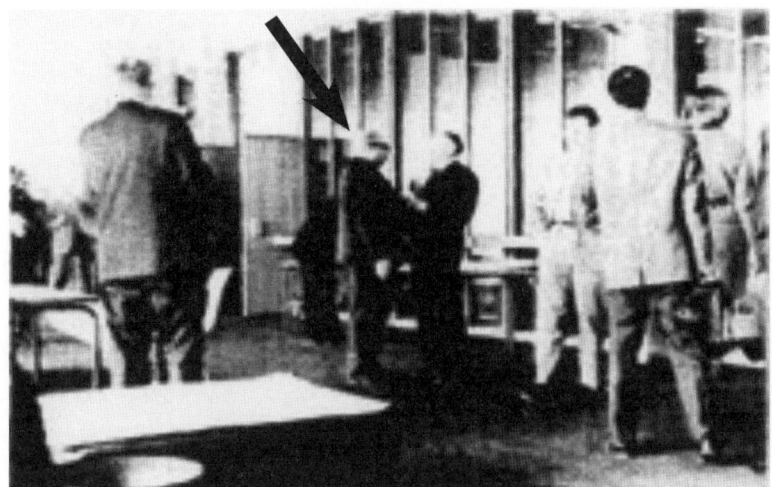

A CIA photo of Radó taken at the Third International Conference on Cartography, 17-21 April 1967, in Amsterdam. He has been marked with a black arrow. Radó's attendance at geography and cartography conferences was of interest to US Intelligence agencies until the end of his life.

Studies in Intelligence Vol. 12 No. 3 (1968)

locker. I suspect Professor Sándor Radó might yet be charming his listeners and telling jokes in six languages in some special place for old spies, a place whose name and location, of course, cannot be revealed.'[132]

Of all the characters the Map Library introduced me to, Radó was one of my favourites as he embodies the importance of mapmakers as collectors and organisers of information and the power this gives them in times of upheaval. His story also evoked a bygone age, one where many geographers 'had a past' that generated both intrigue and also acceptance among colleagues and peers, not just within their own country but around the world.

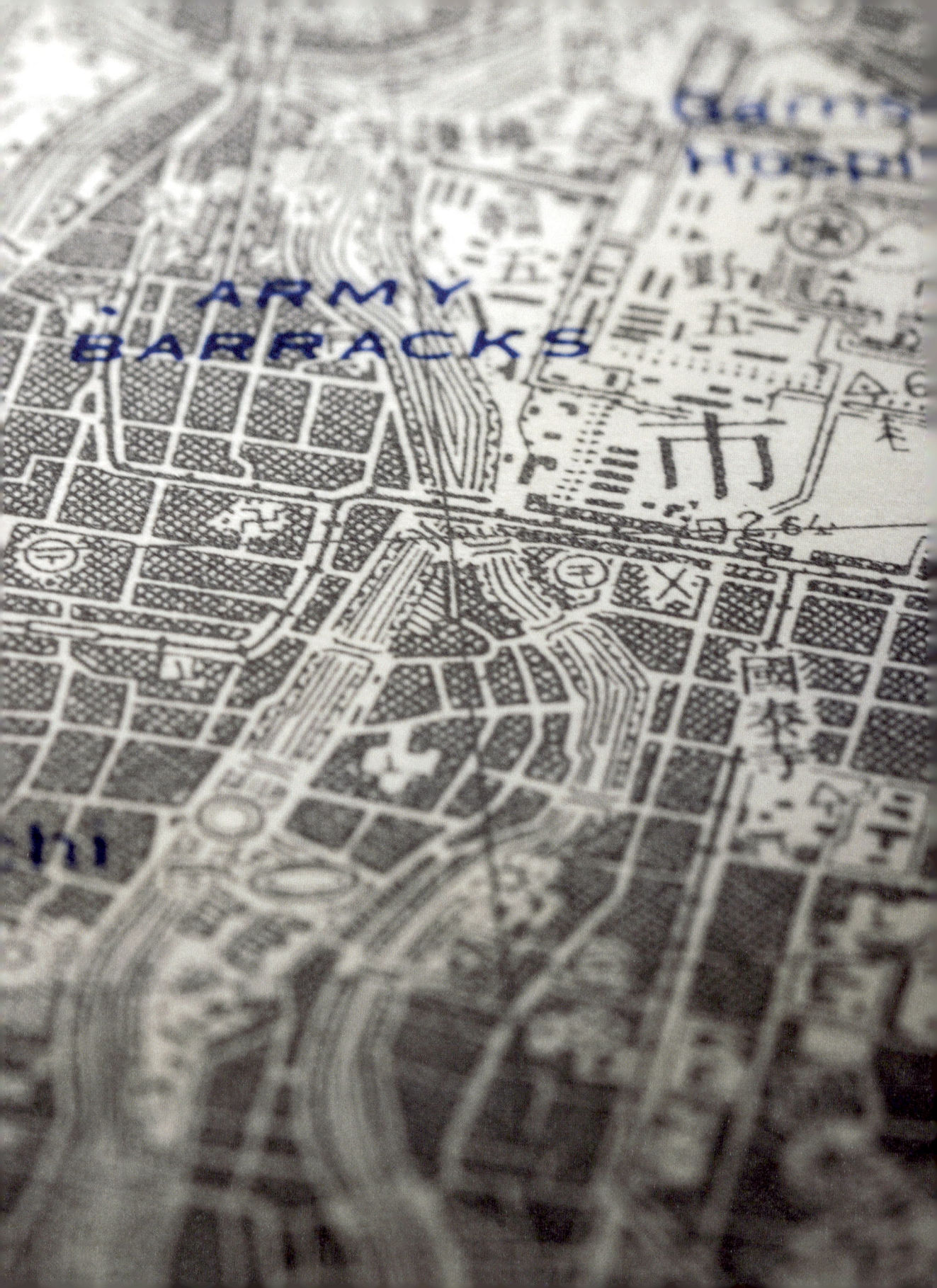

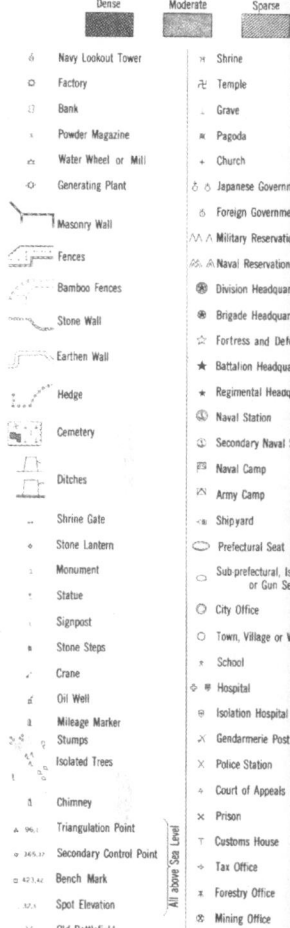

CHAPTER EIGHT
Hiroshima

'The planet is large enough for peace, but too small for war.'
William Bunge[1]

The epic struggle that Alexander Radó was engaged in started with the fight against the horrors of fascism and then took the world to the brink of nuclear annihilation in the Cold War. With so much history, it is easy to take the maps from the era at face value without considering the immense weight of the consequences that they had.

When a Japanese professor I work closely with paid a visit to London I invited him to the Map Library. I wanted to show him a pile of topographic maps of Japan that were crammed into their drawer so tightly that it was almost impossible to get it open. Produced by the US Army Map Service (AMS) in 1944 to support its operations in the Pacific region during the Second World War, these maps, at a scale of 1:50,000, spanned hundreds of sheets and covered the southern half of Japan. To create this series the AMS

Map 78 [Extract]

Hiroshima Sheet 4550-II:
US Army Map Service (1944)
45.5 × 37.5cm

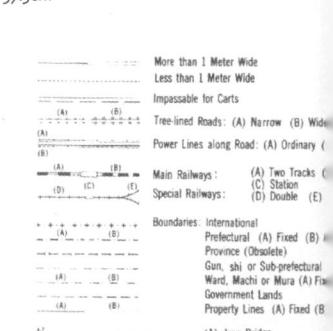

THE LIBRARY OF LOST MAPS

Map 78 [Extract] The Aioi Bridge in Hiroshima was the intended target of the atomic bomb. It had a distinctive T-shape (tricky to see at this map's scale but in focus on the previous page) and was in the centre of the city. In the event, the bomb fell 300 metres away, exploding in the air over Shima Hospital. The bridge was reconstructed and continued in service until a complete replacement was needed in 1983. Today the downstroke of the bridge's T is an important link to the Hiroshima Peace Memorial Park.

Hiroshima Sheet 4550-II: US Army Map Service (1944) 45.5 × 37.5cm

had made copies of the 'Japanese Imperial Land Survey' of 1933 in black and white and then overprinted Americanised labels in dark blue ink. Naturally, these labels emphasised military landmarks, such as army barracks, as well as important infrastructure like reservoirs and railway lines.

It turns out they were quite a rare series, and my colleague requested that they be scanned in order that he could have a record of them to share in Japan. As I flicked through the pile to count how many sheets there were, their titles were zipping by, and I registered that SHEET 4550-II was titled 'HIROSHIMA'. It took a few moments for my subconscious to catch up before I connected the year of the series – 1944 – to historic events. Within a matter of months, the city in the south-eastern corner of 4550-II would be the first in history to be almost erased from the map by an atomic bomb.

This must have been among the last maps to have been printed of Hiroshima as it was before 6 August 1945, when somewhere between 90,000 and 160,000 of its residents were killed.[2]

Jacob Bronowski, a Polish British mathematician and philosopher, was part of a team who visited Japan tasked with establishing the effects of the atomic bombings to help the British government plan for the consequences if a nuclear weapon was used on its shores.[3] Bronowski was profoundly affected by what he saw and in the decades that followed, up until his death in 1974, he became a high-profile communicator, espousing a humanistic approach to science. The peak of his fame as a public academic came with his thirteen-part TV documentary, *The Ascent of Man*.

The series features an extraordinarily powerful monologue by Bronowski from Auschwitz concentration camp, a place where members of his family had been murdered. In it he reflects on the role of science in the atrocities that took place there:

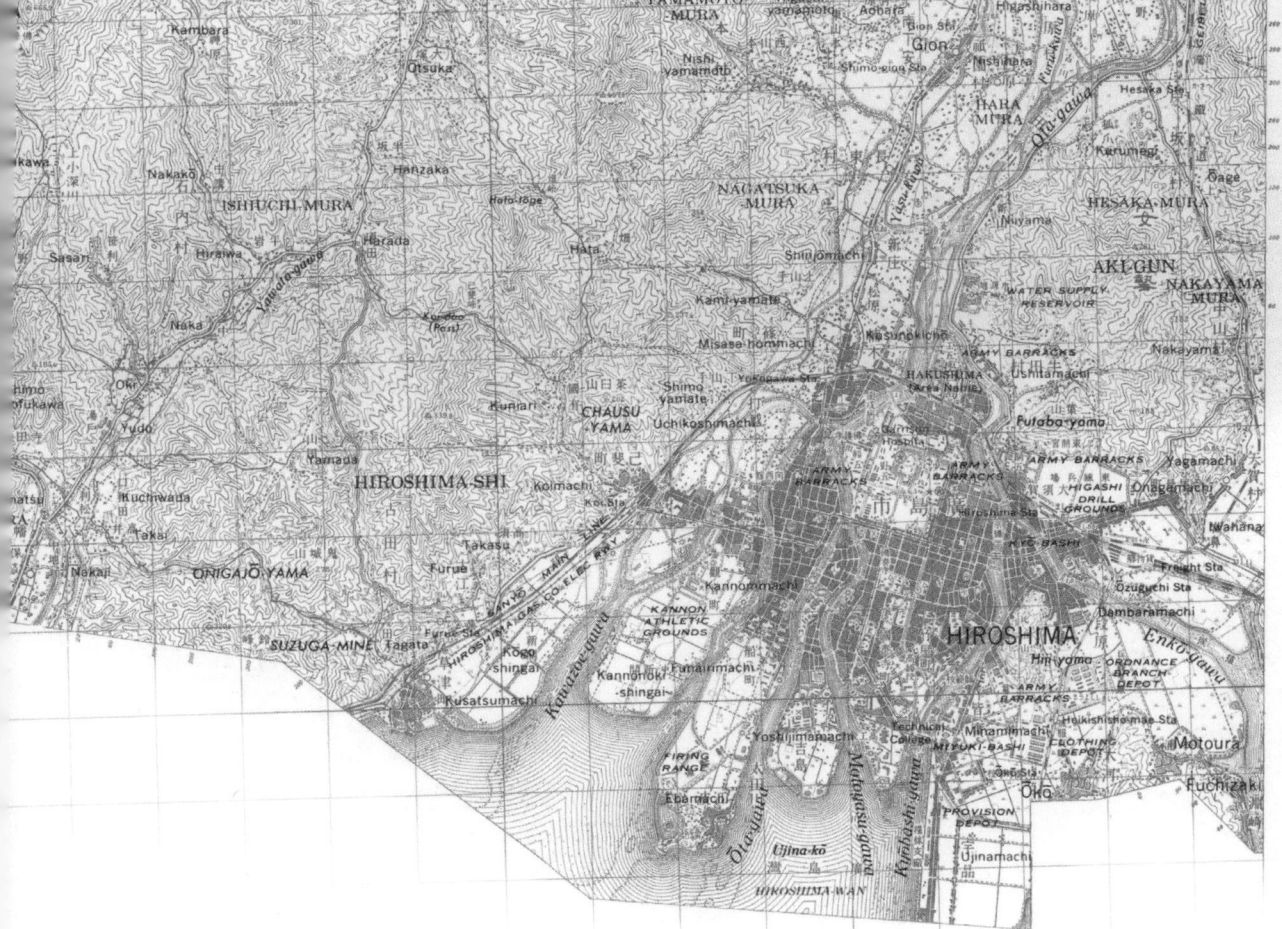

When people believe that they have absolute knowledge with no test in reality, this is how they behave. This is what men do when they aspire to the knowledge of gods. Science is a very human form of knowledge: we are always at the brink of the known, we always feel forward, for what is to be hoped. Every judgement in science stands on the edge of error and is personal.

Science is a tribute to what we can know *although* [my emphasis] we are fallible. In the end, the words were spoken by Oliver Cromwell – 'I beseech you, in the bowels of Christ, think it possible that you may be mistaken.' ... We have to cure ourselves of the itch for absolute knowledge and power. We have to close the distance between the push button order and the human act. We have to touch people.[4]

Bronowski may not have had mapmakers in the front of his mind when he spoke to the camera but, as we now know, they were part of the system that created Auschwitz and indeed countless 'push button orders', including the nuclear destruction of Hiroshima, that reinforced ideologies that resulted in the deaths of millions.

When I now look at some of the maps I have shared in the preceding pages I can't help but hear Bronowski: 'We have to close the distance between the push button order and the human act. We have to touch people ...' But the extraordinary thing about maps is that, although they are complicit in creating distance, we also reach for them to convey the intimacy of our neighbourhoods and communities.

For that reason, it was no surprise to me that one of the exhibits in the Hiroshima Peace Memorial Museum is a map that shows the former Nakajima District of the city before the bomb, now the site of the Peace Memorial Park.[5] Some of those who lived there and survived did not want the spirit of their neighbourhood permanently erased and in 1969 a project was launched to gather the information needed to recreate a detailed map of the area. The final map was preserved as a copperplate to pass on to future generations as a reminder of what was lost. Seeing this map reminded me of the tourist map of Zaporizhzhia I had been given by the Ukrainian academics (see Chapter 1) in their own small act of defiance in the face of the destruction wrought by the Russian invasion of their country.

It is undeniable that maps can inflict terrible damage and there is still a nervousness about this today. Nonetheless, we know that maps can also be a way to rebuild, bring together and hope for better. In 2019 Alex Kent, former president of the British Cartographic Society and a senior member of the International Cartographic Association, wrote an editorial titled 'A Picture and an Argument: Mapping for Peace with a Cartography of Hope' to

HIROSHIMA

observe 100 years since the signing of the Treaty of Versailles:

> One hundred years on, selfish ambition and envy are hardly strangers in this geopolitically volatile world. Maps force definition and certainty onto a real world that resists both of these impostors, and propaganda mapping is alive and well. Yet, the cartographic ideal endures in its hope that, through maps, the world will be a better place.[6]

When I teach my students how to make maps, I now take them to the Map Library and tell them they should never be complacent about the power of their creations. They must believe in what they are doing and own the consequences, intended and unintended, when they release their maps into the world.

The Hiroshima Peace Memorial (Genbaku Dome) was the only building left standing within the vicinity of the atomic bomb's explosion on the 6 August 1945. Designated a World Heritage site in 1996, it serves as a memorial to those who were killed and expresses the hope of world peace and the elimination of all nuclear weapons.

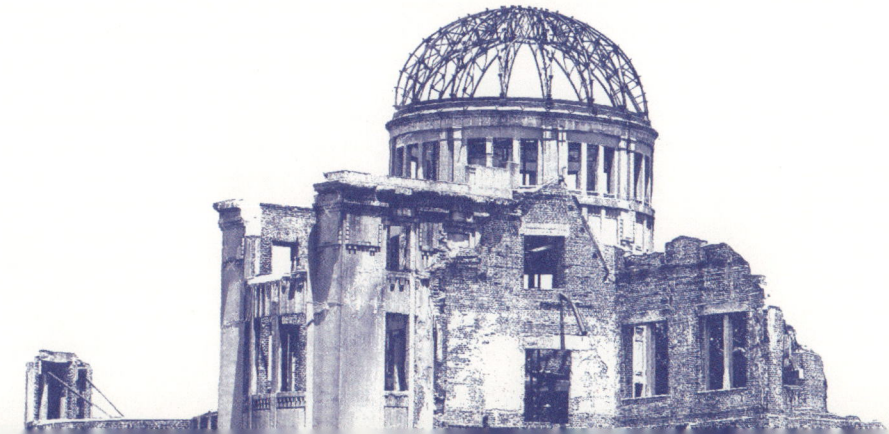

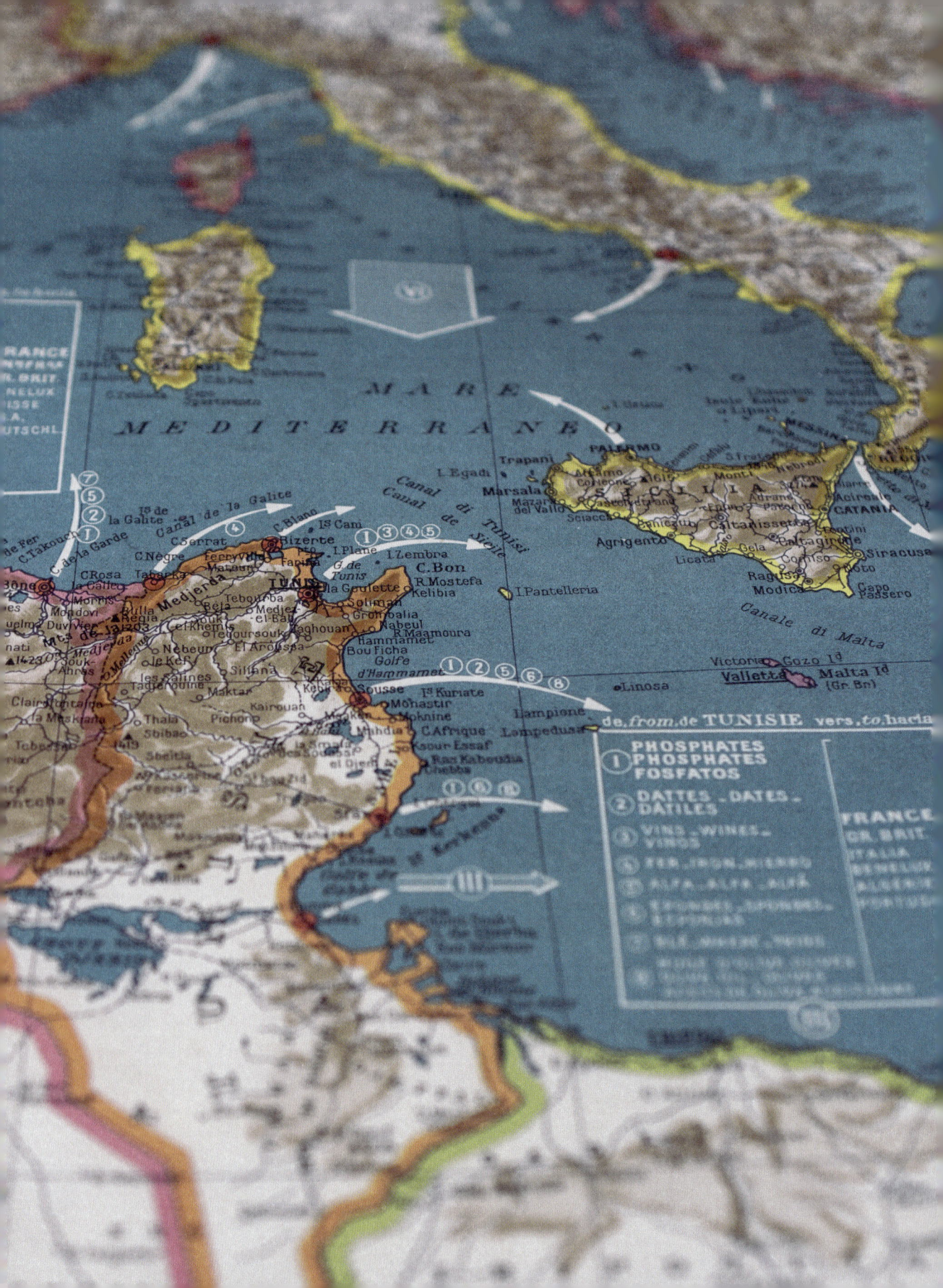

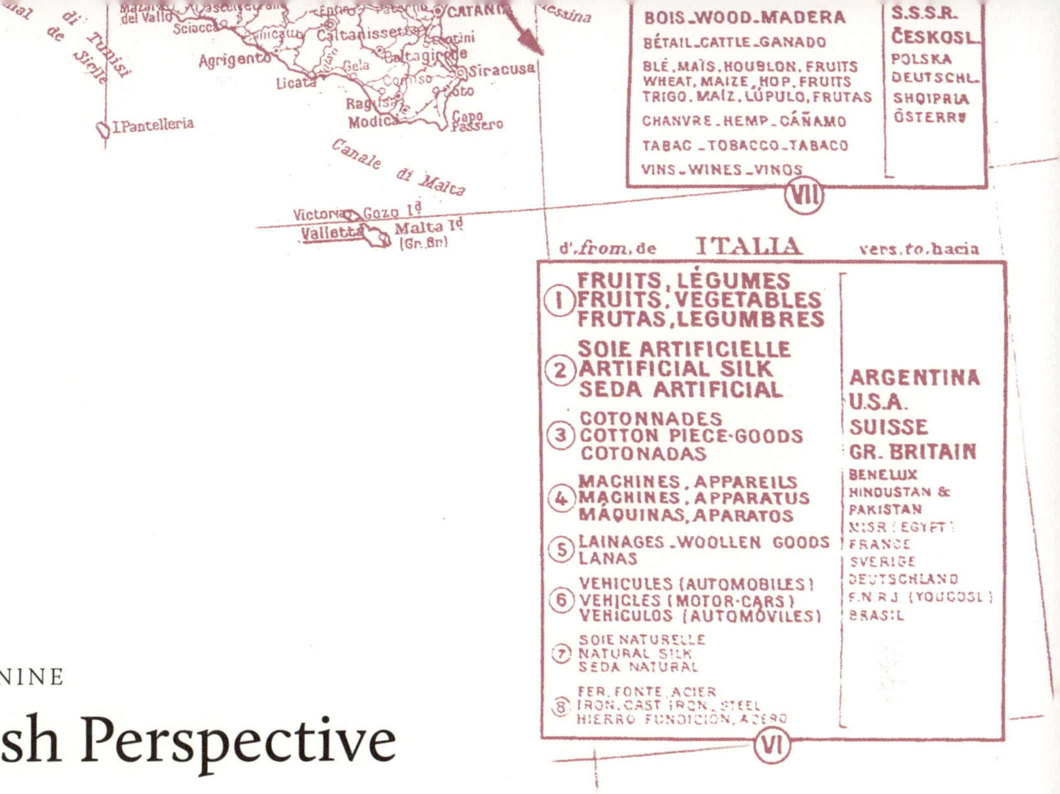

CHAPTER NINE

A Fresh Perspective

'A map is a window not only to the present features of the earth's surface, but to the past as well. Constantly we ask the question: How did these things come to be?'
Armin K. Lobeck[1]

Soon after the Second World War, many mapmakers turned their attention to how maps could be reimagined to reflect a radically different world. As Europe began to rebuild, the USA became the dominant geopolitical force set against the Soviet Union. The European continent lost its status as the centre of the map as it became more peripheral to the prominence of the USA. There was also the acknowledgement that, although geopolitical rivalries were important, maps were no longer just a playing surface for adversaries, but were a basis to forming new connections through economic and technological progress. They were also rediscovered as a window into the latest scientific advances, much as they were in the time of Alexander von Humboldt and

Map 79 [Extract] The bold arrows of trade flows across the Mediterranean as shown in the *Larousse International Atlas* (1950).

Atlas International Larousse (Méditerranée Occidentale): Armand Colin (1950)
59.5 × 40cm

Map 80 [Extract] A giant fold-out map from the 1950 *Larousse Atlas*. It uses a polar projection to show the transport connections between countries by land and sea. These polar projections were an important feature of the atlas, which, despite them erasing the southern hemisphere(!), the editors said, offered 'the advantage of demonstrating more clearly the linking up, through the polar and circumpolar regions of the Arctic, of the political and economic groups of the world'.

Atlas International Larousse (Communications – Hémisphère Nord): Armand Colin (1950)
62.5 × 62.5cm

THE LIBRARY OF LOST MAPS

Heinrich Berghaus a century before.

In 1950 Armand Colin, a French publisher, printed the *Larousse International Atlas*, which evoked many of these themes. It begins with the words: 'The world has emerged from the war in a state of turmoil, and this necessitated, therefore, the embodiment of the new economic and political state of the world in a fresh atlas, founded on new facts and first-hand information.'[2] I was drawn, first, to the rich red cover of the copy I found in the Map Library, then intrigued to see that the atlas was written in French, Spanish and English. The title page features a red and black compass rose set around a dandelion with its seeds dispersing and the words 'JE SÈME À TOUT VENT' (I sow to all winds), which was the metaphor at the heart of Larousse's dissemination of knowledge. Although the atlas was one of the larger volumes in the collection, it expanded further with maps that were folded into four that could be pulled out to create vast vistas of the Northern Hemisphere. This is not an atlas you can enjoy on your lap – it needs to rest on a large table!

In its preface, the renowned French geographer and political scientist André Siegfried set out why such an atlas was needed at the halfway point of the twentieth century. Core to his message was the idea that Earth's dimensions were being 'transformed in a revolutionary manner, under the influence of that fundamental phenomenon of our age, the conquering of distance'.[3] It was thanks to this that the world had entered an 'age of rapid communications … speed is a new vice, invented by the 20th century'.[4]

Harnessing the *Larousse Atlas*, Siegfried was summoning a world of interconnections with countries better thought of as nodes in a network rather than simply distinct entities, which meant that 'there is no longer such a thing as geographical isolation'.[5] He expressed his view that countries needed to work together and that smaller countries especially should form

A FRESH PERSPECTIVE

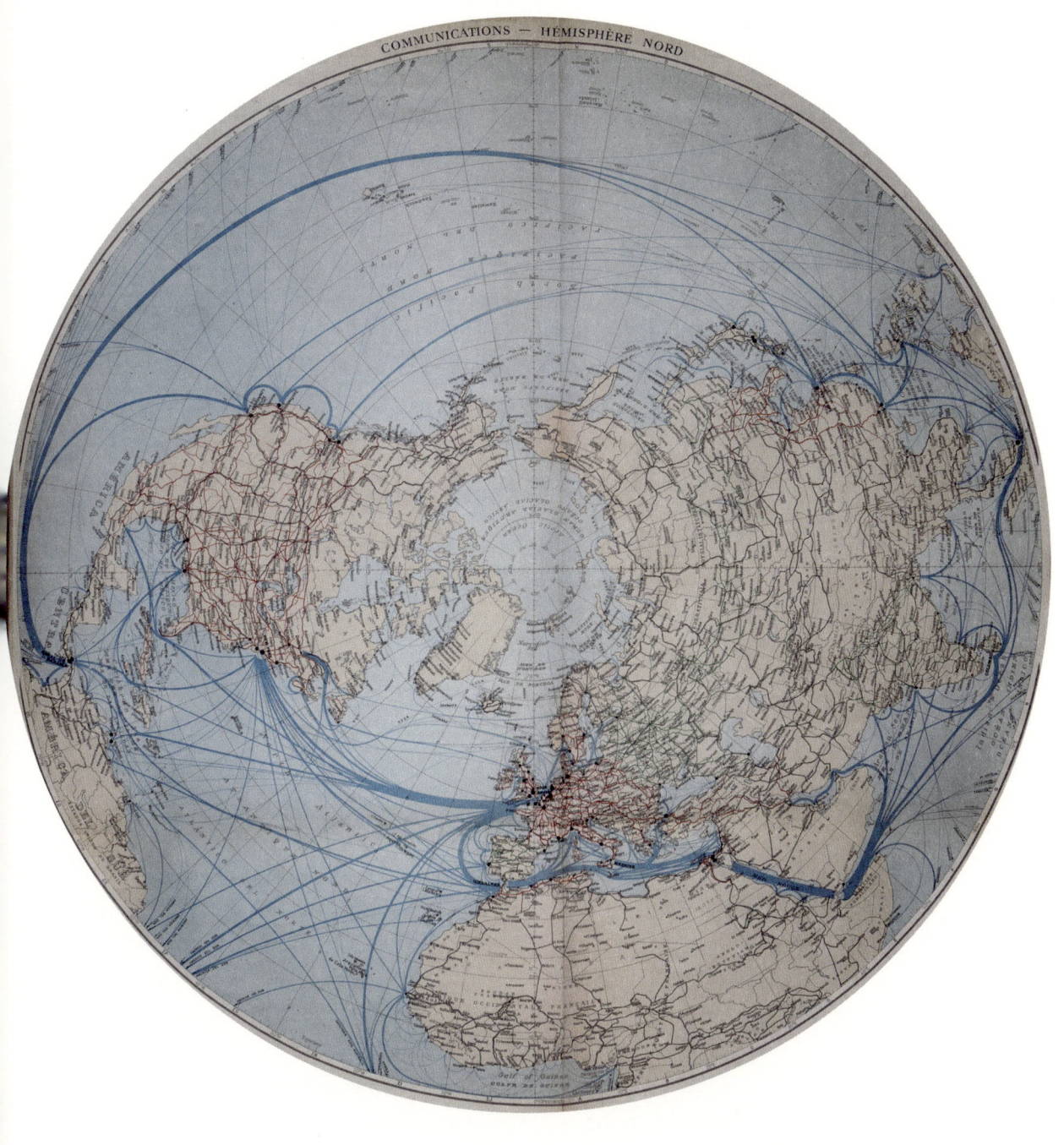

federations: a logic that continues to define global politics to this day. In essence, collaboration, not combat, was the way forward.[6]

To further the sense of coherence and connection, instead of a page to a country/continent, the *Larousse Atlas* chose to extend the maps beyond hard national borders and evoke more fluid collections of states and the linkages that bind them. For example, to map the flow of resources between southern Europe and northern Africa – across the Mediterranean – the atlas does something radical and slices off the top of France to allow more space for North Africa.

These innovations were well received, including by the Royal United Services Institute (RUSI), which commented: 'This sumptuous publication is something new among the greater atlases, for it presents the World in a bolder and more coherent way than does the page to a Continent or a Country system to which we have become accustomed.'[7]

When I was leafing through the atlas, I also noticed that bold arrows were often used that would not have been out of place on a map produced by Karl Haushofer or Alexander Radó in the 1930s (the indefatigable Radó did in fact become an editor of a later edition of the atlas in the 1960s). But the emphasis was different: these were arrows for interconnectivity, not advancing armies or national ambitions.

It would, of course, be naive to see the early 1950s as a utopian period where peace prevailed around the world, but Siegfried's preface ends with a cautiously optimistic statement that 'Despite the sinister rivalries in politics, the marvellous technical progress invites us to proceed without too much pessimism.'[8] It is this sense of optimism that beams out of many of the most innovative maps in the library created in the second half of the twentieth century.

Maps were still being taken into war and there was a boom in the development of national atlases of the kind that the *Atlas de*

A FRESH PERSPECTIVE

Finlande had pioneered (see Chapter 5), which was in part fuelled by newly created countries, particularly in Africa, as colonised regions gained independence. Of course, there were still uncharted lands to be explored with a spirit of adventure and 'discovery' that differed little from the Victorian surveyors a century before.

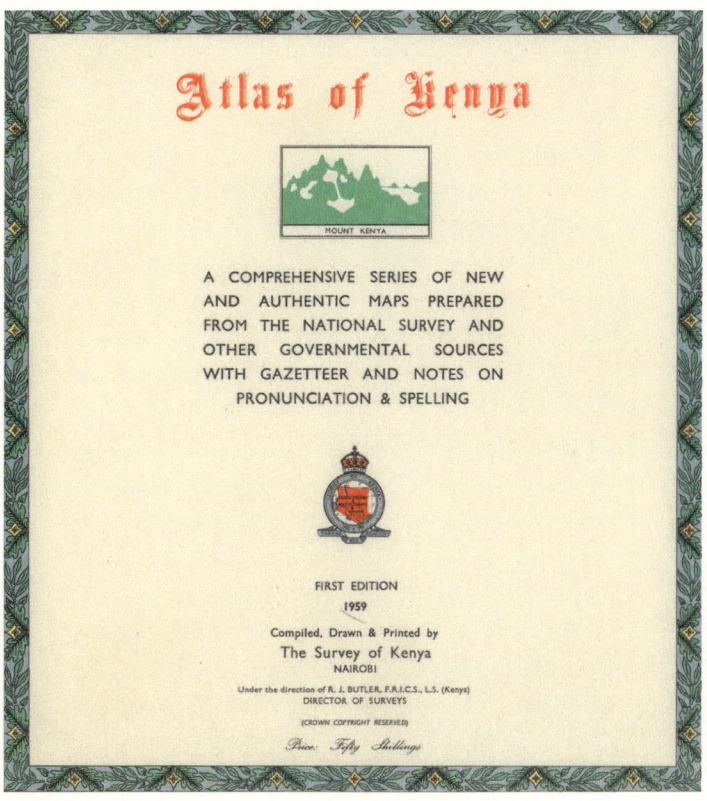

The orderly frontispiece of the *Atlas of Kenya* belies the bloodshed of nearly a decade of unrest in the country from the Mau Mau Uprising. Published in 1959, the atlas pre-dates Kenyan independence by five years but it was the first in a series that spanned the independence era, culminating with the publication of the *National Atlas of Kenya* in 1970.

Atlas of Kenya : The Survey of Kenya (1959)

I was to discover, too, that in this age of optimism the beauty and sophistication of maps literally reached new heights as (mostly Western) climbers were drawn to the Himalayas in the wake of Edmund Hillary and Tenzing Norgay's summit of Everest in 1953.

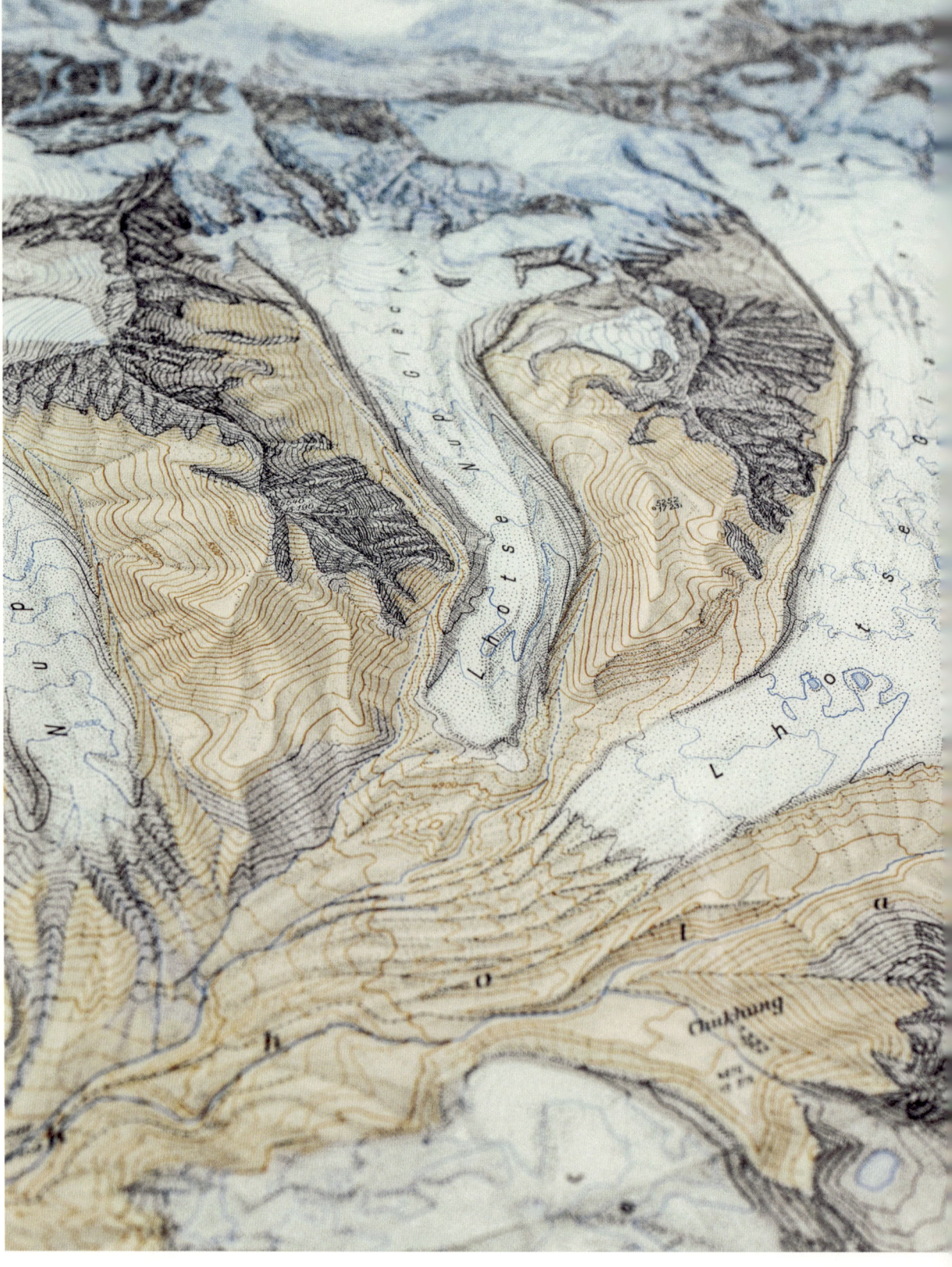

A FRESH PERSPECTIVE

Everest

The Map Library has only one detailed map of Mount Everest and its surrounding peaks, but it ranks among the finest ever published. It represents the pinnacle of human endurance, pioneering technology and unsurpassed cartographic skill. More than seventy years after its creation, it remains hugely important today. The map is titled 'Mahalangur Himal Chomolongma [sic] – Mount Everest' and when it was completed in 1957 it was the most comprehensive map of this famous part of the Himalayas.

    I first pulled it from a drawer that was crammed full of maps of Nepal and northern India. I had to be careful not to tear the fragile paper which was partly folded, with frayed edges stained from the dull yellow of ageing tape that had become detached and brittle. But as soon as I glimpsed the top of a glacier buttressing the paper's edge, the shabby exterior of the map was forgotten.

    I laid it out on the large table in the centre of the Map Library and immediately felt the cold blast of the torrents of ice flowing down from the world's highest mountains into deep valleys and melting into roaring streams. The vastness of the map appeared to dwarf Mount Everest itself, with its understated label battling to be seen against the craggy ice clinging to the mountain's eastern face. It was astonishing to me how the cartographers had made such a vertigo-inducing map, as if levitating above the Earth's surface on a clear day.

    Far below me was crystal-white snow, the deep blues of the thickest ice, the rugged browns of the rocks and the greys of the

Map 81 [Extract] Himalayan glaciers flow down the mountains of this exceptional map, which, as we shall see, was surveyed in the harshest conditions.

Mahalangur Himal – Chomolongma – Mount Everest: Kartographische Anstalt Freytag-Berndt und Artaria (1957)
77.5 × 83cm

cliff faces poking out on some of the highest peaks in the world. I began imagining ribbons of mountaineers, dwarfed against this landscape, threading to the summit of Everest, 8,849 m above sea level. I could see from the stamps at the bottom of the map that it was given away for free to the Map Library as a 'cancelled' map – probably because it was deemed out of date or a duplicate by the Directorate of Military Survey, which had picked it up in December 1961.

By directing my attention to Erwin Schneider, whose name appears several times in the map's right-hand margin, I discovered that he ranks among the greatest cartographers of the twentieth century and is particularly renowned for his maps of the Himalayas. An accomplished climber, he was able to operate in the most extreme mountain environments, remarkably despite frostbite causing such damage to his toes that he had all ten amputated in 1939.[9] He continued to map the region for two decades after the completion of the Everest map before me, and versions of his maps, known simply as the 'Schneider Maps', are still available to climbers today.

The map was created as part of the International Himalayan Expedition of 1955, which was organised by the 'High Priest of the Himalayas',[10] Norman G. Dyhrenfurth, a mountaineer and filmmaker born in Germany in 1918 who moved to the USA (via Austria then Switzerland) just before the Second World War. Dyhrenfurth's military service gained him US citizenship, and he is now best known for leading the expedition that placed the first American, Jim Whittaker, on Everest in 1963. The fame that followed, which included a *Life* magazine cover story and being presented with the National Geographic Society's Hubbard Medal by President John F. Kennedy,[11] is credited with establishing mountaineering as a popular sport in the United States.[12]

Eight years before his success on Everest (and two years

after Hillary and Tenzing Norgay made it to the summit), Dyhrenfurth was granted permission to make the first ascent of Lhotse, a neighbouring mountain rising to 8,516 m above sea level (fourth highest in the world). Beyond the prospect of bagging an unclimbed peak, it was a rare chance to access the Nepalese side of Everest and so he assembled a team that could film the area and map it. As part of this work, he recruited Schneider to undertake the first comprehensive survey of the region.

Dyhrenfurth's accounts of the expedition give a sense of the ambition of the undertaking and the scale of the logistical challenges. In his report to the American Alpine Club, he wrote:

> In view of the time element involved in producing several documentary films, and in making the first truly professional map ... of the entire Everest region, it was decided to leave for Nepal in two groups: Dipl. Ing. Erwin Schneider, outstanding cartographer and well-known Austrian mountaineer, Ernst Senn, one of Austria's leading 'extreme' climbers, who had been to Broad Peak with Dr. Herrligkoffer in the Fall of 1954, and myself left Europe on March 30 by ship, taking with us the bulk of the expedition's baggage ... The second group, consisting of the three Americans, George I. Bell, Fred Beckey, and Richard McGowan, and the two Swiss, Dr. Bruno Spirig and Arthur Spöhel – the latter had been to Everest with me in 1952 – were to leave Europe on July 30.[13]

The first group congregated in Mumbai (called Bombay at the time) and loaded 167 pieces of luggage (weighing 6,500 kg) onto its own coach, which was then towed for four days and five nights by a slow-moving passenger train – with the expedition team aboard – across India to the Nepalese border.

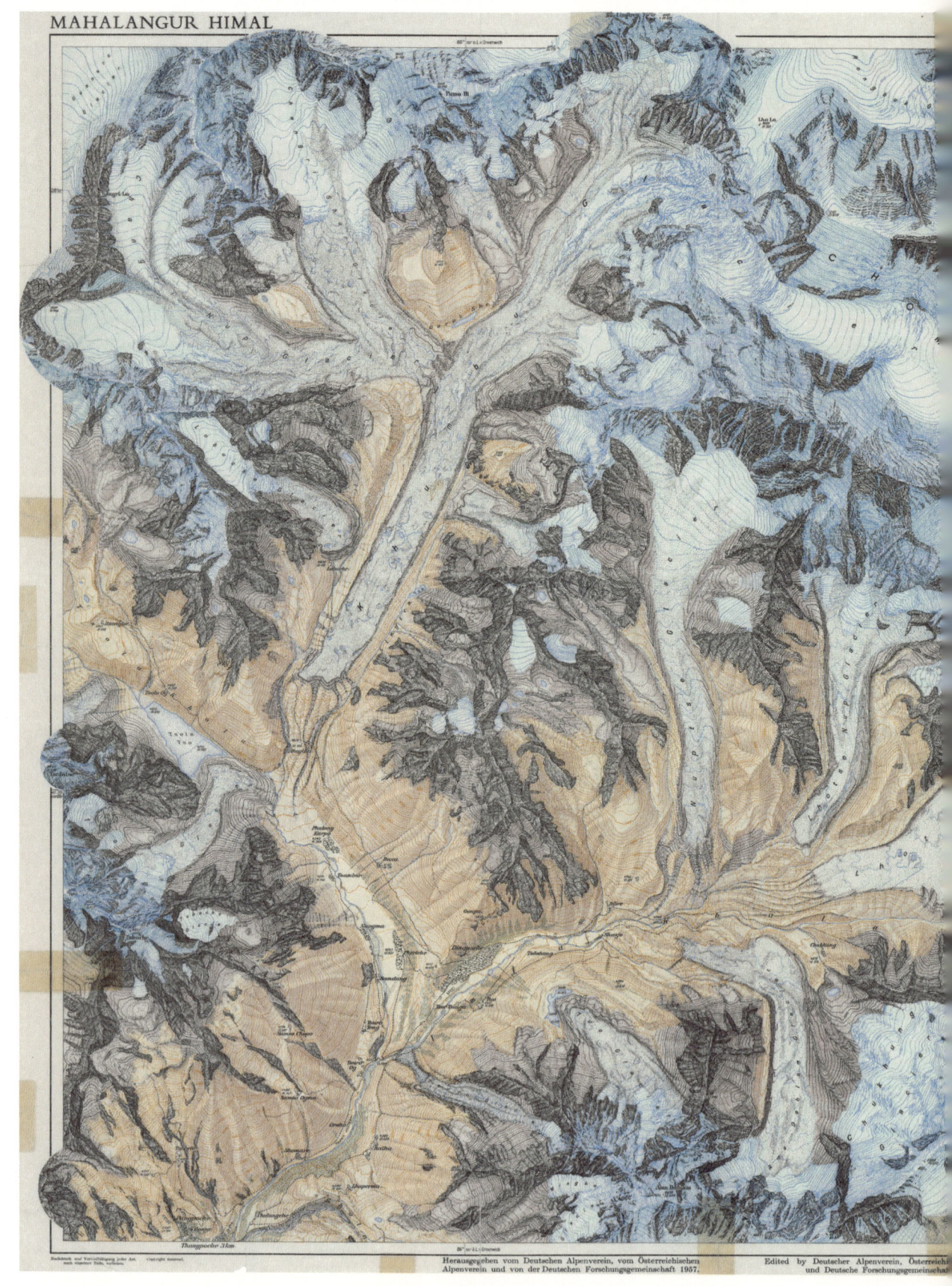

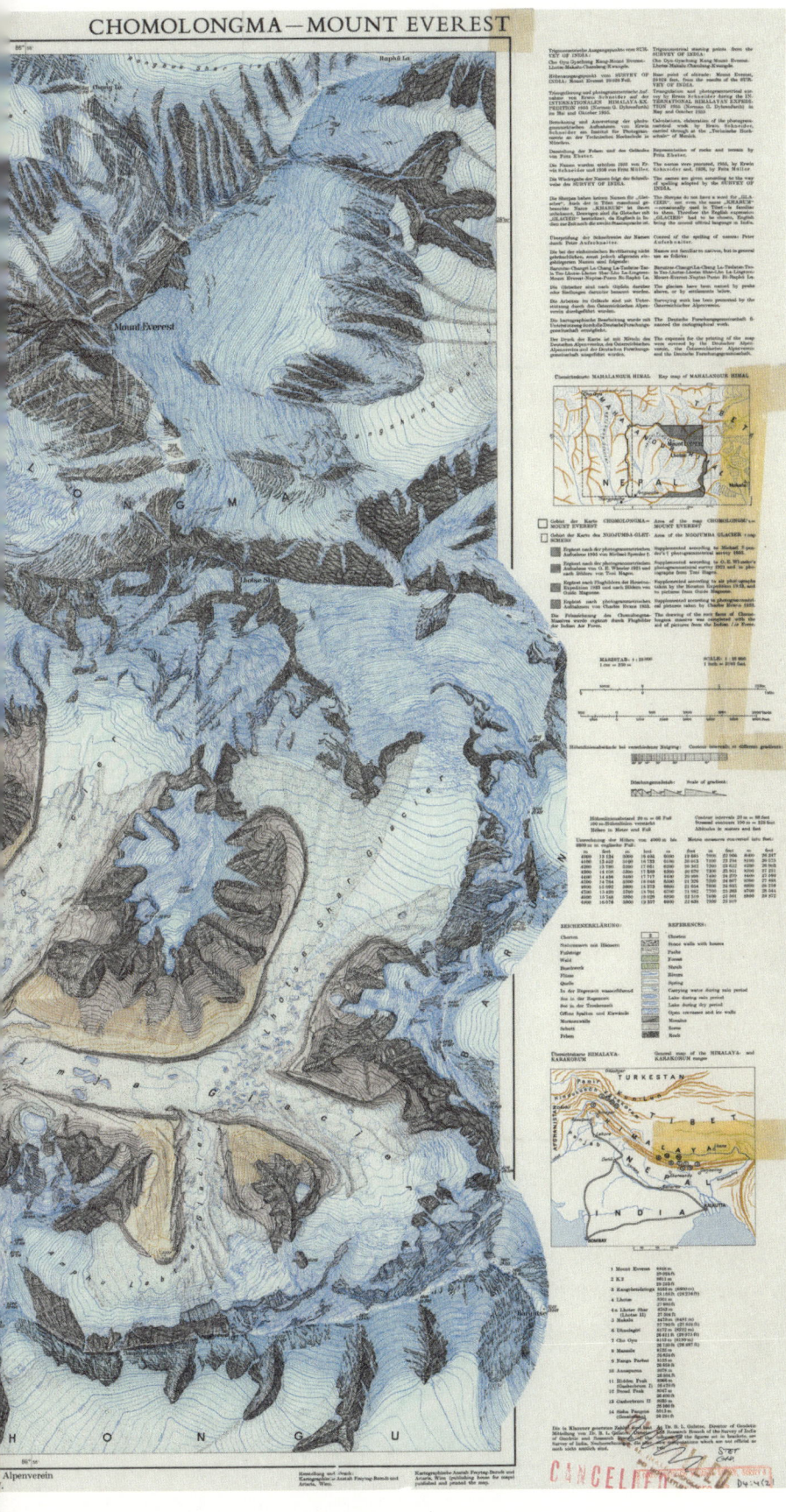

Map 81 With access restricted and only an overview survey completed in 1920s, the Nepalese side of Mount Everest was a 'big enticing secret' to western mountaineers. The immensity of the terrain was hard to comprehend; Schneider wrote 'Whether you look at the Lhotse from the valley floor (4500m) or from a point of view 1000m higher – you do not have the impression of coming closer to the summit.'

Mahalangur Himal – Chomolongma – Mount Everest: Kartographische Anstalt Freytag-Berndt und Artaria (1957)
77.5 × 83cm

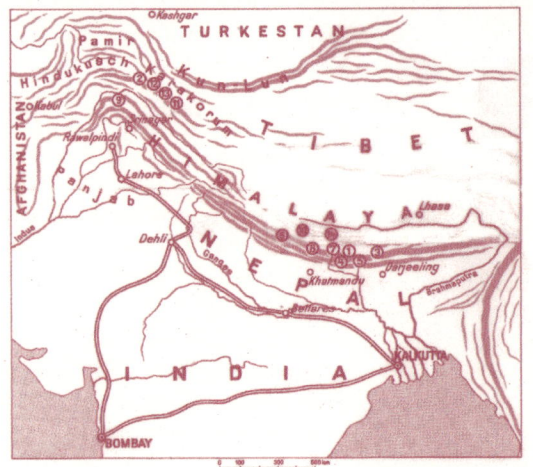

THE LIBRARY OF LOST MAPS

Despite the luggage coach twice becoming detached from the train, once as it rounded a long corner and once being left in a railway siding, the vast quantity of supplies and the expeditioners who depended on them arrived safely in the border town of Jogbani, where they had to clear Nepalese customs and repack for the journey by road to the foothills of the Himalayas. The approach march started on foot from the city of Dharan, with a team that had swelled with the employment of 196 carriers (each carrying around 40 kg) and ten sherpas. The salary costs alone were 800 Indian rupees a day (around £1,500 at today's rates, a high total but modest per individual employed).

It was a gruelling sixteen-day hike to the village of Namche Bazar, nowadays known as the gateway to the high Himalayas, where the team stopped to change carrier teams and recharge. Understandably, Schneider's lack of toes meant he struggled with the long hike over consecutive days, but Dyhrenfurth's expedition report shows his passion for the landscape they were passing through: '[Schneider] who did not feel very comfortable during the approach is like a newborn. He assures us again and again that this is the most beautiful mountain landscape that he has ever seen and Erwin Schneider has seen a lot.'[14]

Two days later – on 9 May 1955 – they arrived in the village of Dingboche, where they based themselves for the next couple of months. By the beginning of June, Schneider had made rapid progress on the southern section of the photogrammetry survey, which Dyhrenfurth refers to as the 'Imja Khola area'. Frustratingly, poor weather brought on by the arrival of the monsoon put an end to this work and so on 7 June they took an excursion north-westwards beyond what would become the top left corner of the map[15] in search of clearer days.

Delays to the photogrammetry work were serious because it was integral to the success of the expedition. The technique was a relatively new approach to surveying, but one that Schneider had mastered to generate the data he needed to make the detailed map of the Everest region. Rather than undertake the labour-intensive work of detailed triangulation – where the location of a point of interest (such as a mountain peak) is calculated by measuring the angles to it from known points, which can then be converted into a distance using Pythagoras' theorem, and then sketching in the terrain between the measured points – several overlapping photographs could be taken from different angles and the points within them triangulated.[16] This removed the need for Schneider to make as many manual measurements to achieve the level of precision he needed, so he could cover much more ground in more detail, assuming the conditions were clear enough for the photography.

Upon returning from Imja Khola, still in the midst of the monsoon season, they moved the main camp to a yak pasture called Lobuje (labelled Lobuche on the map) at about 5,000 m above sea level and on the western side of the Khumbu Valley, with its mighty glacier flowing through it. Dyhrenfurth's explanation for the move was that there was a local superstition that prevented strangers staying more than two months, 'otherwise nothing would grow any longer'.

Respecting this local custom meant they were now living in tents, surrounded by 100 yaks and in the worst of the monsoon rain. Dyhrenfurth soon grew fond of their hairy companions, describing them as 'magnificent and quite peaceful animals' that suddenly appear from the monsoon fog with 'a hum and a grunt'. But the weather started to get to them: 'The next few weeks were about as depressing and demoralizing as any I have ever lived through. Even Schneider was feeling mighty low, since there wasn't a thing he could do in the way of photogrammetry.'[17]

By August, things started to improve. The expedition had kept morale high by reading mail from home (they benefited from two pairs of postmen running mail back and forth to Kathmandu, a journey of ten to fifteen days) and had celebrated Swiss National Day on 1 August by reading 'mountains of letters' and drinking rum long into the night. The team then started to move provisions to their higher base camp and from there Dyhrenfurth was turning his attention to preparing for the Lhotse summit to the east. The second group arrived on 16 September and by 10 October the summit team were ready to start their attempt.

Despite good weather conditions initially, before reaching the summit the climbers (who did not include Schneider) became caught in a terrifying 'hurricane-like' storm that stopped them in their tracks. It battered them and their tents for seventeen days before they could stagger back to the base camp dejected and in very poor shape.

While the main part of the expedition was being defeated by the elements Schneider kept working away on his survey. As Dyhrenfurth notes: 'Schneider has done the most fruitful work of all of us. In the weeks we spent in the western basin and on the Lhotse flank, he continued to measure tirelessly: Khumbu glaciers, Chola Khola and the entire catchment area of the Dudh Kosi and its wonderful glacial phenomena.'[18]

Schneider was someone with a rare combination of the technical abilities required to undertake a cutting-edge survey and being comfortable in the mountain landscape. It meant he could climb some of the highest peaks in the world (he summited 6,000 m peaks as a matter of course) to get the camera angles required for the photogrammetry and then rapidly descend them to move on to the next viewpoint. In one of his reports, Dyhrenfurth notes with pride seeing Schneider and the expedition medic Spirig ski down the infamous Khumbu Icefall 'taking their

A FRESH PERSPECTIVE

skis off only twice for short stretches'.

At the end of October the expedition began its slow descent from the high mountains, and by 8 December, more than eight months since the first group set out, everyone had headed home. 'Twenty full working days' had been spent doing the surveying, work that Schneider acknowledges was assisted by 'Gaja Nanda Vaidya from Kathmandu ... five Sherpas from Namche Bazar, Khumjung and Pangboche, and two Sherpani'.[19]

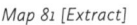

Map 81 [Extract]

The travelling might have finished but the production of the map still had a long way to go: it had to be drafted and then printed. Credit is given on the maps to Fritz Ebster for the 'representation of rocks and terrain'. While he was not on the expedition, Ebster was a close collaborator with Schneider and they took five months to bring the survey measurements, notes, sketches and photographs to life in the map. The total time the work took was estimated to be 600 hours.[20] The costs of printing were covered by the German and Austrian Alpine Clubs as well as the German Research Foundation.

Without achieving the stated aim of summiting Lhotse, the 1955 expedition has not gained the iconic status it deserves. For example, it gets no real mention in any of Dyhrenfurth's obituaries following his death in 2017, aged ninety-nine. Nonetheless, it gave something hugely valuable to mountaineers: a map. This was not lost on Dyhrenfurth himself at the time:

> At the end of such an expedition one is bound to ask oneself: Was it worth it? Must it be called a failure, since we did not reach the summit of Lhotse? I don't think so. The ascent of Lhotse was not our only goal: the mapping, photography, and documentary films were of no lesser importance.[21]

When he offered this reflection, Dyhrenfurth could not have anticipated that Schneider's map would grow further in importance. The impact that the climate crisis is having on the Himalayas is benchmarked against such maps. They serve as antidotes to what is known as shifting baseline syndrome[22] – a phenomenon in which each generation perceives the environmental conditions they grow up with as 'normal', leading to a gradual acceptance of environmental degradation over time. Historic maps can jolt us out of this mindset by giving us the baselines of generations before.

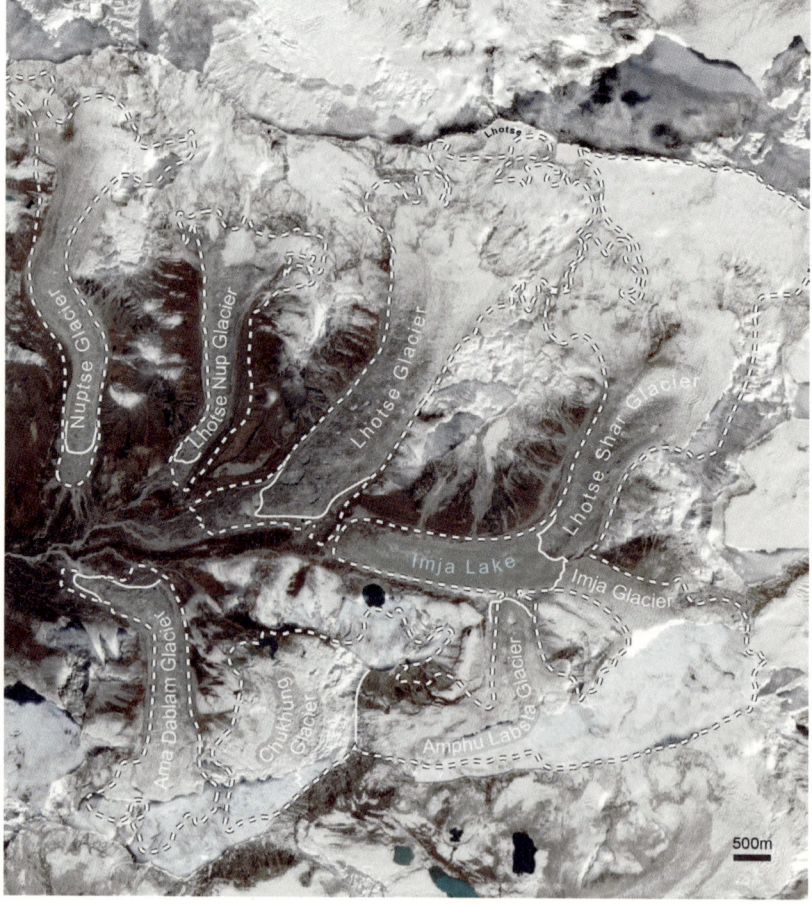

In dashed white lines I have traced the outlines of eight Himalayan glaciers as they were mapped by Schneider in 1955. These are set against a satellite image taken in November 2024, with solid lines added to show the furthest extent of each glacier today. There has been a drastic retreat of the Imja Glacier, leaving a lake in its place. Imja Lake was one of the fastest growing in the Himalayas and risked breaching the barrier of moraine holding the meltwater back until the Nepalese army was able to drain over 4 million cubic metres of water from it in 2016.[12]

A FRESH PERSPECTIVE

In the decades since Schneider completed his survey, the glaciers around Mount Everest have thinned rapidly[23] and even those at the highest altitudes are at risk of disappearing entirely by the middle of this century.[24] We know this in part thanks to him, since his map and photographs offer a snapshot in time to compare to present-day surveys.[25] If all Dyhrenfurth had done was summit Lhotse, there'd be nothing else to show for the months he and his team spent in one of the most inhospitable places on Earth. There'd be no map to pave the way for other climbers, no map to chart a changing planet and no map for me to discover seventy years later.

Dyhrenfurth and Schneider's quest to map the Himalayas epitomised the best of map-making in the 1950s. It is steeped in innovative surveying techniques and bolstered by the immense skill of formal cartographic draughtsmanship. But while these traditional-looking maps and atlases flourished, the idea that there was a single, methodical perspective that was privileged for cartography had disappeared during the Second World War. What's more, the awareness of the utility of making maps had increased in other fields, not least to those without formal cartographic training. This created opportunities for experimentation and for producing maps that were unlike any others that had come before.

The unprecedented demand for maps during the war had also meant that map-making talent was sought beyond formally trained cartographers and geographers and from the world of artists and graphic designers. These people had the skills to create maps that were unshackled from the cartographic rules imposed by the geographers. Designers and illustrators thought more about how the maps were being used and how they might engage with their audiences better. Harry Beck, with his London Underground map, was an early pioneer of this in the 1930s, but many more cartographic outsiders began making maps during the war years.

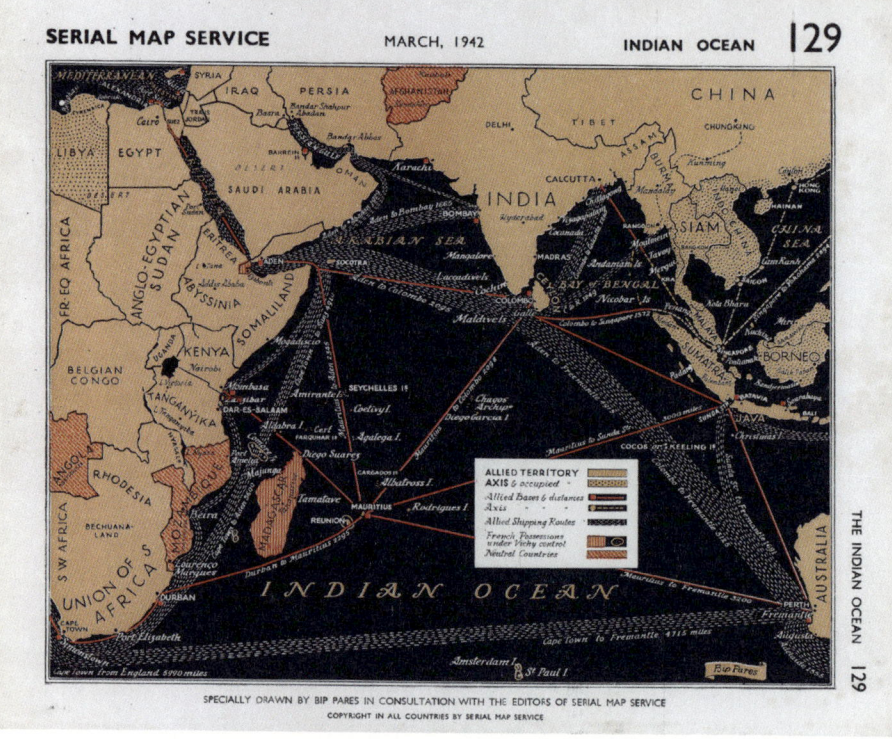

Map 82 A Bip Pares map for the March 1942 edition of the *Serial Map Service* (SMS). Like Frank Horrabin, she was not a conventional cartographer, but an illustrator and storyteller. This gave her maps a very different appearance to the more traditional-looking maps the *SMS* were printing prior to her and Horrabin's arrival at the publication.

The Serial Map Service (The Indian Ocean) (1941) 24 × 19cm

For example, I was delighted to discover that an art deco designer by the name of Bip (Ethel) Pares, who, like Beck, produced designs for London Transport, worked alongside J. F. Horrabin (TV's 'map man') on the *Serial Map Service* (SMS) maps. After my discovery of the Horrabin map (see Chapter 7), I went on a quest to find more of the SMS and found a couple of folders of maps online. Pares's maps are the best among them.

After the war Bip Pares became well enough known to generate her own bylines at the *Observer* newspaper, which also led to a global profile as her maps were picked up by publications as far away as Australia.[26] The accompanying articles would begin with 'A new Bip Pares map showing ...'[27] and would command the front page. Pares died in 1977 at the age of seventy-two and her obituary highlighted her skills as a mapmaker above her other work.[28]

But not everyone felt that the arrival of designers and illustrators with a journalistic bent onto the cartographic scene was a positive development.

A FRESH PERSPECTIVE

In the archives of the American Geographical Society (AGS), I came across a document[29] written by one of their cartographers named Stanley Smith. Smith worked at the AGS from 1927 up until his death in 1949. He was the perfect example of a cartographer from the first half of the twentieth century, employed on hugely expensive mapping projects or tasked with repurposing maps from one format to another. I'm not sure if the document was a letter, a speech or something for publication; it was written on 14 February 1948 and addressed to John K. Wright, the Director of the AGS. It is excoriating about some of the maps created in the war that amounted to a 'bastard product which has emerged from the drawing boards of commercial artists and advertising copy boys … with inflated wages and egos to match their speedily acquired titles …' He goes on to say, 'Undoubtedly a fair number of these people will emerge as competent workers but for the majority it would be doing the trade of map-making a great service if they would return to their advertising copy, fabric designing, ballet dancing, stage managing, teaching and spring ploughing. (The world needs food!)'[30]

But Smith (and the AGS) missed a trick here. The audience for these 'bastard products' was growing while the very traditional form of cartography that the AGS produced was less in demand, especially their brand of generalist maps. New printing technology also made it economical to publish complex graphics in full colour, which grew the public's appetite to catch sight of the latest scientific developments of the age. This extended to maps, which were often used to reveal everything from how global flyways were developing to the detailed demographics of cities.

When this all came together, it led to some of the most iconic maps of the century.

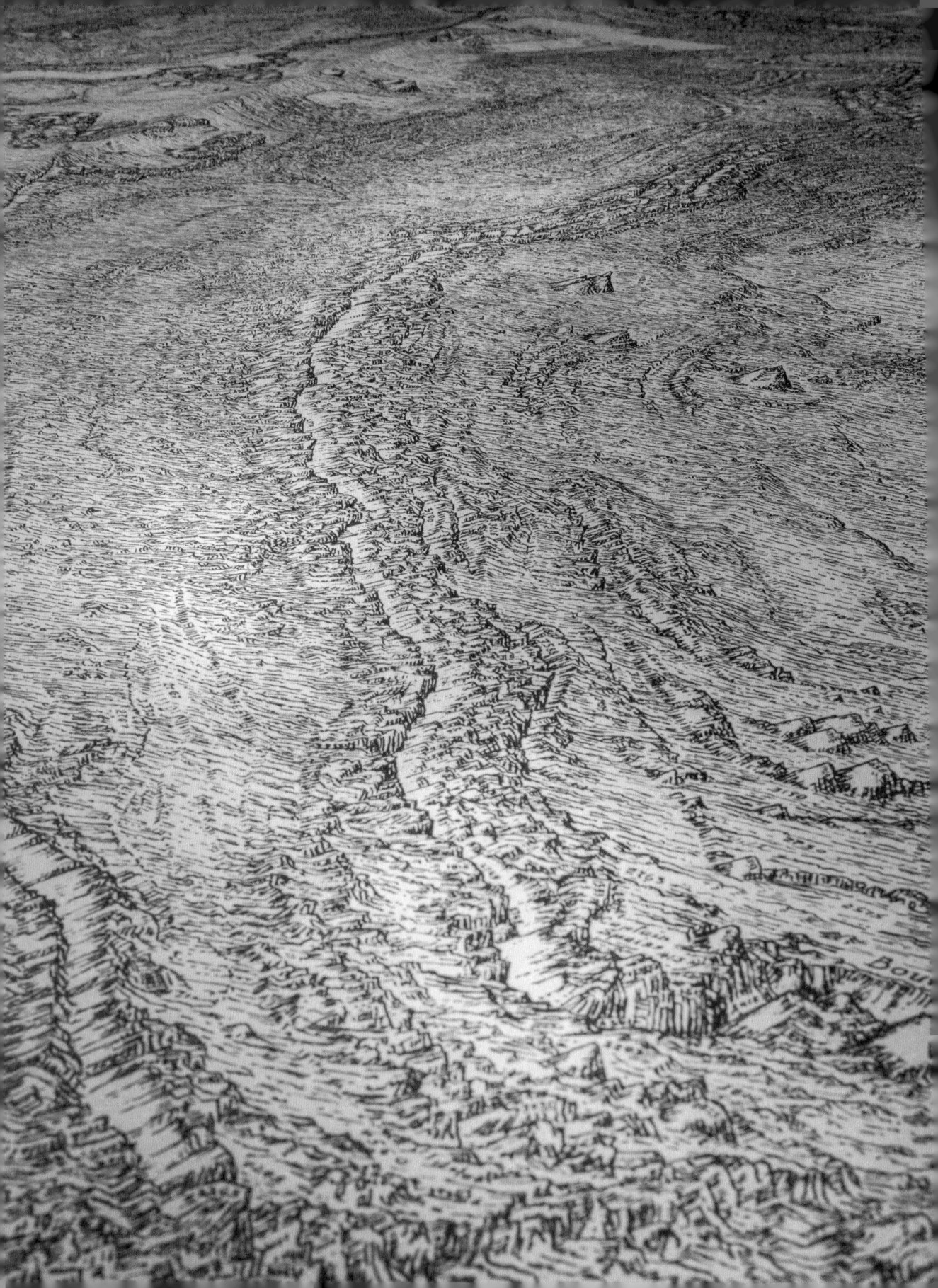

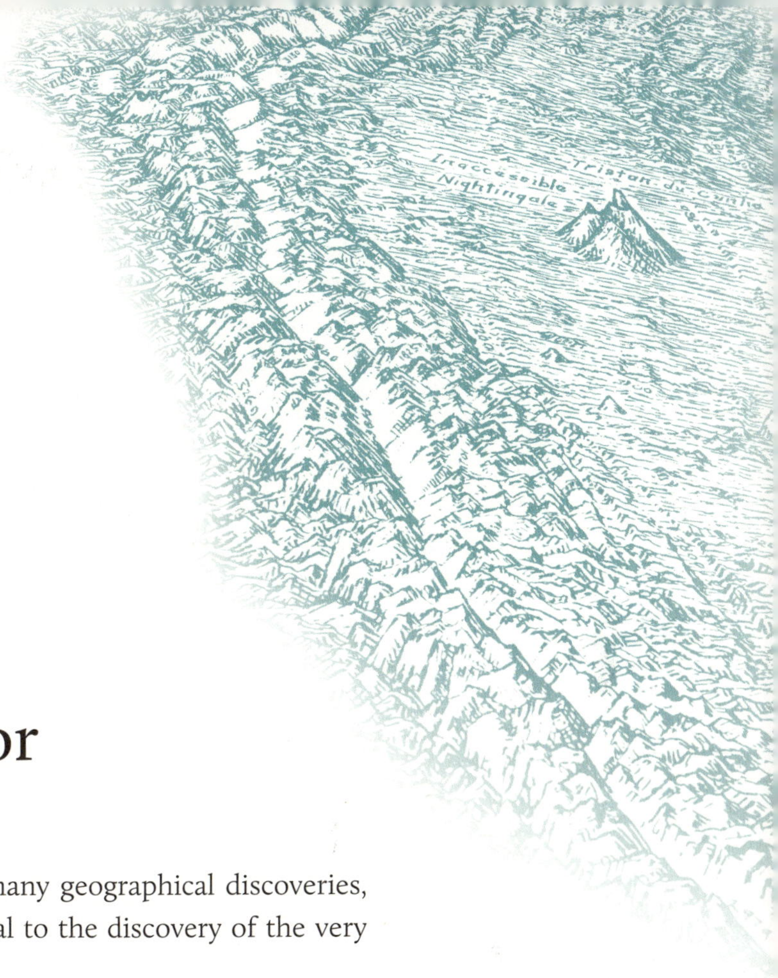

CHAPTER TEN
The Ocean Floor

'Maps have been instrumental in many geographical discoveries, but rarely have maps been so crucial to the discovery of the very nature of the Earth.'
John Noble Wilford[1]

At over a metre wide, and nearly a metre and a half tall, the first map I unfolded from the 'Oceans' drawer was suitably vast. When I laid it out on the table, it transported me to the depths of the Atlantic Ocean and a rugged landscape with more mountains and valleys than the roughest terrain on land.

Coming up from the south of the map was the tongue of the Antarctic Peninsula that directed my eyes to the east where they met the vast ridge that runs up the centre of the Atlantic. There were many conical mountains standing out, like termite mounds on the blue surface but infinitely larger, as atop their peaks teetered tiny islands each with handwritten names adjacent to them. Nestled among the ridges and crags were numbers showing

Map 83 [Extract] The Mid-Atlantic Ridge as drawn by Marie Tharp using the distinctive 'physiographic' mapping technique.

Physiographic Diagram of the South Atlantic Ocean: The Geological Society of America (1962)
118 × 142cm

Map 83 Heezen and Tharp's stunning physiographic map of the South Atlantic Ocean floor, which was first published in 1961.

Physiographic Diagram of the South Atlantic Ocean: The Geological Society of America (1962)
118 × 142cm

the depth of the ocean where measurements had been taken. It was a map unlike any other I had seen in the Map Library.

To my delight this map of the South Atlantic was accompanied by equally astounding maps of the Indian Ocean and the South Pacific. All three were credited to a Bruce Heezen and Marie Tharp and were created between 1961 and 1971. Interspersed among the giant sea-floor charts were other much smaller maps published as posters for *National Geographic* magazine. Again, Heezen and Tharp were credited, and so, too, was an artist called Heinrich Berann. As I was to discover, this was a small collection of maps that totally changed how we saw the world.

I immediately started to research Heezen and Tharp, and realised that a huge amount has already been written about their work (Google even featured Tharp on its homepage in 2022), so I was a little embarrassed not to have been very familiar with them as individuals even though I had seen their maps before online.

To find out more, I went to the Library of Congress in Washington DC and the Heezen–Tharp Archive. In 1995 Tharp donated her papers, containing over 40,000 items,[2] to the library and until her death in 2006 helped arrange them for the benefit of future researchers.[3] The collection is the largest held by the Library of Congress Map Division and spans 255 boxes, which require over 100 m of shelving.[4] It is an extraordinary trove of information that contains not just the drafts and workings behind the maps but also pages and pages of transcripts from recordings that Tharp made of Heezen before his unexpected death in 1977 in preparation for a biography that never materialised, as well as draft transcripts corrected by Tharp from interviews conducted later in her life. There are reams and reams of correspondence from Tharp, too, as she told and retold her story to those who wrote to her. Rarely does one get such a privileged insight into a scientific, creative and personal partnership, especially one that

THE OCEAN FLOOR

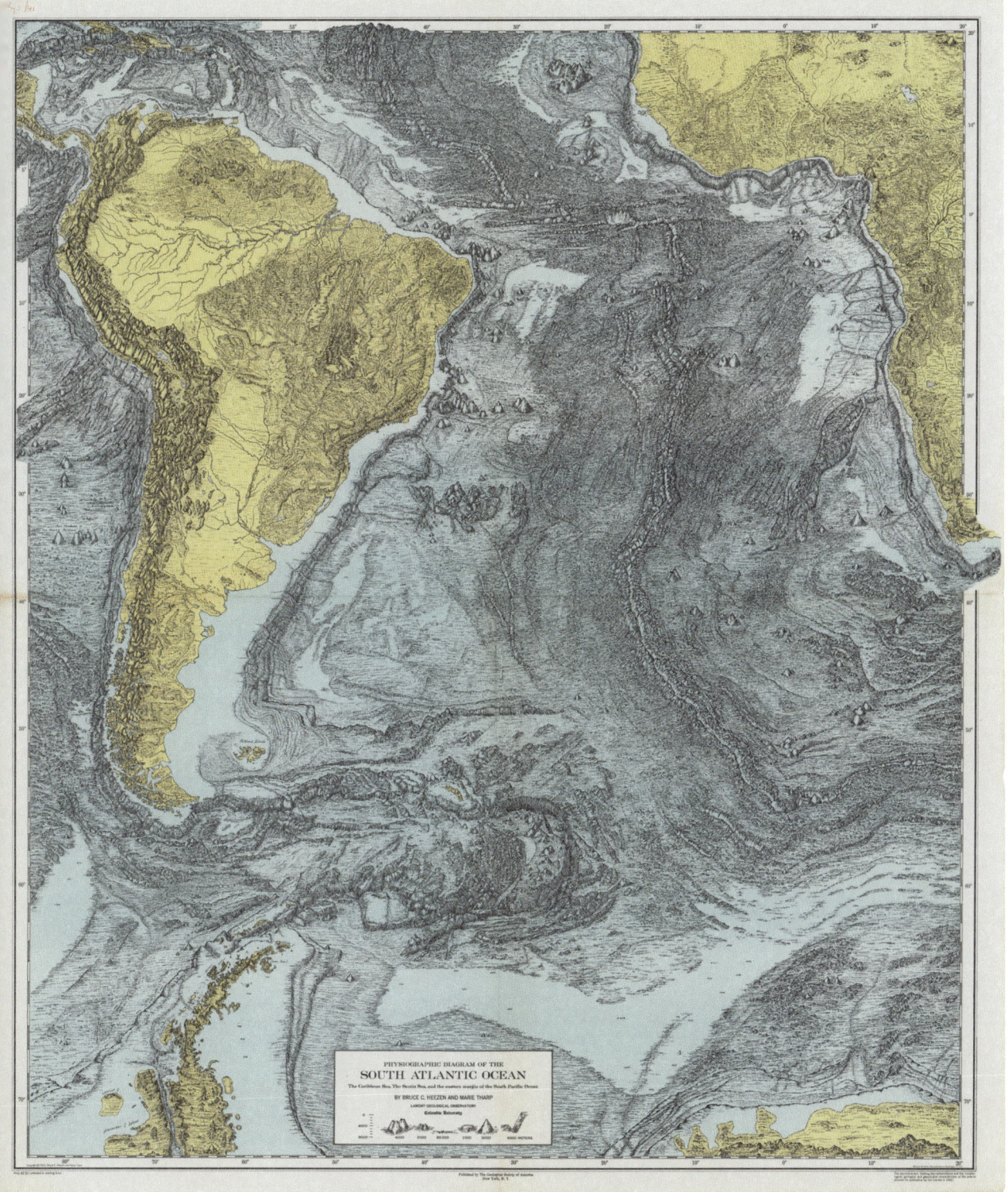

would come to define our vision of Earth.

I had only five days to sift through as much as I could and was left in no doubt that between the early 1950s and the late 1970s the quest to map the ocean floor brought together the best artists, cartographers and scientists who coalesced around Heezen and Tharp. Some of these characters had trained as far back as the Paris Peace Conference of 1919 and were immersed in the kinds of maps that appealed to Stanley Smith of the American Geographical Society (AGS), while others were the innovators who reformed maps for the second half of the twentieth century. All played their part in this extraordinary cartographic endeavour.

Before we get to peek into a world of cartographic giants mixing with competitive scientists seeking to make their mark, we first need to add to our geological knowledge and delve into some of the debates that surrounded one of the biggest shifts in scientific thinking to have occurred in the twentieth century: the acknowledgement that the Earth's crust is fractured into different tectonic plates that are on the move.[5]

There were many who worked on this issue and the arguments that bubbled up between the various personalities involved were just as impassioned (and at times cantankerous) as those that George Bellas Greenough became embroiled in over a century before. Perhaps it's no surprise as many of Greenough's ideas, including some that were refuted, were part of the intellectual inheritance of the geologists and sea-floor mappers of the 1950s. I can't cover it all here so, as with the rest of this book, I shall focus on the stories that are present in the Map Library and leave the rest to others.[6]

As soon as the first cartographers began sketching detailed outlines of the continents, it was clear that these landmasses might just have been connected like the pieces of a giant jigsaw. The cartographer Ortelius noticed it in the sixteenth century, while in the seventeenth century Francis Bacon observed that the fit between the continents was too close to have occurred by chance. In the nineteenth century Alexander von Humboldt saw similar rock formations in South America and Africa that indicated that the two continents most obviously fitted together.[7]

You would think, then, that the map of the world is all the evidence you'd need to prove that the continents had indeed been one large landmass and over billions of years have moved apart from one another. But this idea was not universally accepted until the end of the 1960s. I remember first learning about it in geography classes at school in the early 2000s and being puzzled how something so obvious could have taken so long to have been confirmed. The reason was that geologists could not explain *how* it could be that the Earth's crust – almost entirely made of solid rock – was moving.

It was a German geologist named Alfred Wegener who made the first serious attempt at settling the argument in 1912 by proposing the idea of 'continental drift'. He suggested that Earth's continents were indeed once part of a single landmass called Pangaea, which over the course of millions of years had fractured into the shapes we recognise today. He backed up his theory not just by making the jigsaw puzzle argument but also by aligning the similar rock formations and fossil species found on continents now separated by oceans. Wegener's continental drift hypothesis, however, still did not offer a compelling explanation for how the continents could move. His suggestion that they were ploughing through the oceanic crust – like an icebreaker across a frozen lake – was deemed too implausible by the geologists.

Wegener's continental drift hypothesis was not universally dismissed, however, and geologists started to join one of two camps: the 'fixists' and the 'mobilists/drifters'. The fixists were united in their view that the continents were stationary, but

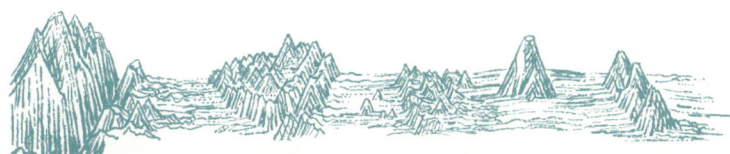

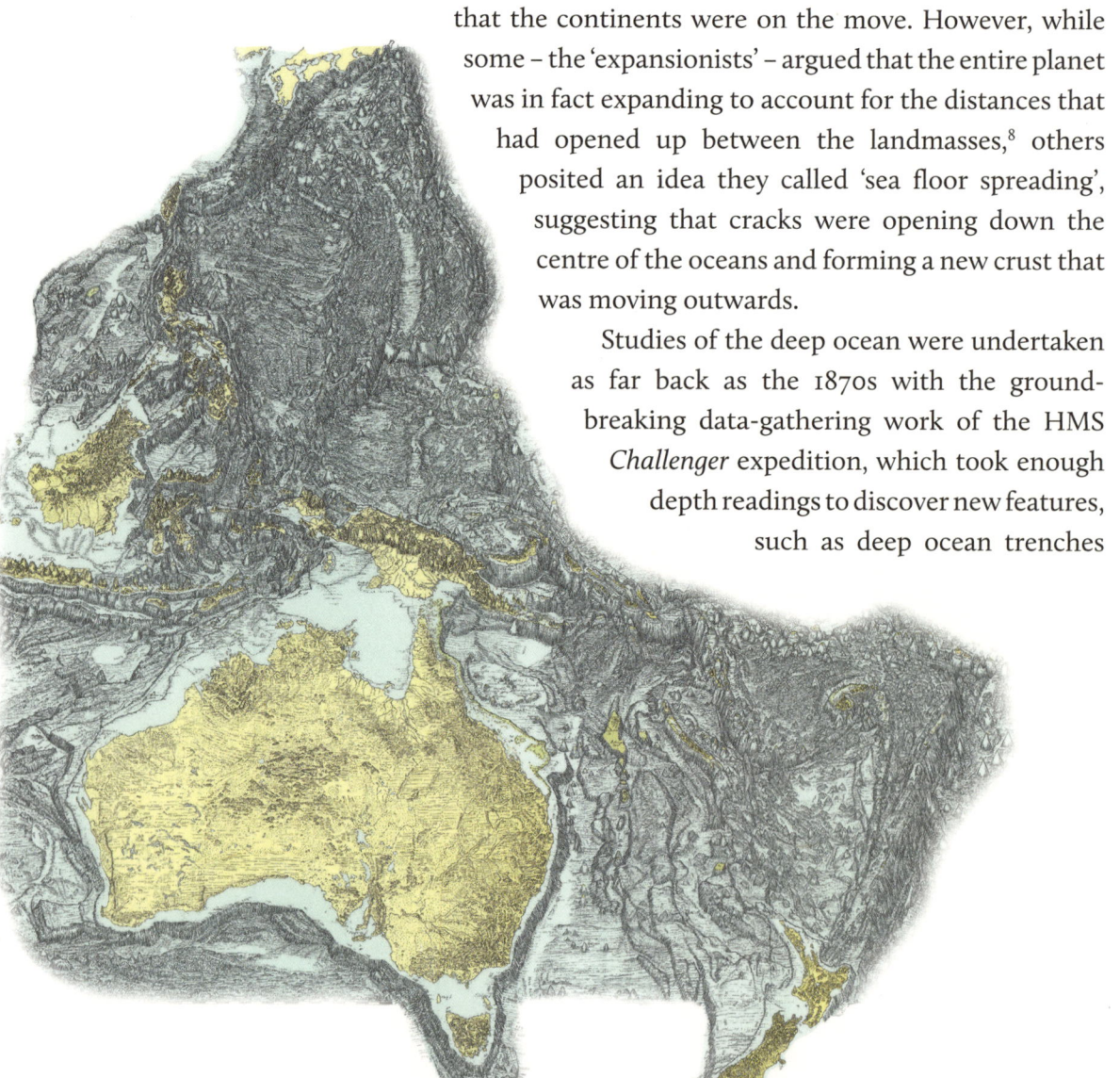

Map 84 [Extract]
Physiographic Diagram of the Western Pacific Ocean: The Geological Society of America (1962)
118 × 142cm

some believed that while the continents' relative positions stayed fixed, their positions relative to the poles had moved. There was also disagreement among the group about how the continents formed in the first place, not least because they acknowledged there were geological and biological connections between them. Some geologists therefore argued for the existence of ancient land bridges, while others opted for 'island hopping' to explain migration paths for common species of flora and fauna.

On the drifters' side there was, of course, agreement that the continents were on the move. However, while some – the 'expansionists' – argued that the entire planet was in fact expanding to account for the distances that had opened up between the landmasses,[8] others posited an idea they called 'sea floor spreading', suggesting that cracks were opening down the centre of the oceans and forming a new crust that was moving outwards.

Studies of the deep ocean were undertaken as far back as the 1870s with the groundbreaking data-gathering work of the HMS *Challenger* expedition, which took enough depth readings to discover new features, such as deep ocean trenches

THE OCEAN FLOOR

and mountain ranges.[9] Nonetheless, early approaches to recording ocean floor depths were extremely rudimentary and amounted to little more than measuring the lengths of weighted ropes dropped overboard. In the vastness of the oceans these single, not terribly accurate, data points were never going to fill out a map.

The arrival of echo-sounding technology in the 1920s and then the development of sonar for submarine warfare unlocked the power of sound waves to measure the depths of the oceans by timing how long sound pulses took to travel to the sea floor and then bounce back. Unlike the weighted rope method, this, for the first time, offered more continuous readings of ocean depths and gave the finer details needed to realise the sea floor was as rugged and varied as the terrain on dry land.[10] This exciting development meant that, just as the space race was heating up in the 1950s so, too, was the quest for data about the oceans, and huge investments were made in data gathering, especially by the USA and USSR, who took a surprisingly collaborative approach to sharing data and ideas.[11]

Theories developed rapidly during this time thanks to the sudden abundance of data to help prove or disprove them. The key purpose of much of the funding for data collection, however, was not in and of itself scientific curiosity. A better understanding was needed to help military submarines to navigate the murky depths and for stringing communications between continents. Cables were snapping and submarines risked crashing into the seabed (a problem that exists even to this day, as the $3 billion USS *Connecticut* discovered in 2021 when it collided with an unmapped undersea mountain[12]).

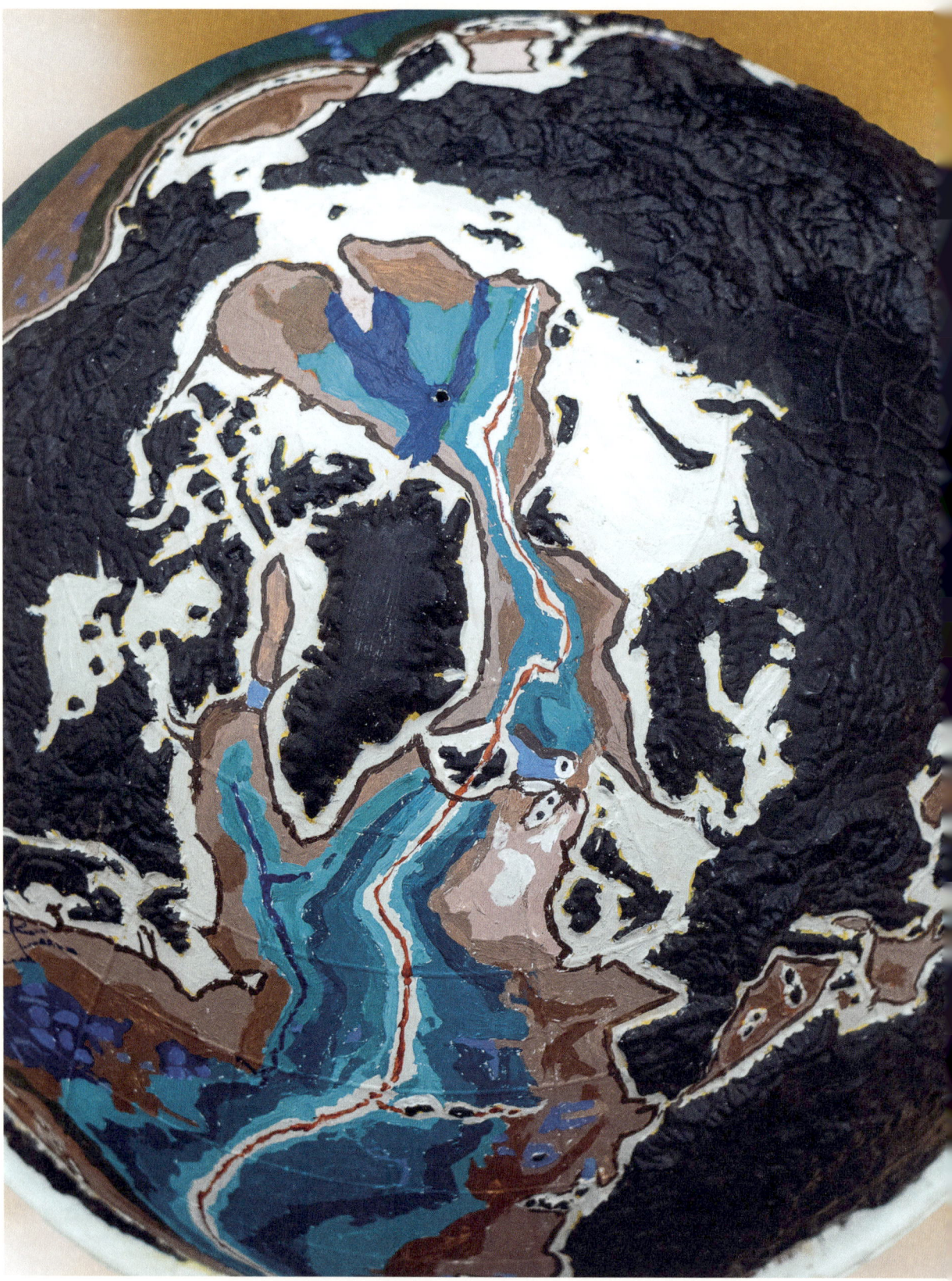

Heezen and Tharp

Bruce Heezen – one of the names on the maps I was so excited to discover – was among the scientists commissioned to work on the challenge of where to lay undersea cables without them snapping. In September 1957 he wrote a paper for the *Bell System Technical Journal* titled 'Oceanographic Information for Engineering Submarine Cable Systems'.[13] It was a specialist paper for engineers rather than for oceanographers and geologists, but the fifty-page paper offers a primer on the sea floor. What's more, it was a hot-off-the-press rundown of what would become one of the most significant scientific discoveries of a generation: a canyon running down the centre of the Mid-Atlantic Ridge. The paper was illustrated through a series of conventional graphics, and tucked in the back was an extraordinary map of the North Atlantic Ocean.

Instead of a relatively flat surface with the occasional contour – which was how the best available maps up until that point had appeared – here was a map that showed a sea floor packed with details that made it resemble a fantasy map of sharp peaks and open plains. This map, credited to Heezen and Marie Tharp, was a first glimpse of an entirely new view of the world, and the first in the series that included the vast maps I pulled from the 'Oceans' drawer.

Tharp was raised in a world of data collection and mapping. She spent her childhood trailing after her father who was a soil surveyor criss-crossing the USA, gathering much-needed

Not all pioneering maps demonstrate the finesse of a skilled cartographic hand. This crudely painted globe was completed by Bruce Heezen and must have been the first to show many of the recently discovered undersea boundaries that were so important to the theory of plate tectonics.

Ocean Floor Globe (1957)

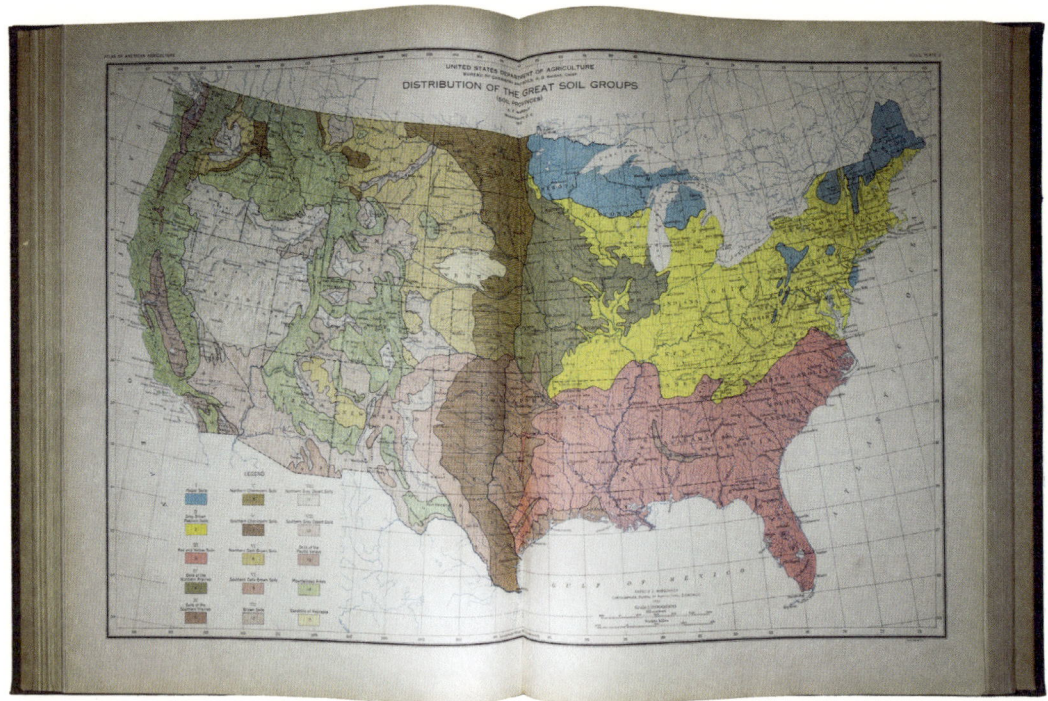

Map 85 A double-page spread from the 1936 *Atlas of American Agriculture* that Marie Tharp's father contributed data to. It is an extraordinary atlas in terms of the volume of information it contains and the range and diversity of the maps, which tackle everything a farmer might need to plan for their crops and livestock. This map shows the major soil groups of the USA.

Atlas of American Agriculture (Distribution of the Great Soil Groups); United States Department of Agriculture (1936)
59 × 39cm

information at a time when parts of America were turning into a dustbowl. When I learned this, I went to the atlas cabinet in the Map Library containing the 1936 *Atlas of American Agriculture*[14] – a beautiful atlas of extraordinarily detailed maps and diagrams of climate, geology, land use and soil – and was delighted to see the name W. E. Tharp credited for collecting the data in three locations.[15] Tharp was an eager field assistant for her father and so this may in fact have been her first data publication.

After a tumultuous time at seventeen schools across four states, she graduated from Ohio University (English and music) in 1943 and also earned degrees from the University of Michigan (petroleum geology) and the University of Tulsa (mathematics). Tharp started work at an oil company in Tulsa but 'found it unsatisfying as she was confined to an office where she collected maps and data for the men going into the field'.[16] So in 1948 she went for a job at the Lamont Geological Observatory, Columbia University (New York). According to Tharp, Maurice Ewing, the imposing boss of the research institute, was left wondering who had shown up to the interview: 'And he wanted to know what my background was, and so I told him. And since I had gotten my education in such a funny order ... his final question was, "Well, can you draft?"... He was so bum fuzzled at this funny collection of courses, and in my taking them in such a back-ass way.'[17]

Tharp got the job and her drafting skills were in demand right away. She was sent work from researchers across the institute, but was soon overwhelmed by so many demands that she left. She was talked back by Ewing and agreed to continue working, but only under the management of Heezen who was a graduate student at the time. They had first met when he had not long returned from a research trip soon after she started.[18] 'So that was that.'[19]

Tharp and Heezen's relationship was known for its intensity, with stories of 'ink bottles and erasers being thrown around the room and maps being torn up',[20] but the pair had a deep affection for one other. The first drawing Tharp made for Heezen was a contour map of the Hudson Submarine Canyon: they were 'so happy to get it recorded in its correct manner'.[21] Buoyed by their success, they then tackled something much more ambitious: the North Atlantic Ocean.

Until I saw the metres and metres of paper on my visit to the Library of Congress, I did not appreciate what a manual process this was. Heezen and Tharp went through reams of tracing paper, using it to draw everything from the routes the data-gathering ships took, to the sounding profiles they generated, the epicentres of earthquakes, the gentle undulations of contour lines and then the jagged mountains and valleys.

One of Tharp's gifts was her ability to get all these different layers processed to convert the raw data into usable information.

THE LIBRARY OF LOST MAPS

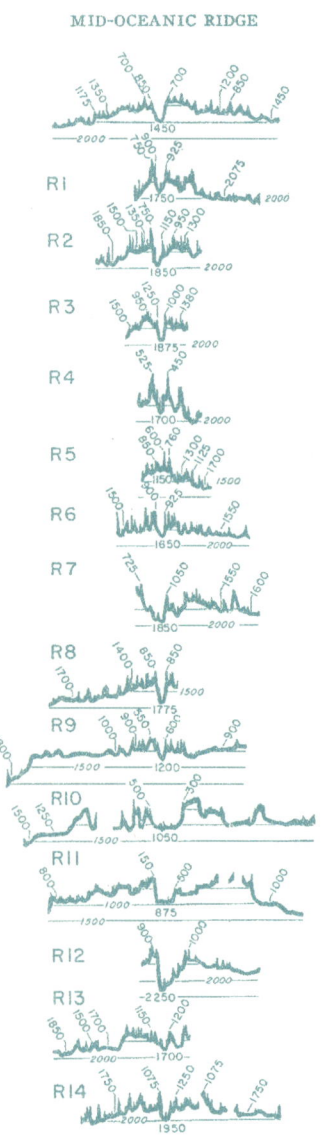

MID-OCEANIC RIDGE

It was this ability that led to her making the profoundly important discovery of the rift valley running down the centre of the Mid-Atlantic Ridge. The ridge itself was already known, but the new knowledge of a vast canyon was of huge significance to the geologists as it meant it could be an active feature and perhaps the location of where new oceanic crust was being created. As Tharp remembers, she had 'a hodgepodge of disjointed and disconnected profiles of sections of the North Atlantic floor' that took six weeks to arrange in order from west to east into six parallel profiles spanning the North Atlantic. She noticed immediately the general similarity in the shape of the ridge in each profile, with a consistent V-shaped indentation sitting at the centre of them. This, Tharp thought, might just be a rift valley.[22]

John Tukey, who is considered one of the pioneers of what we now call data science, wrote in 1977 that **'The greatest value of a picture** is when it *forces* us to notice **what we never expected to see.'**[23] This is precisely what happened the instant that Tharp lined the profiles up and caught sight of the same dip occurring in all of them.[24] Fixist geologists, in particular, had denied the existence of such a valley because it pointed to the growth and movement of the Earth's crust, but Tharp's plots forced them to notice its presence, even if they didn't want to see it. She recalls that the fixists 'not only said it wasn't fair they said it was a bunch of lies'.[25] On its own this was not enough to settle debates about the drifting of continents, but it was a huge leap forward for the drifters.

What I find so powerful about this moment is the colossal effort needed to create it. If we think of all that Erwin Schneider put in to mapping the peaks around Everest, it doesn't even come close to what was needed to gather these few data points from the sea floor. Schneider was part of just one expedition with a small team (although I appreciate he inherited past maps to work with, too), but here we are talking about tens of expeditions, buffeted by ocean storms and costing millions of dollars to put the ships and their crews to sea.

Tharp and Heezen then spent thousands of hours collating the data that was sent back – which had to be distilled again and again to a point of unequivocal clarity. No more than six jagged lines on a chart, but all indicating a rift valley, the existence of which would help to settle one of the greatest scientific debates in history.

Tharp excitedly showed what she had found to Heezen, but he groaned and said, 'It cannot be. It looks too much like continental drift.' He was an expansionist and she recalls that:

> at the time, believing in the theory of continental drift was almost a form of scientific heresy. Almost everyone in the United States thought continental drift was impossible. Bruce initially dismissed my interpretation of the profiles as 'girl talk'.[26]
>
> But I thought the rift valley was real and kept looking for it in all the data I could get. If there were such a thing as continental drift, it seemed logical that something like a mid-ocean rift valley might be involved. The valley would form where new material came up from deep inside the Earth, splitting the mid-ocean ridge in two and pushing the sides apart.[27]

The rift valley discovery occurred in 1952, but it was not until late 1956 that Heezen and Lamont director Ewing published a paper[28] that announced its presence as a major feature on the ocean floor. Clues had been published before, but they had not hit the headlines in the same way. Newspapers ran sensational headlines like 'World is Cracking Up ... And Here's the Proof'[29] and 'Gigantic

Tharp aligned depth profiles collected by multiple research cruises across the Atlantic. She spotted that they shared a V-shaped indentation at similar points, which suggested to her that there was a rift valley canyon running the length of the Mid-Atlantic Ridge. Here profiles are shown of the canyon from a publication Tharp co-authored with Heezen and Ewing that was published to give more information on their pioneering map of the North Atlantic.

Text to Accompany the Physiographic Diagram of the North Atlantic; The Geological Society of America (1959)

THE LIBRARY OF LOST MAPS

Undersea Crack in Earth, Says Scientist'.[30] Heezen had no doubt it existed, but, given his expansionist perspective, remained unconvinced about its origins[31] so there was still work to do, not least more detailed mapping, which is something they lobbied for as part of the International Geophysical Year (1957/8) that their paper had coincided with.

Alongside the sensational announcement, Heezen and Tharp had another trick up their sleeve: the map of the North Atlantic tucked at the back of the 1957 *Bell System Technical Journal*. This transformed the sea floor into the rugged arrangement of mountains and valleys that jolted people out of thinking it was a flat, featureless plain. Suddenly there was a whole new world to discover, and they wrote it up in more detail for the Geological Society of America,[32] as they did for all their subsequent maps. The maps, worth thousands of dollars today, were for sale 'unfolded in a mailing tube' for just $2.

Map 86 [Extract] This is the first physiographic sea-floor map Heezen and Tharp completed and the first to show the central rift valley running through the centre of the ocean.

Physiographic Diagram Atlantic Ocean: The Geological Society of America (1957)
137.5 × 68.5cm

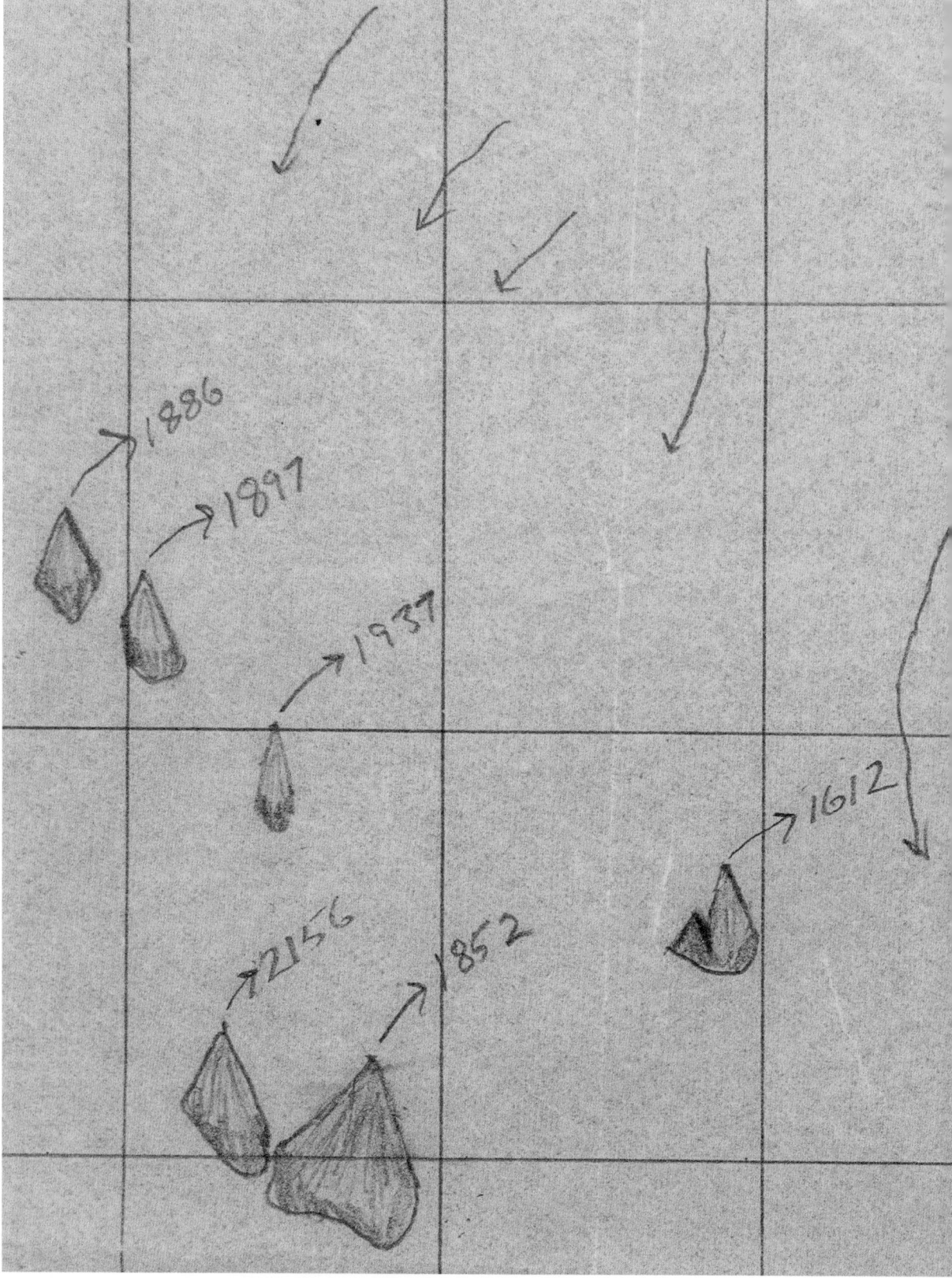

An Ocean of Data

Bruce Heezen and Marie Tharp took three crucial decisions early on in their work together that enabled them to make such an important discovery and then go on to find fame by producing such compelling maps. The first decision was a technical one that accelerated Tharp's breakthrough moment of seeing the rift valley that runs down the Mid-Atlantic Ridge. Every sheet of paper, whether it was a vast profile that spanned half an ocean or a tiny scrap with a couple of fragmentary contours, was drawn to scale with neat gridlines that meant it could be aligned with every other map and layered on a light table. Being able to layer and re-layer in this way meant that any relationships between features from different data sources could be seen.

This, as Tharp recalls, was crucial to confirm her discovery:

> Bruce hired Howard Foster, a young, deaf graduate of the Boston School of Fine Arts ... to plot [by hand] the location of recorded earthquakes in the Atlantic and other parts of the world. He plotted tens of thousands ...
>
> While I was at my map table plotting the position of the Mid-Atlantic Ridge and the alleged valley, a map of the same scale as mine, showing the position of the earthquakes, sat on an adjoining table ... Because all our data were on maps of the same scale, the locations of the epicenters within the valley showed up when we superimposed the maps on a table lighted from below. At that point, I was completely convinced that the valley was real.[33]

A glimpse of Tharp's process from her archive in the Library of Congress. These are rough drawings of individual sea mounts rising from the floor in the North Atlantic, just south of Newfoundland, that have been drawn on tracing paper. Such sketches would be layered with others on a light table and guided Tharp as she filled out the final map.

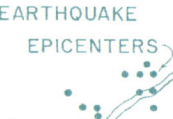

THE LIBRARY OF LOST MAPS

It was this alignment between the earthquakes and the undersea terrain that made Heezen wonder if the lines of earthquakes measured elsewhere in the Arctic Ocean, Indian Ocean and Eastern Pacific might also be pointing to similar rifts. He surmised that these rifts might well connect into a colossal 'world-girdling'[34] 40,000-mile-long Mid-Oceanic Ridge system.[35]

The second decision they made was a pragmatic response to restrictions placed on certain maps by the military. Publishing detailed contours of the ocean floor was banned so Heezen and Tharp alighted on a 'physiographic approach' which focused on the textures of the Earth's surface rather than its elevation (or depth in the case of the ocean floor). This more artistic rendering meant there was enough ambiguity in the positions of the features on the map that the censors were happy for the maps to be published.

This way of mapping was developed by a cartographer named Armin Lobeck. Lobeck honed this approach for the Paris Peace Conference at the end of the First World War where he prepared physiographic maps for the decision makers[36] who, as we saw in Chapter 6, sometimes struggled to understand the more technical-looking cartographic outputs. Or as Heezen put it:

Text to Accompany the Physiographic Diagram of the North Atlantic; The Geological Society of America (1959)

> [Lobeck] found all these politicians couldn't tell a mountain from a mole hill or a river from a valley or anything from a shoreline. So this technique of sketching was developed so that in their discussions of European boundary questions, the politicians could have some idea what the topographic maps were trying to say. [The] Technique worked so well he went on to use it mostly for high school and graduates and young students to understand what the mountains looked like.[37]

THE OCEAN FLOOR

Erwin Raisz (whose map of Cuba we saw in Chapter 2) was a fan of this technique, too, since he had shared a PhD adviser with Lobeck.[38] In his cartography textbook he set out that 'the advantage of the physiographic method is that the map appeals to the average man. It suggests actual country and ... It works on his imagination.'[39]

This decision to use the physiographic technique resulted in maps that were a radical departure from those that came before, and, as Raisz points out, they were maps that worked on people's imaginations. They might be tricky to read precise depths from, but they conjured an image of the ocean landscape that people could immerse themselves in.

Tharp recalls the moment in 1952 that work on the first physiographic map of the ocean began: '[Bruce] did the first drawing, and then he handed it to me and said, "Well, why don't you fill in the rest." I didn't quite realize that it would take me the rest of my life ...'[40]

'By 1956 we finished our first detailed physiographic diagram of the North Atlantic. And like the cartographers of old, we put a large legend in the space where we had no data. I also wanted to include mermaids and shipwrecks, but Bruce would have none of it.'[41]

Tharp had taken an early draft of her map to Lobeck to get advice and encouragement, but by the time the map was finished he was gravely ill, so she was not able to share her final creation with him in person. Heezen and Tharp still sent Lobeck a copy, who saw it just before he died. His wife wrote to them to say how much he liked it.[42]

After 1961, the US declassified the use of contours on marine maps so the original justification for taking the physiographic approach no longer applied. However, Tharp and Heezen continued to use it because, as Tharp told interviewers in the 1980s, 'we kept on using that sketchy method ... because we liked it.

THE LIBRARY OF LOST MAPS

Map 87 [Extract] Lobeck's inspirational map of North America. During the Second World War, his maps were used for military planning, not least for the Allied invasion of Normandy on 6 June 1944.[13]

Physiographic Diagram of North America: The Geographical Press, Columbia University (1948)
57 × 79.5cm

It was a very demanding technique where you had data. You could show everything. And it allowed you to invent and extrapolate where you had no data.'[43]

The brilliance of Tharp was her ability to take very limited information, no more than a few transects and soundings across an ocean, that pointed to the existence of ridges and mountain ranges and then expand – or extrapolate – this out to an entire ocean floor. Tharp saw deep sea soundings obtained along a ship's track as 'a ribbon of light where all was darkness on either side, until the next ship's tracks which might be 5, 10 or 500 miles away'.[44] To illuminate the darkness, she would work 'from the known to the unknown', by starting with the well-surveyed shorelines and shelf breaks and moving on to the continental slopes and abyssal plains before inking in the crests of the ridge.[45] She created beautiful details, but much of it was a combination of imagination and scientific data. This is similar to the approach that William Smith took in his geological map of Britain and that George Bellas Greenough had initially pooh-poohed (see Chapter 3). Tharp and Smith teach us that sometimes you need to take a leap of faith to drive science forward.

The third decision Heezen and Tharp took was to be uncompromising in their commitment to communicating their science. This was one of the motivations for adopting the physiographic technique, since they could have used contours and kept the maps under lock and key for use only by approved scientists and the military. Heezen said that they 'only had one client all these years. That is the scientific public.'[46] I think by this he meant the scientific community at large, but from his and Tharp's actions it was clear they wanted to share their maps with as many people as they could beyond the scientists.

This remains a pioneering perspective since good scientific communication continues to be one of the most overlooked

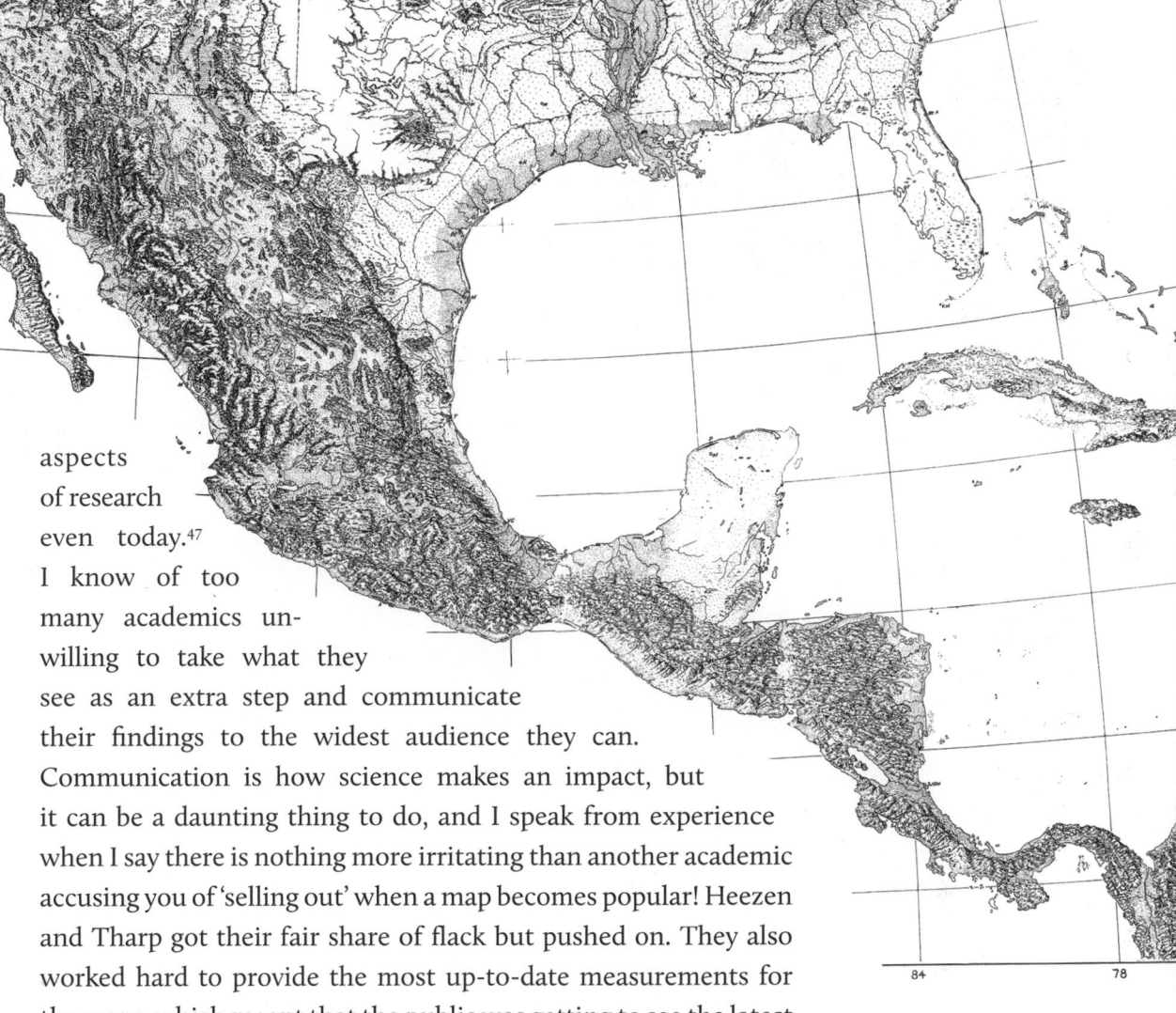

aspects of research even today.⁴⁷ I know of too many academics unwilling to take what they see as an extra step and communicate their findings to the widest audience they can. Communication is how science makes an impact, but it can be a daunting thing to do, and I speak from experience when I say there is nothing more irritating than another academic accusing you of 'selling out' when a map becomes popular! Heezen and Tharp got their fair share of flack but pushed on. They also worked hard to provide the most up-to-date measurements for the maps, which meant that the public was getting to see the latest depictions of the ocean floor at the same time as other scientists.

The headlines of the discovery of the Mid-Atlantic canyon, as well as their articulation of the vast ridge in the centre of which it sat, attracted the attention of some of the most iconic magazines, and their skilled illustrators, of the day. And Heezen and Tharp were only too happy to work with them.⁴⁸

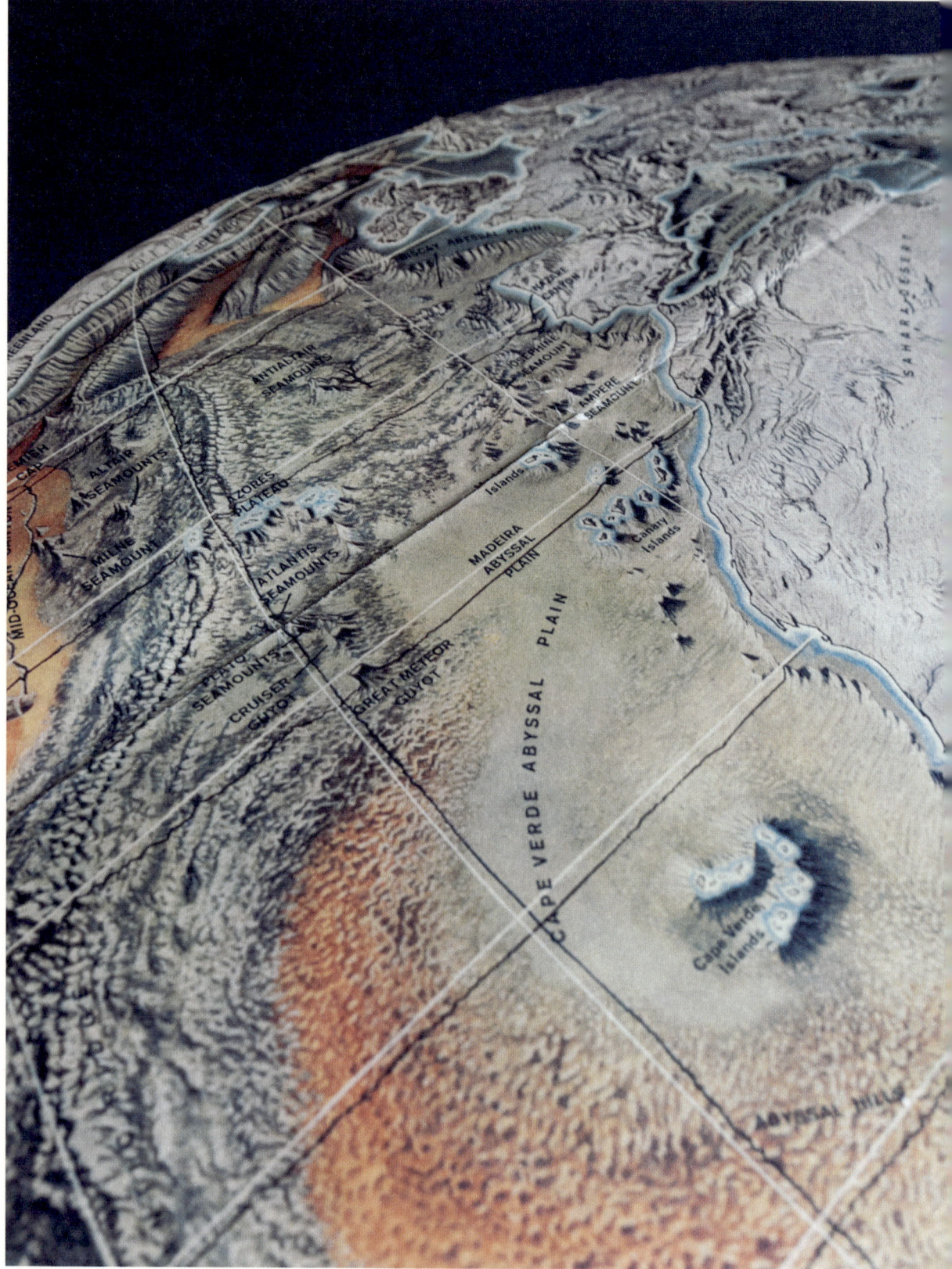

THE OCEAN FLOOR

Draining the Oceans

One of the first illustrators to come to Heezen and Tharp was Richard Edes Harrison. As Susan Schulten, the great historian of American cartography, writes: 'Harrison was the person most responsible for sensitizing the public to geography in the 1940s. He drew dozens of maps for *Fortune* magazine, so distinctive that they literally re-created the look of the world and set a standard that others would emulate.'[49]

Despite this impact, Harrison may well have been one of the mapmakers that the AGS cartographer Stanley Smith was so irked by. As Schulten says, Harrison disowned the label 'cartographer', preferring 'artist' instead, since it meant he could liberate himself from rules the likes of Smith wanted mapmakers to follow. Seeing the profession as mired in 'a static condition bordering on senility',[50] Harrison wanted radically to change people's perspective of the world by offering maps that broke with conventional map projection choices.

For his map of the sea floor, Harrison sought out one of Smith's colleagues, William Briesemeister, who was Chief Cartographer at the AGS and who, like Armin Lobeck, had spent his formative years supporting the Paris Peace Conference at the end of the First World War. Harrison wanted an interesting map projection that could offer a seamless view of the oceans and had spotted that Briesemeister had developed his own projection for a hugely successful disease atlas that the AGS produced in the 1950s. As it happens, this was an alternative atlas by Jacques May to the German

Map 88 [Extract] Published as a giant poster, this is only a section of *Life* magazine's map of the ocean floor. The text accompanying the map explains how the colours give Earth a Mars-like appearance: 'Land above sea level is outlined in aquamarine. Reddish tones are red clay, greenish are debris from land, and grayish and yellowish sediments are plankton oozes.'

The New Portrait of Our Planet: Life (1960)
128.5 × 76cm

THE LIBRARY OF LOST MAPS

Map 89 This is a small inset from a much larger map of global diets produced by a pioneering medical geographer named Jacques May for the American Geographical Society (AGS). He asked the AGS cartographer William Briesemeister for an innovative map projection for his *Atlas of Disease*.

Study in Human Starvation 2: Diets and Deficiency Diseases: American Geographical Society (1953) 33 × 19cm

offerings from the same time and was made by the American research team that the US Navy had requested collaborate with the former Nazi scientists mentioned in Chapter 7.

Harrison realised that he could take the Briesemeister projection and centre a world map along the westerly coast of Australia.[51] In so doing he pulled the landmasses to the edge of the map and pushed the oceans towards the centre. His extraordinary image, published in 1959, was the first global view of the ocean floor that contained Bruce Heezen and Marie Tharp's processed data. Harrison went on to refine this for a larger map he produced for the Association of American Geographers (AAG) in 1961.

Surprisingly, given Harrison's high profile, his maps have been overlooked in the history of Heezen and Tharp's efforts. There was also an amazing map created soon after by another artist for *Life* magazine, but that Heezen was less keen on for its 'using slightly

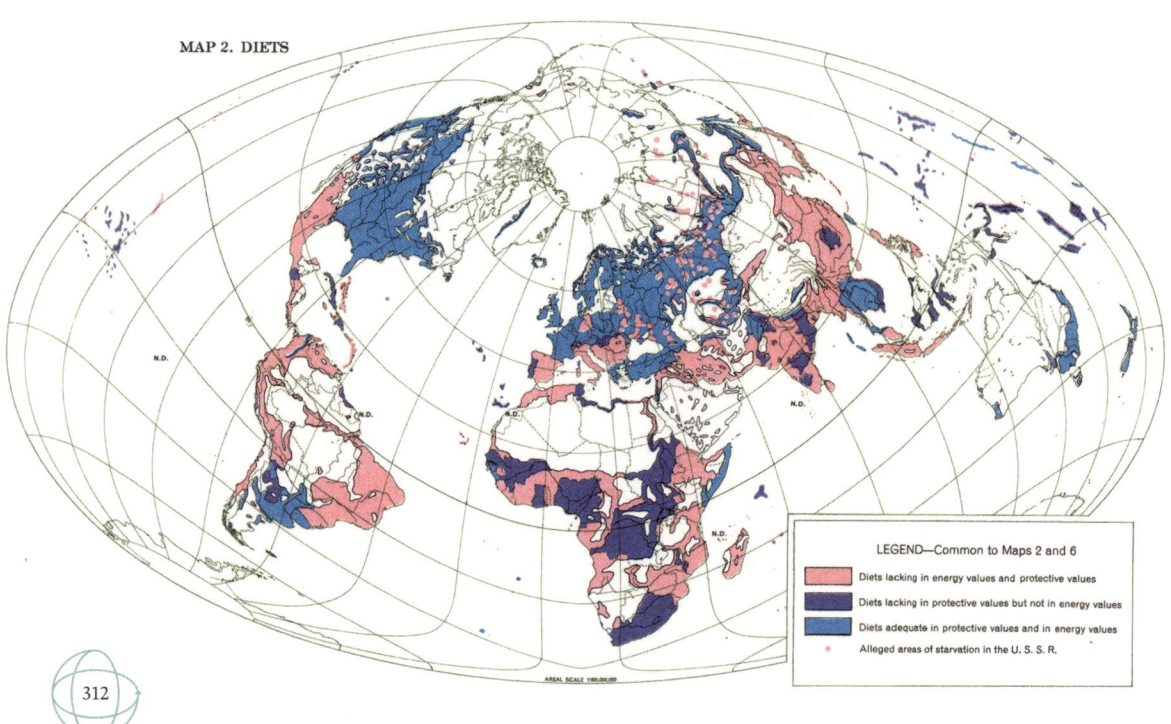

THE OCEAN FLOOR

too bold a color for the sediments.'⁵² The reason for this is that all maps that came before were eclipsed by a relatively unknown (at the time) Austrian named Heinrich Berann. Through his artistry, Heezen and Tharp's ocean floor mapping became a cartographic blockbuster.

I had encountered Berann's work before in my explorations of the Map Library, when I pulled from the Switzerland drawer one of the most remarkable tourist maps of Europe I had ever seen. It was a painting of the entire arc of the Alps sweeping round from Austria along into Spain and France, and then hooking down the spine of Italy with the remainder of the country fading into the horizon. Berann, like Harrison, was a master of perspective. Not only is this map looking south, rather than the usual north, but it evokes the curvature of the Earth and draws you in.

Map 90 The second of two maps of the ocean floor that Richard Edes Harrison created. It's an update on a map he produced for *Fortune* magazine in 1959 in which he asked readers to imagine that 'The surface of the earth has been cut, like the skin of an orange, along the American continents (at longitude 70° west) and pressed flat in such a way that the relative sizes of all areas are free from distortion.' ¹⁴

The Floor of the World Ocean: Association of American Geographers (1961) 68 × 40cm

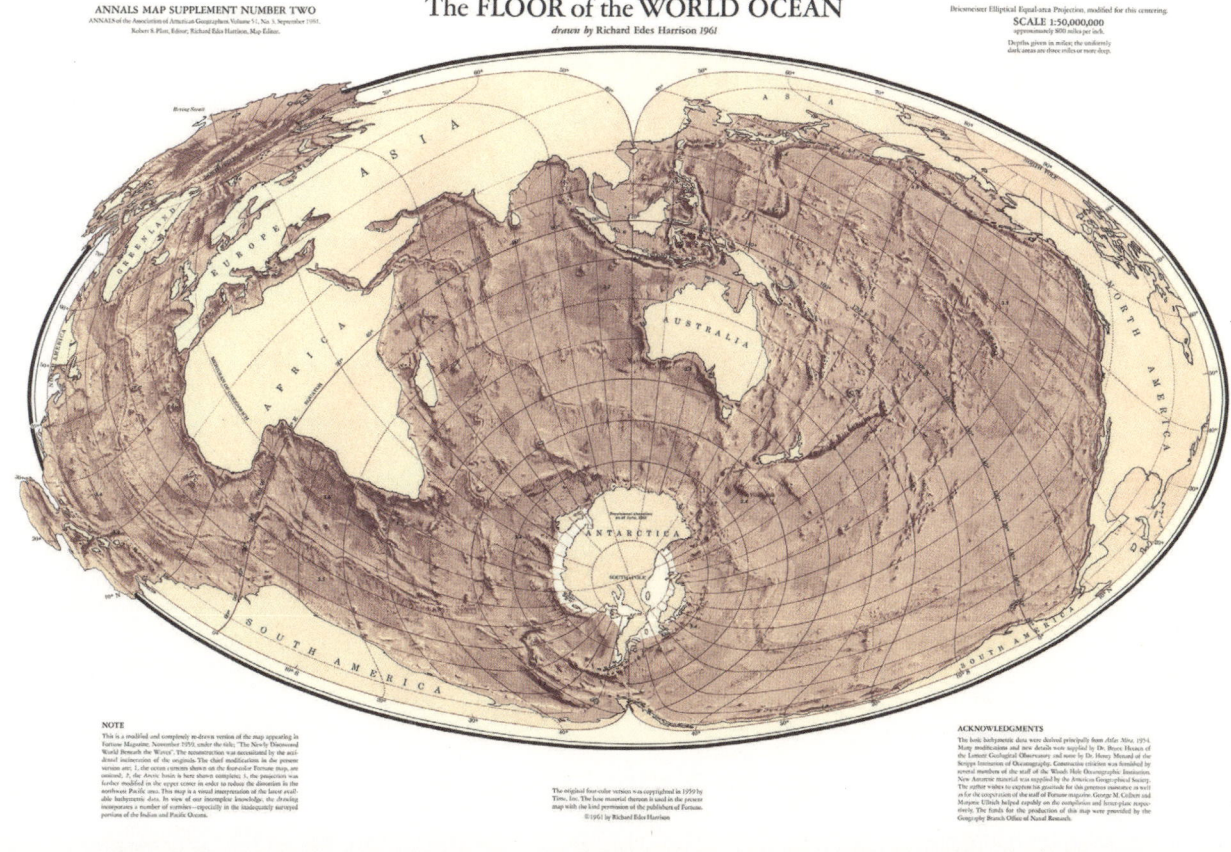

THE LIBRARY OF LOST MAPS

Map 91 [Extract] Berann's stunning vista of the Alps, which was published in the 1960s to entice tourists to the mountain range (perhaps after they'd watched *The Sound of Music*, which was released in 1965).

The Alps – Europe's top attraction all year round: Joint Publicity Commission of the Alpine Countries (1968) 80 × 47.5cm

Berann's map immerses us in the great lakes of Como, Maggiore and Garda spilling out into the plains of northern Italy. We can ride a train through the great valleys that run along the Alps to Innsbruck, to Andermatt, and run our fingers along to the gateways to the Alps such as the Swiss capital of Bern or the big cities of Zurich and Geneva. Follow the base of the mountain range and we come across Salzburg and Vienna in the bottom left corner just above the signature of Berann written in his distinctive style. I reflected how this part of Europe has changed a lot since the 1960s, when Berann painted it, with more development and larger resorts: the gondola lifts that Berann had painstakingly

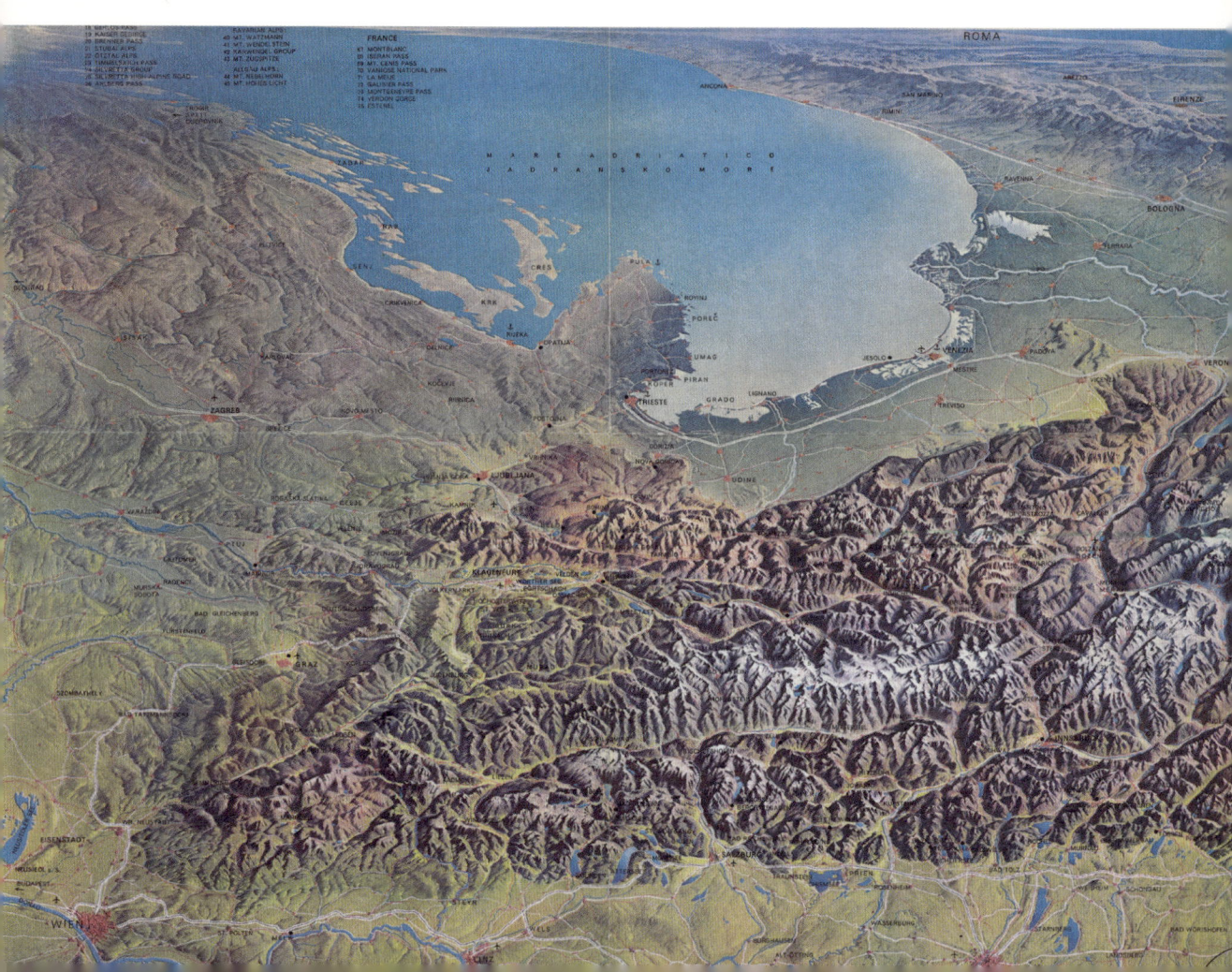

slung from the valleys to the mountain peaks with delicate red lines have now multiplied and the mountain towns that support them have sprawled.

Berann's plunge from the High Alps into the deep ocean was facilitated by *National Geographic,* which first paired him with Tharp and Heezen to create a map of the Indian Ocean. Heezen had shown Tharp's physiographic map of the Indian Ocean to the magazine, but as she recalls[53] 'they liked the map, but they thought they'd like it in a different media'. So the magazine's chief cartographer, Albert Bumstead, got them working up a draft map that could be presented to the editor.

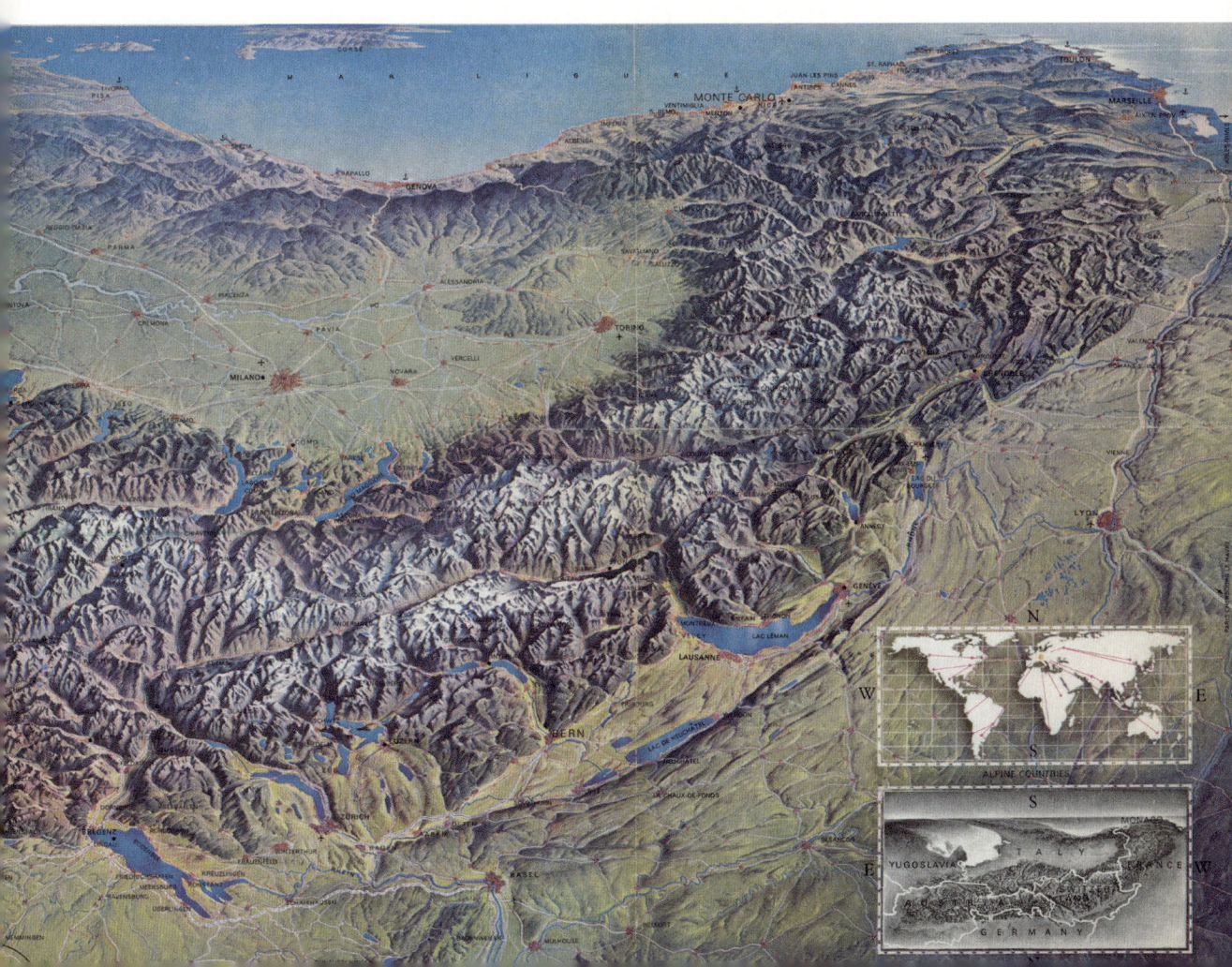

Map 92 [Extract] Berann's first ocean-floor map was produced for *National Geographic* magazine. Published in 1967, it amazed readers with this unfamiliar perspective and shocked them with the sight of great fissures carving up the Indian Ocean.

Indian Ocean Floor: National Geographic magazine (1967) 63.2 × 48.2cm

According to Tharp, Berann had come to be employed by *National Geographic* thanks to 'a letter from a little girl who said, "I've been looking at your maps, and my father can paint better than you can".' Bumstead was sent to Berann's home in Lans, on the outskirts of Innsbruck 'to case the joint'. There 'he found this fabulous man, Heinrich Berann ... He had a beautiful home and a separate studio, and he did serious paintings there. Very serious paintings.' These paintings 'in the style of Leonardo da Vinci', were not how Berann earned a living, however; his stock-in-trade was creating beautiful mountain panoramas for resorts and attractions as well as sweeping vistas of entire mountain ranges – like the map of the Alps I was so pleased to find – for national tourism agencies. Berann's first commission for *National Geographic* was a stunning view of Mount Everest for the October 1963 issue, which celebrated the USA's successful summit of the mountain (under the leadership of Norman G. Dyhrenfurth).

The Indian Ocean map was published in October 1967, with *National Geographic* inviting its readers to:

> Suppose all the water could be drained from an ocean! What would the bottom look like? The remarkable map-painting that accompanies this issue, Indian Ocean Floor, combines scientific discoveries and an artist's skills to answer that intriguing question ... Heinrich Berann shows the round Earth as flat from east to west, but rolling gently away northward, so that most of the vast Asian landmass is visible ... Towering undersea mountains and plunging slopes of this waterless ocean are highlighted as if at sunrise.[54]

This brief article was accompanied by a photograph of Heezen and Tharp with a physiographic diagram (probably of the Pacific,

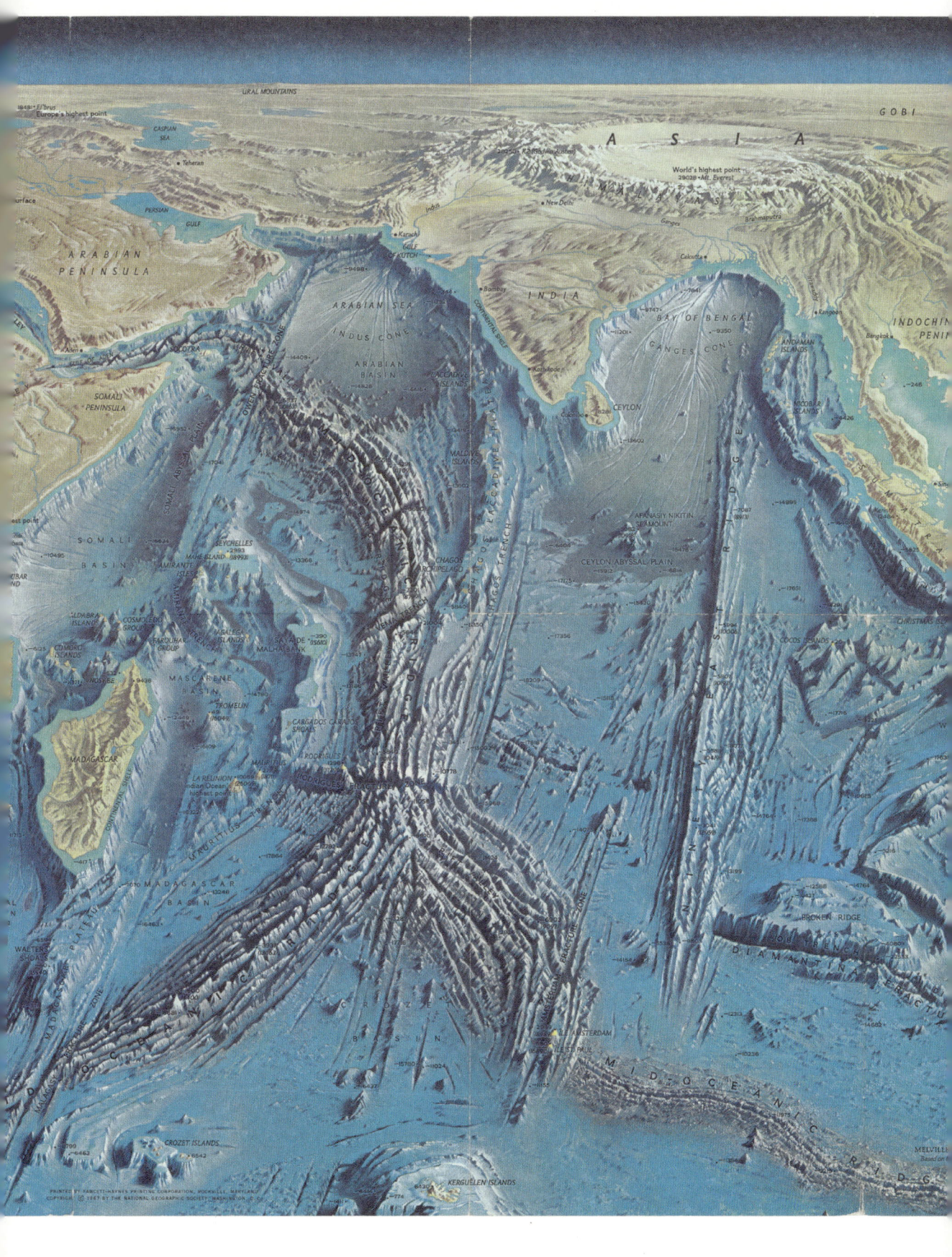

which is what they'd have been working on at the time). Heezen's account of how the map of the Indian Ocean came to be published reveals how close it came to falling onto the cutting-room floor in favour of more newsworthy events that the magazine's editor at the time, Melville Bell Grosvenor, felt should take precedence.

> Grosvenor said: 'Oh, let's dump that ocean floor map and put in a new map of the Holy Lands.' It would be so topical because of the [Six Day] war. For a few weeks our map was dumped from the schedule. The carto-art department of the geographic said to Grosvenor 'Where do we draw the international boundaries now?' They simply couldn't decide so inserted our ocean floor map at the latest moment they could in order to get it in the October issue ... Printing 6 million maps in a matter of weeks.[55]

After publishing the map of the Indian Ocean, the Atlantic followed in June 1968. This pull-out map was accompanied by a much longer article that hints at the developing ideas of sea floor spreading and doesn't hold back on the drama with an arresting image of a giant fracture excising the ocean between the Americas and Europe and Africa.

> Here the S-shaped trough of the Atlantic lies unveiled, drained of its 85,000,000 cubic miles of water. As if suspended high above the earth, the viewer gazes down on an infinitely varied suboceanic landscape on blue-black trenches plunging as far as five miles down; on violet abyssal plains averaging three miles below the surface; and on lighter blue continental slopes and shelves rimming the buff-hued continents.[56]

These unprecedented maps were taking the ocean floor into the homes of the millions of *National Geographic* subscribers, who could piece them together by purchasing any of the maps they didn't have for just $1. After the Atlantic came the Pacific in 1969 (overtaking Tharp's output of physiographic diagrams – her Pacific was published in 1971) and finally the Arctic in October 1971. This was to be the last in the series as, according to Heezen: 'Grosvenor had saturated with physical geography and wanted to go back to pretty girls and plants and calico skirts and things.'[57] Nonetheless, other maps made by Heezen and Tharp did feature in *National Geographic* atlases and desk globes, the last of which was a map of the region around Antarctica in 1975.[58]

While the *National Geographic* maps were being published, debate continued to rage between the fixists and the drifters about the movement of tectonic plates. In 1962, a geologist at Princeton University named Harry Hess pieced the newly mapped features of the sea floor together to propose the idea that became known as 'sea floor spreading'. He spotted that the middle of the oceans were generally shallower, with the greatest depths achieved closer to their edges, for example the 8 km deep Peru–Chile Trench that runs down the western coast of South America. In Hess's view it made sense that the oceans were growing with molten rock oozing from the ridges running along their centre lines and pushing outwards. Over time the older ocean floor became cooler and denser in its journey to the margins of the oceans where it then collided with the lighter continental shelf[59] and was pushed beneath it – forming deep trenches – and back into the mantle where it became molten again.[60]

New features, such as transform faults, were beginning to appear on the maps that further bolstered this idea. These were identified in 1965[61] as smaller faults perpendicular to the

Map 93 [Extract] The tiny island of Surtsey, which lies to the south of Iceland, emerged from a volcanic eruption along the Mid-Atlantic Ridge that took place between 1964 and 1967. It was still growing when this map was drawn in 1964.

Surtsey: Landmælingar Islands (1964)
25 × 25cm

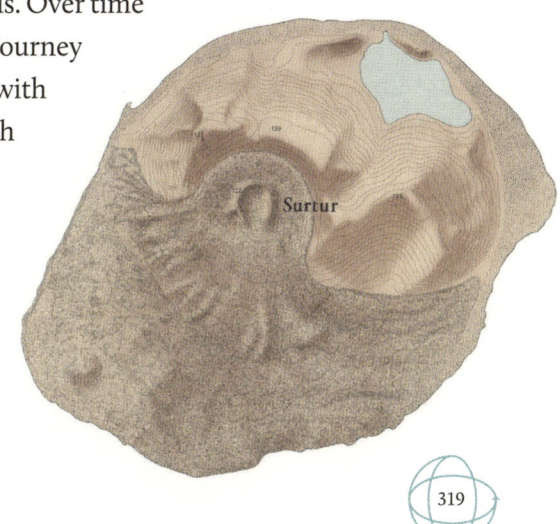

Map 94 This map, produced in 1974 by the Lamont-Doherty Geological Observatory, shows the age of the ocean floor. In one image it offers compelling proof of the idea of sea floor spreading as the youngest rock (in red) sits at the centre of the ocean with the oldest (in purple) approaching the coastal margins.

The Age of the Ocean Basins: The Geological Society of America (1974)
113 × 56.5cm

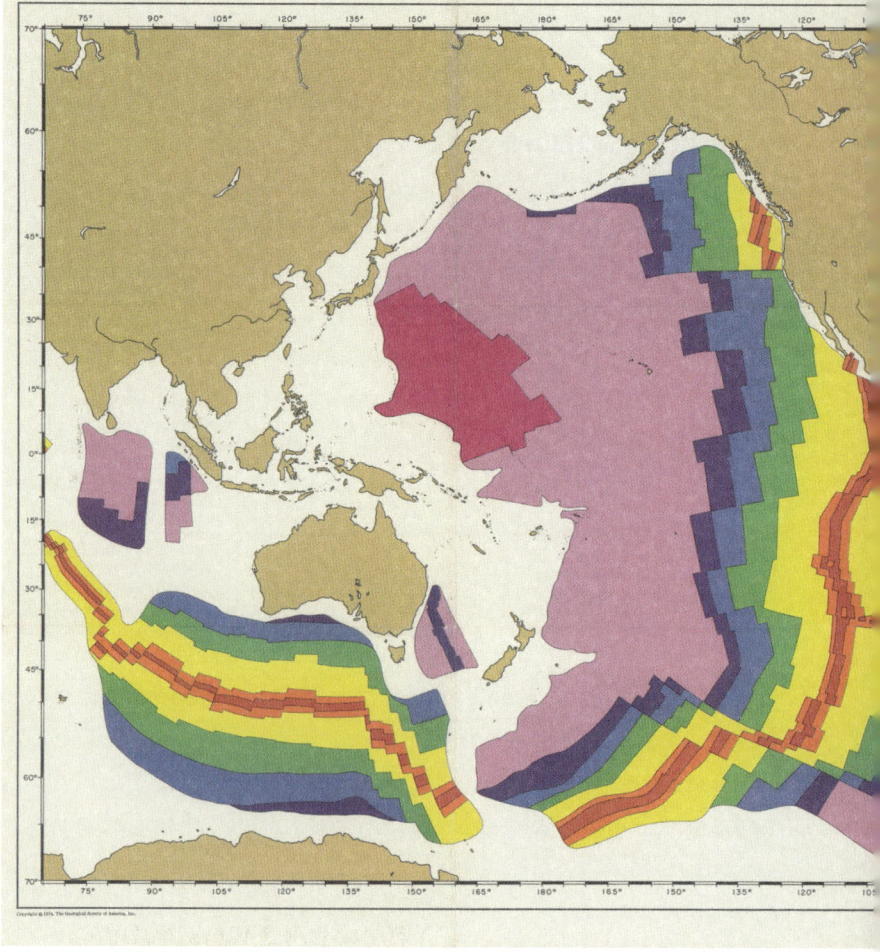

main fault line caused by the Earth's crust moving at different speeds along the length of the fault and fracturing where some parts are moving more quickly than others. This gives the Mid-Atlantic Ridge a step-like appearance, which became much more obvious in the later maps to which Tharp and Heezen contributed data from the early 1960s onwards. The *National Geographic* article accompanying the Atlantic maps offers an evocative description of 'scores of east-west fractures, as if slashed by a gigantic cleaver'.[62] Knowledge of these faults was combined with maps of the age of the Earth's crust, which in another technical

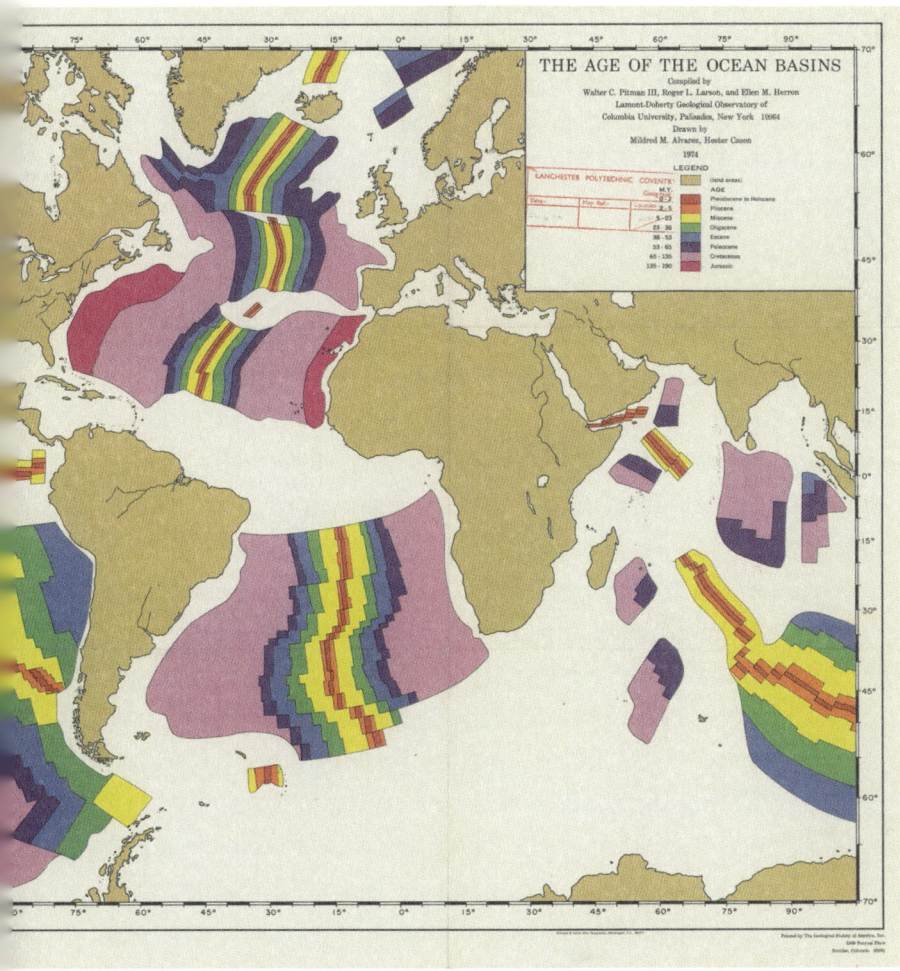

advance could be determined by the magnetic characteristics of the rock,⁶³ to show how it aged the further from the central ridge it was.

By the late 1960s, the integration of evidence from sea floor spreading, paleomagnetism, seismic activity and geology led to the formalisation of the theory of plate tectonics. This confirmed that the Earth's crust was fractured into distinct plates that were bobbing around on its mantle. Plate tectonics explained not only the movement of continents but also the formation of mountain ranges, oceanic trenches, earthquakes and volcanoes.

A decisive paper that left the drifters in no doubt that the fixists had lost their argument was published in 1968[64] by Xavier Le Pichon, a French geologist working at Lamont alongside Heezen and Tharp. Le Pichon was able to put the layers of evidence into a geometrical model that mathematically proved how the movements of the plates were not just possible but had resulted in the features of the Earth's surface that were being mapped. This then allowed him to go back in time and reconstruct the historic movements of the tectonic plates. Le Pichon's 'kinematic model' completed the slew of evidence that was needed to prove Alfred Wegener's original idea of a once great continent – Pangaea – that had shattered into the shapes we now see on the world map.

When Le Pichon was writing his paper he knew that he was going to move from the Lamont Observatory back to France to set up his own research group[65] and he needed a way to raise the profile of oceanography in his home country to help attract research investment. Seeing the success of the Heezen and Tharp maps of individual oceans, he thought it a good idea to produce a global map of the ocean floor – something that hadn't been published since Richard Edes Harrison's 1961 update of his pioneering 1959 *Fortune* magazine map. So in late 1968/early 1969 he commissioned an illustrator named Tanguy de Rémur to do just that, and gave him the available maps including the physiographic maps that Heezen and Tharp had completed up to that point.

Le Pichon recalled to me that he left de Rémur to combine the various maps without much supervision, save for the way that the Southwest Indian Ridge was to be depicted. He was concerned that Heezen and Tharp had exaggerated the transform faults on their map of the region, so he told de Rémur to tone them down, a recommendation that Le Pichon notes he implemented in 'a drastic way, which makes this ridge look quite bizarre'.[66] The map was published in French in 1973 and, to the untrained eye, there

THE OCEAN FLOOR

is nothing amiss: it was a fabulous depiction of the oceans that ignited the interest of the French in just the way Le Pichon had hoped.

In the back catalogue of Heezen and Tharp's maps, de Rémur's map is often mistaken for one that they had a direct hand in. This is understandable because it has clearly drawn inspiration from the *National Geographic* maps and, what's more, a later edition was published in English under the banner of the American Geographical Society that is credited to Heezen and Tharp. I found a version of this map in the Map Library at UCL. In fact, they were unhappy with the lack of credit on the original, which resulted in them gaining permission to publish the map under their own name and donate the proceeds to the AGS. I couldn't find any reference to the map in the Heezen-Tharp Archive and the AGS only has a brief letter from the director to Heezen confirming the delivery of 1,000 copies that were to be sold at $10 each.[67]

I suspect that Heezen and Tharp were holding out to complete their ocean-by-ocean run of maps with *National Geographic* before turning their attentions to a single, unified view of the world.

(Overleaf)

Map 95 The American Geographical Society edition of the Tanguy de Rémur ocean-floor map, which was first published in French in 1973.

The Floor of the Oceans: American Geographical Society (1977)
96 × 54.5cm

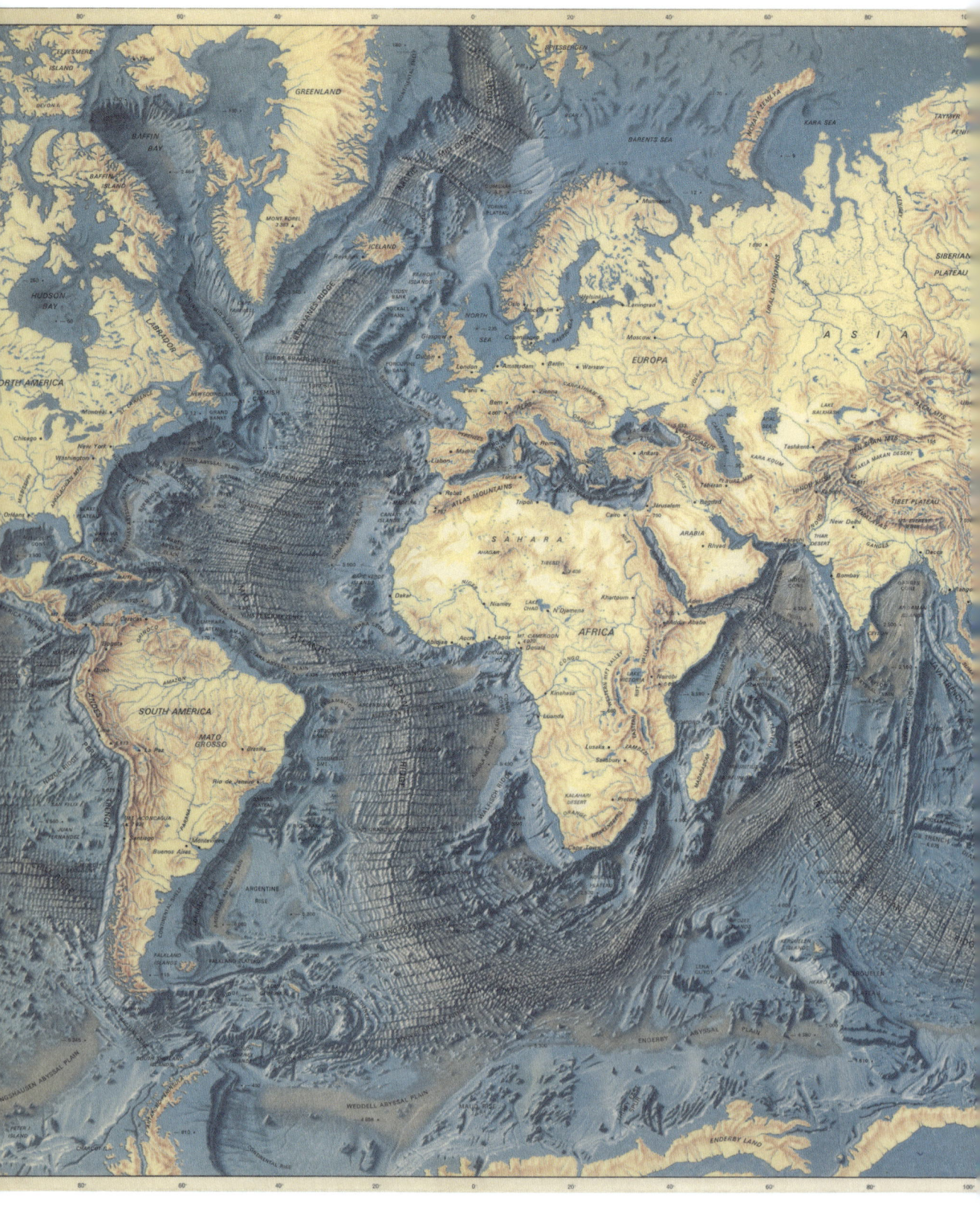

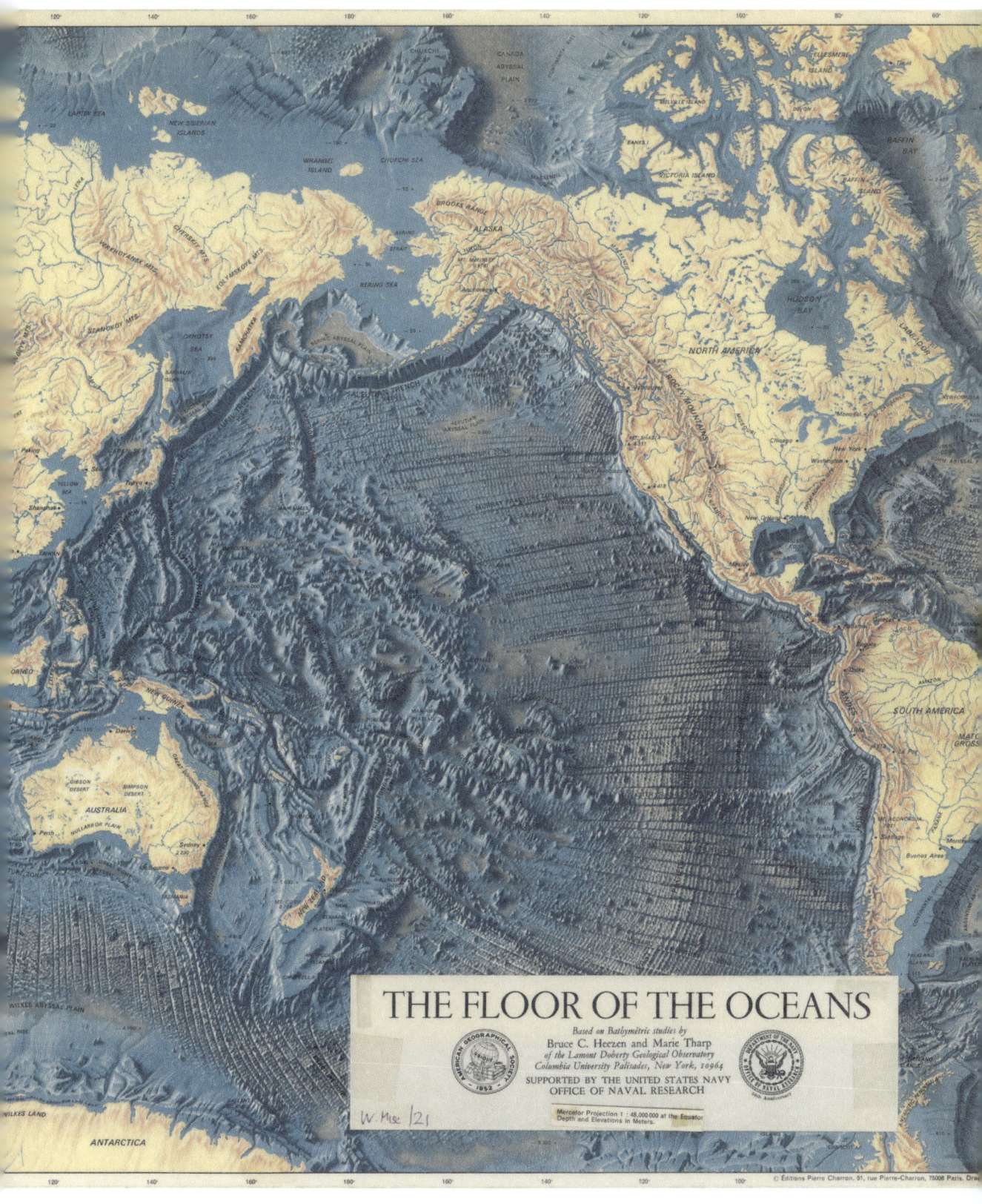

THE FLOOR OF THE OCEANS

Based on Bathymétric studies by
Bruce C. Heezen and Marie Tharp
of the Lamont Doherty Geological Observatory
Columbia University Palisades, New York, 10964

SUPPORTED BY THE UNITED STATES NAVY
OFFICE OF NAVAL RESEARCH

Mercator Projection 1 : 48,000,000 at the Equator
Depth and Elevations in Meters.

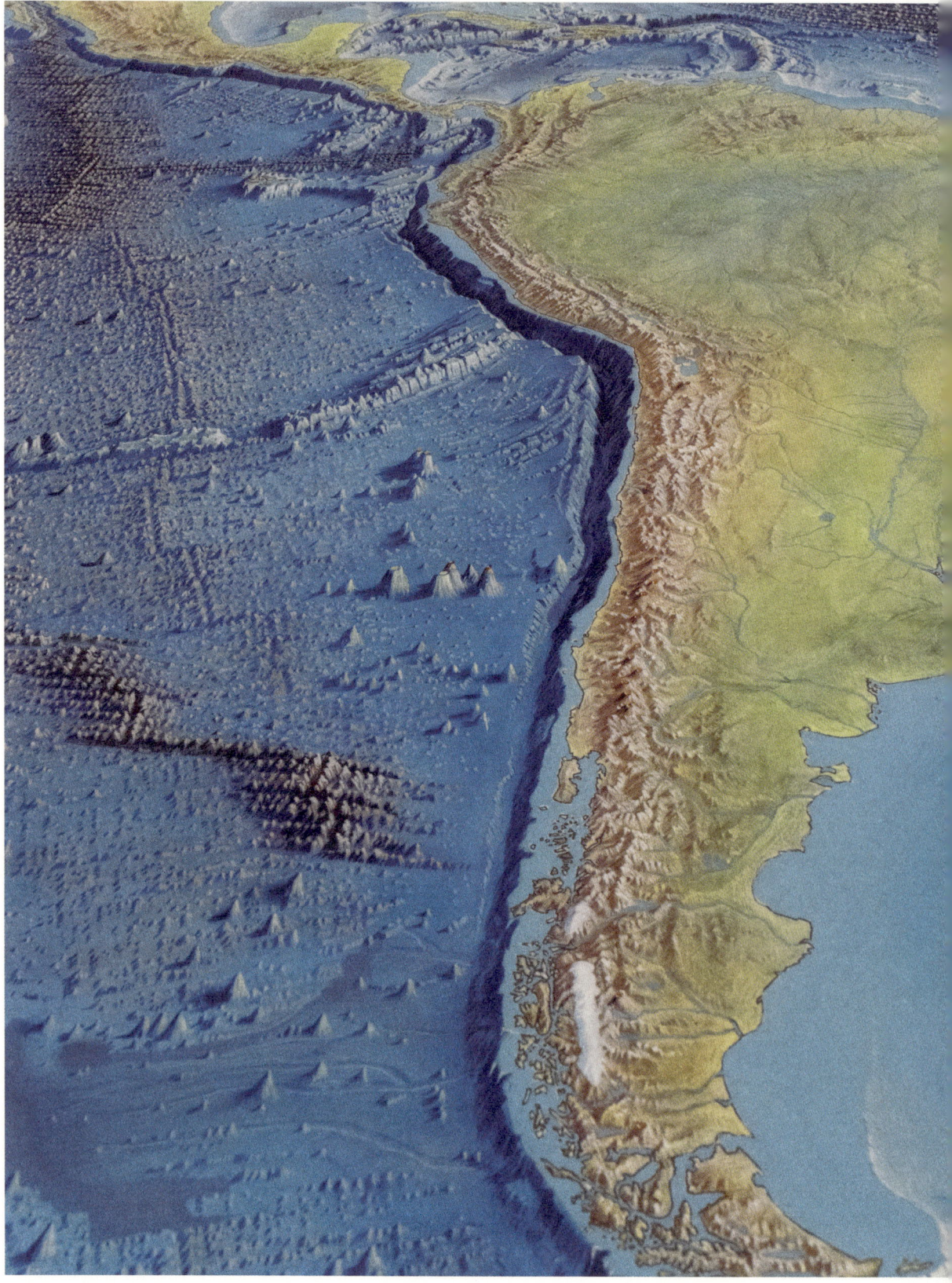

THE OCEAN FLOOR

The World Ocean Floor Panorama

A map of the entirety of the ocean floor was a vision that Bruce Heezen and Marie Tharp would have long held, as it was the only way to portray the continuous set of connecting rift valleys that span the planet, which Heezen had inferred from Tharp's plot of earthquakes way back in 1952. But without the backing of *National Geographic*, they were short of funds, especially to pay Heinrich Berann, with whom they had loved working. So they turned to the US Office of Naval Research, an organisation with which they had previously had links, and in 1974 it fortunately agreed to fund their project.

The ambition of the map should not be underestimated: it was much more labour-intensive than Richard Edes Harrison's pioneering efforts, or the more recent map by Tanguy de Rémur. Put crudely, those maps were compositions and copies of other maps, and so did not consider the underlying data. By contrast, Tharp and Heezen would go back to first principles and the basic soundings to further improve their depiction of the ocean floor. No map of theirs was ever finished and they did not take the maps produced by others at face value. They sought the original soundings where they could.

Berann, who at this point was well used to working with them, threw himself into the project. He constructed a specialist table to hold this giant map, and he and his assistant, Heinz Vielkind,

Map 96 [Extract] A section of the exquisite hand-painted *World Ocean Floor* that now resides in the Library of Congress. Berann was a master of drawing mountains either above sea level or below and it's hard to believe this is a painting and not a three-dimensional model.

World Ocean Floor (1977)
191.5 × 106cm

THE LIBRARY OF LOST MAPS

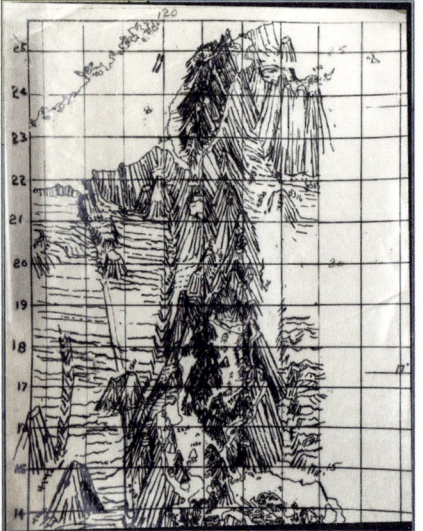 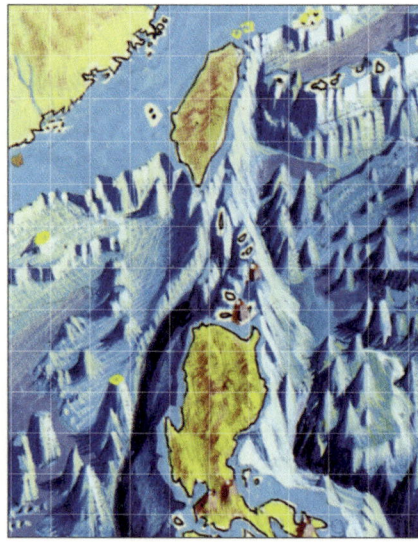 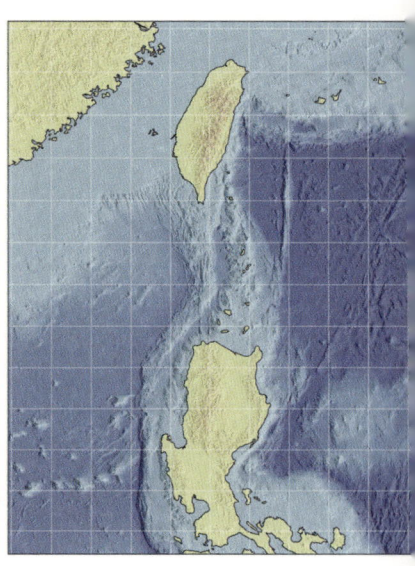

On the left is one of Tharp's physiographic sketches of a part of the ocean around Taiwan and the northern tip of the Philippines. In the centre is Berann's interpretation of it and on the right is a map I created from the latest available data for the region.

began blocking out the shapes of the continents to frame the ocean that was to come. Next came the task of transferring over the sketches that Tharp had produced, working through them in tiny sections to slowly fill the map. Berann and Vielkind would toil on this for long periods at a time, and then, over the course of a number of visits to their studio near Innsbruck, Heezen and Tharp would peer over the map and offer feedback.

Photographs from the time show the work underway, with one captioned 'Bruce discussing a fine point with Heinz Vielkind during the preparation of the World Ocean Floor Panorama'. The work was intense and there were changes that needed to be made that often resulted in redrafting whole sections of the ocean. But since they were all perfectionists and entirely committed to the project, they seemed extremely relaxed in each other's company. One photograph shows Berann spooning an enormous serving of gateau into Tharp's mouth; he nicknamed her 'map frog' after she laid out her charts on the studio floor and then 'jumped around them'.[68] This was a huge contrast to the stress that Tharp and Heezen felt working back home at Lamont.[69]

In May 1977, the map was finally finished and Heezen and Tharp flew with it back to the USA. Because the map had been funded by the US Office of Naval Research and not a magazine, they did not benefit from the printing expertise that came with their *National Geographic* maps. It therefore fell to them to arrange every step in the production process, and they had the odd stumble along the way. Perhaps the largest was an over-estimation of the size of the biggest printing plate available so they could not print it the full size they had intended. Rather than crop the edge of the map, they chose to reduce the overall size of the map, resulting in the slightly awkward scale of 1: 23,230,300.[70]

The first proof of the map arrived just in time for Heezen and Tharp to see it before they headed out on different research cruises around Iceland. Heezen offered some initial feedback, saying that the 'colours were too red and too limey green on the land, too turgid on the seas'.[71] Despite necessary tweaks, this proof was a major step towards the completion of their most ambitious project together.

Tragically, it was also to be their last. Heezen died from a heart attack aged just fifty-three while in a submarine. He was exploring the Reykjanes Ridge, part of the Mid-Atlantic Ridge system, to which he and Tharp had dedicated so much of their lives. Tharp was given the news over the radio as she was still at sea on a different vessel, studying the ridge from above, when it happened.[72]

She then faced the harrowing task of handling Heezen's estate, which included settling the costs of the repatriation of his body, as well as dealing with his vast (and disorganised) archive.[73] But this did not deflect her from her work, and she pushed on with the map. After all, it was now so much more than a depiction of the sea floor: it was an epitaph to Heezen. Labels had to be added, and more colours corrected, but Tharp got there. The map went into print on 17 May 1978, a year after Berann had handed it over to Tharp and Heezen.[74]

Heezen was honoured under the map's title with the words 'Published by the United States Navy as a memorial to Dr. Bruce C. Heezen in recognition of his contributions to man's knowledge of the world ocean floor'. The map itself would go on to be printed in various editions and Tharp sold it under her company Marie Tharp Maps LLC. It commonly appears in the UK as a special smaller edition shared by the Open University in the mid-1980s. The map's iconic status means that even its later editions can attract surprisingly high valuations by antique map sellers.[75] Thanks to the Library of Congress posting a digital copy of the Berann map (without the labels) online, there are also hundreds of print-on-demand retailers offering it to those wishing to have their own version of the map for display.

In October 1978, Tharp and Heezen were honoured with the Hubbard Medal of the National Geographic Society, its highest award, in recognition of their pioneering work. But, despite this, at the back of Tharp's mind was the pressing concern of how the work could be continued in the absence of any further funding. Heezen's mother was still alive, and Tharp wrote to her a couple of weeks after the award ceremony requesting a fund of $100,000 to enable her to continue to revise the World Ocean Floor map over the next three to five years and to finish all the physiographic diagrams she had started with Heezen.[76]

Very sadly, no funding was offered, and a revised World Ocean Floor map never appeared. Hali Felt, who wrote a biography of Tharp, recalls: 'As several people told me, no one was willing to place a large bet on a Bruceless Marie. It didn't matter how important her (or their) work had once been, her time it seemed had passed.'[77]

Knowing this, it felt bittersweet as I waited excitedly for Berann's hand-painted map to be brought up from the vaults of the Library of Congress. It was the last map I saw in my time there and it was one hell of a finale. I entered the manuscript room as the librarian was gently unfurling it, carefully weighting down the leading edge before unspooling the map from its acetate cover. The grey, windowless room suddenly lit up with the blues of the ocean as Berann's iconic colours were illuminated by the neon strip lighting.

The map itself is roughly 191 × 106 cm and painted onto thick paper, with pin holes in the corners from when it was anchored to Berann's special drawing board. It has a slight tear halfway down its left side, but well away from the map itself. I spent three joyful hours pretending I could fly over the Earth's surface, recognising parts of the ocean that I'd encountered through Tharp and Heezen's eyes in the records over the previous days. A chunk of the Caribbean that was a mass of sheets of messy, smudged contours in their notes is on the map in glorious clarity, the warm sea beckoning me in, then over to the patch of 'rough terrain' that Tharp had scrawled on one of her drafts and Berann had rendered in vivid high definition.

I thought of the thousands of hours that went into creating this glorious map – from the researchers and crews of the survey vessels who spent months in rough seas testing the depths, to the data processors and plotters creating the profiles, to Tharp herself working along each one imagining the sea floor and then transferring it into her own underwater mountain ranges. Endless back and forth and discussion before handing over to Berann to begin the cycle again, in his studio surrounded by the Alps. I had Tharp's words ringing in my ears:

(Overleaf)

Map 96 [Extract] The 'World Ocean Floor' painting by Berann. This is a scan from the Library of Congress and lacks the labels and title that Tharp added to the final print version. I prefer it this way!

World Ocean Floor (1977)
191.5 × 106cm

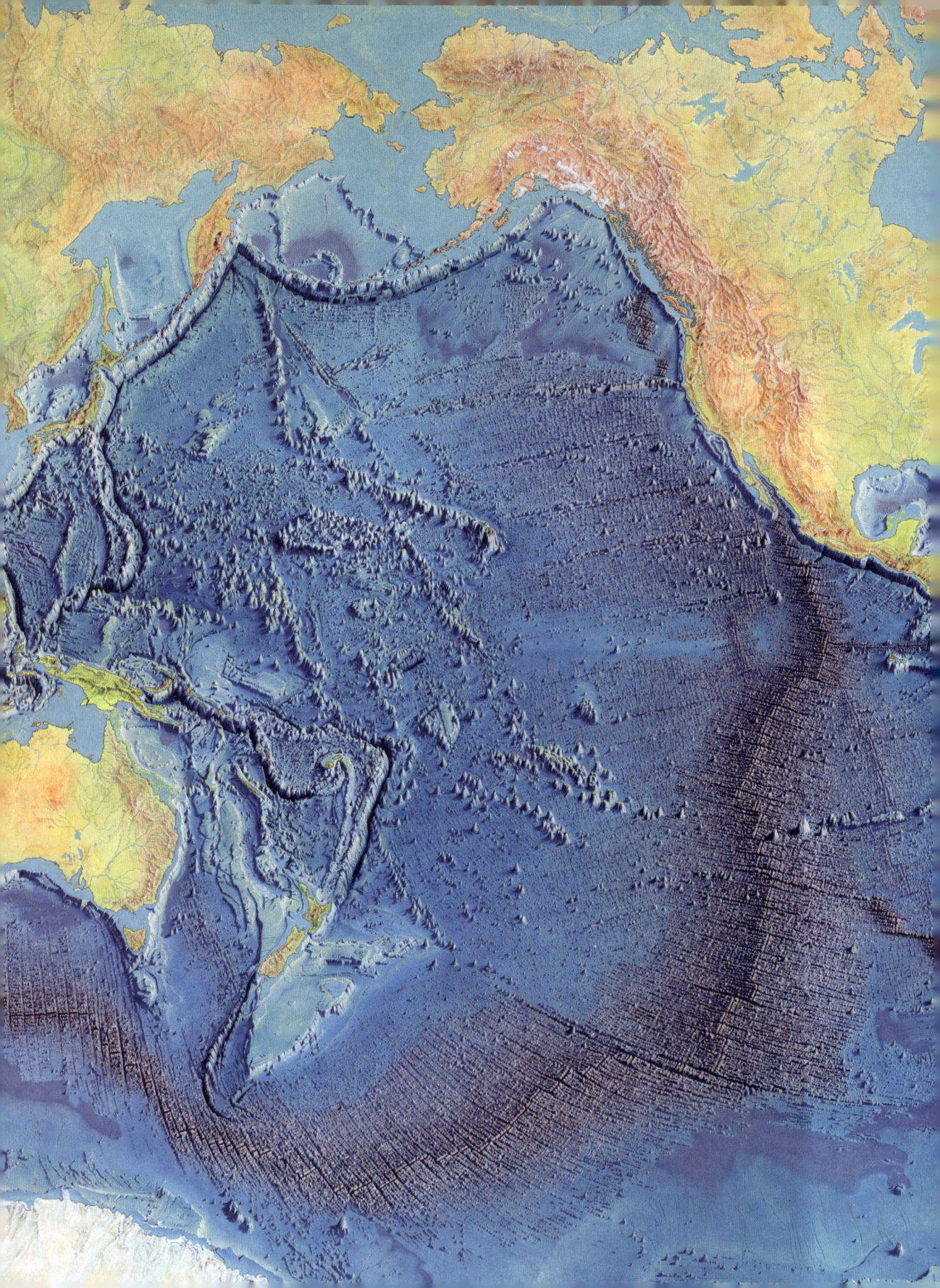

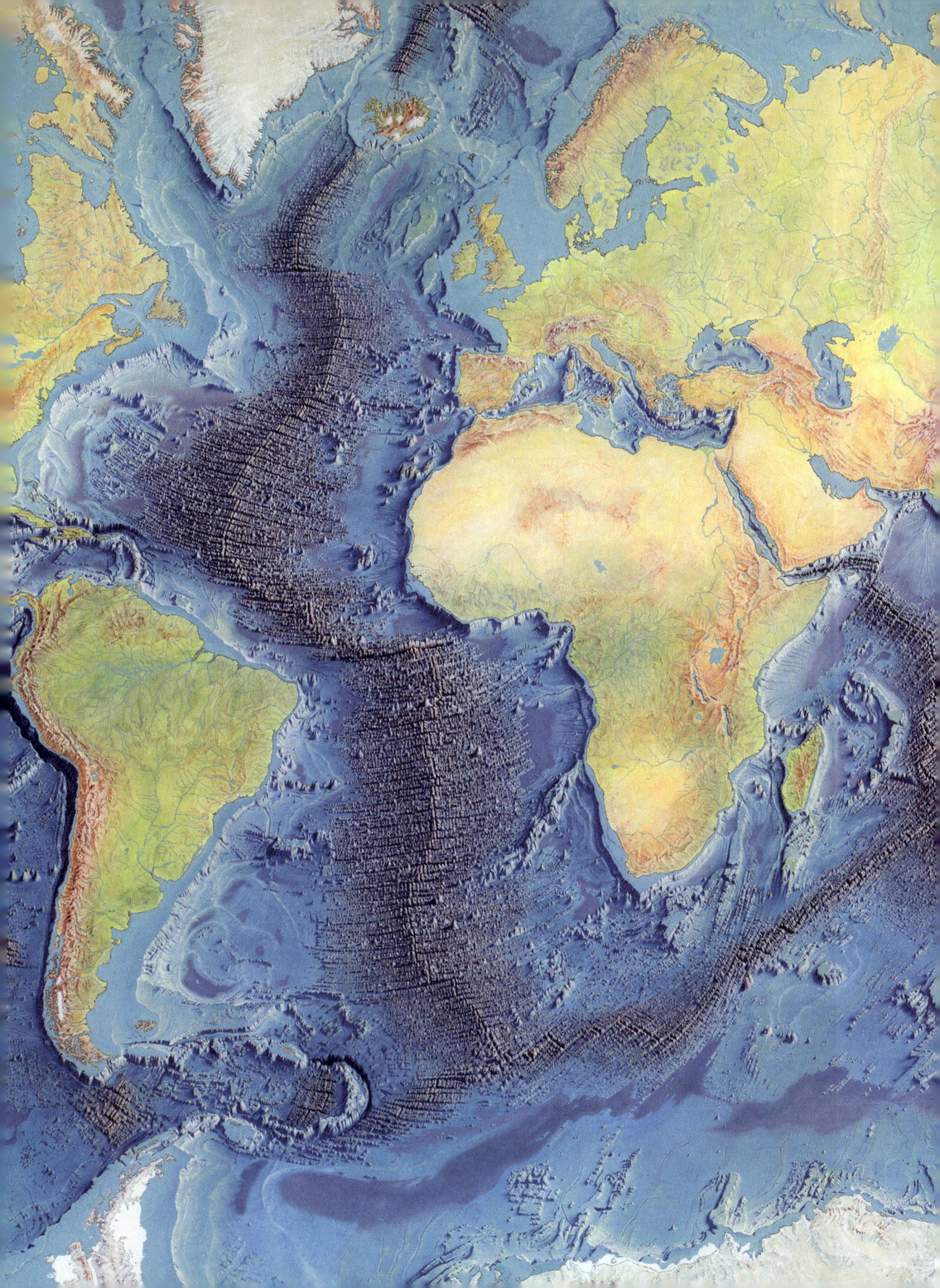

I worked in the background for most of my career as a scientist, but I have absolutely no resentments. I thought I was lucky to have a job that was so interesting. Establishing the rift valley and the mid-ocean ridge that went all the way around the world for 40,000 miles – that was something important. You could only do that once. You can't find anything bigger than that, at least on this planet.[78]

The texture of the paint intensifies the remarkable 3D effect that Berann was able to achieve and the minuscule details are incredible – looking closely, I could see the embers of volcanoes atop sea mounts barely larger than a pinhead. Stepping back from its tiny glow, I could follow the ridges of the plate boundaries resembling the joints of a skull. On land, the topography is just as intricate, and so subtle in its colours that it was as if the entire planet was bathed in intense sunshine. The bright white snowcapped peaks feed crystal-clear rivers that flow into the depths of the oceans, alive with terrain. I felt like I was seeing the Earth reborn.

Heezen said: 'What people think the bottom of the ocean looks like, that is what most scientists and most informed laymen think it looks like, is what Marie thinks it looks like.'[79]

What's so extraordinary is that this is still the case today. Despite the maps being created half a century ago, the World Ocean Floor map is the top of the Google search listings for ocean-floor maps and even the most modern digital outputs can't compete with its beauty. Before this map was even created, Tharp and Heezen estimated that there were more than 50 million maps based on their efforts in circulation.[80] Today, billions of people must have seen the world through Tharp's eyes.

I came away from my time in the archive poring over Heezen and Tharp's notes feeling an appreciation of where their brilliance lay. They had an immense love and respect for one another – camaraderie that not only spurs creativity but also allows for resilience in what can be a petty and sniping world of science and a clear sense of self-belief. But, most importantly, they were able to fill their heads with literally oceans of data, process it and then, through thousands of drawings across hundreds of metres of paper, convert that into a visual representation. They were uncompromising on data quality and constantly drafting and redrafting. They could see patterns missed by others only because they worked so intimately with the data. They were not outstanding artistic cartographers, but they didn't have to be as those who were, like Harrison and most impressively Berann, found their way to them. Berann (and his assistant Vielkind) transformed their data into something beautiful so that it no longer spoke just to geologists but also to the public at large.

Heezen and Tharp's records are special because they reveal how the maps I was so astounded to find in the Map Library came to be. They are also poignant as they are intensely personal. One transcript I came across notes the 'woof woofs' of their dog Inky II[81] in the background; Heezen offers reflections on the geology of a beach while 'half stoned on rum'.[82] Other transcripts I encountered might open with the note 'music in the background' as thoughts were gathered after they had had dinner together, or were rendered unintelligible by Heezen recording rambling thoughts while driving.

This was an important reminder of the human beings behind the maps, a reminder that is even more significant because these maps of the ocean floor were the last high-profile scientific maps to have been produced entirely by hand (save for some data processing): as computer mapping technology was already beginning to take over as the first World Ocean Floor map rolled off the printing press.

THE LIBRARY OF LOST MAPS

It was a period of history that, to me, was the pinnacle of what could be achieved with maps that were, as the introduction of the *Larousse Atlas* set out, different from those that came before. We tend to think of maps as useful or as beautiful, but rarely as revolutionising science. The quest to map the ocean floor shows how map-making can have huge real-life scientific implications and the way that it only takes a few remarkable people – and one remarkable woman in particular – to utterly transform how we see the world.

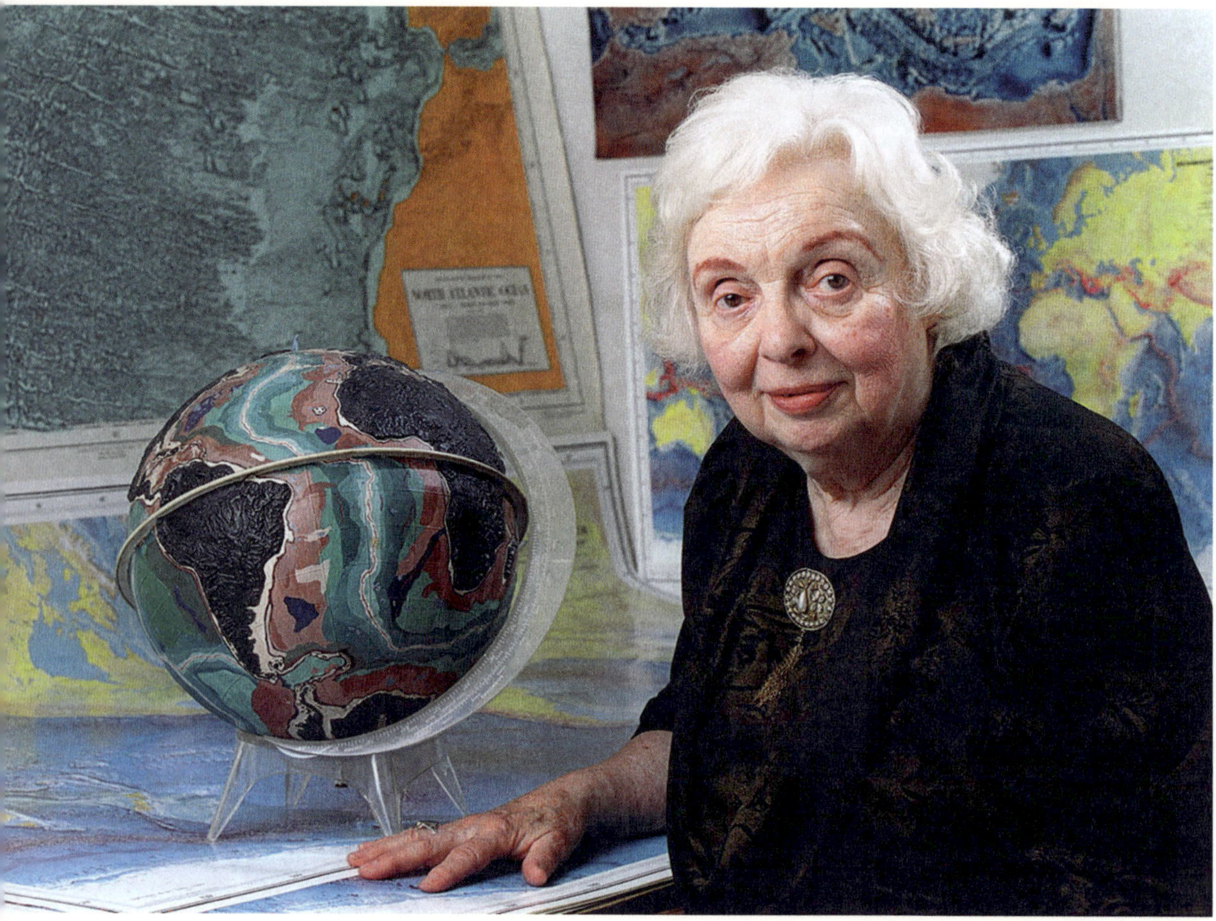

Tharp sat alongside Heezen's painted globe and surrounded by the maps of her extraordinary life.

I'll leave the final words on this to Marie Tharp, who articulates this moment better than I ever could because she helped to create it:

> Not too many people can say this about their lives: The whole world was spread out before me (or at least, the seventy percent of it covered by oceans). I had a blank canvas to fill with extraordinary possibilities, a fascinating jigsaw puzzle to piece together: mapping the world's vast hidden seafloor. It was a once-in-a-lifetime – a once-in-the-history-of-the-world – opportunity for anyone, but especially for a woman in the 1940s. The nature of the times, the state of the science, and events large and small, logical and illogical, combined to make it all happen.[83]

Persons in private households: percentage aged 15-44

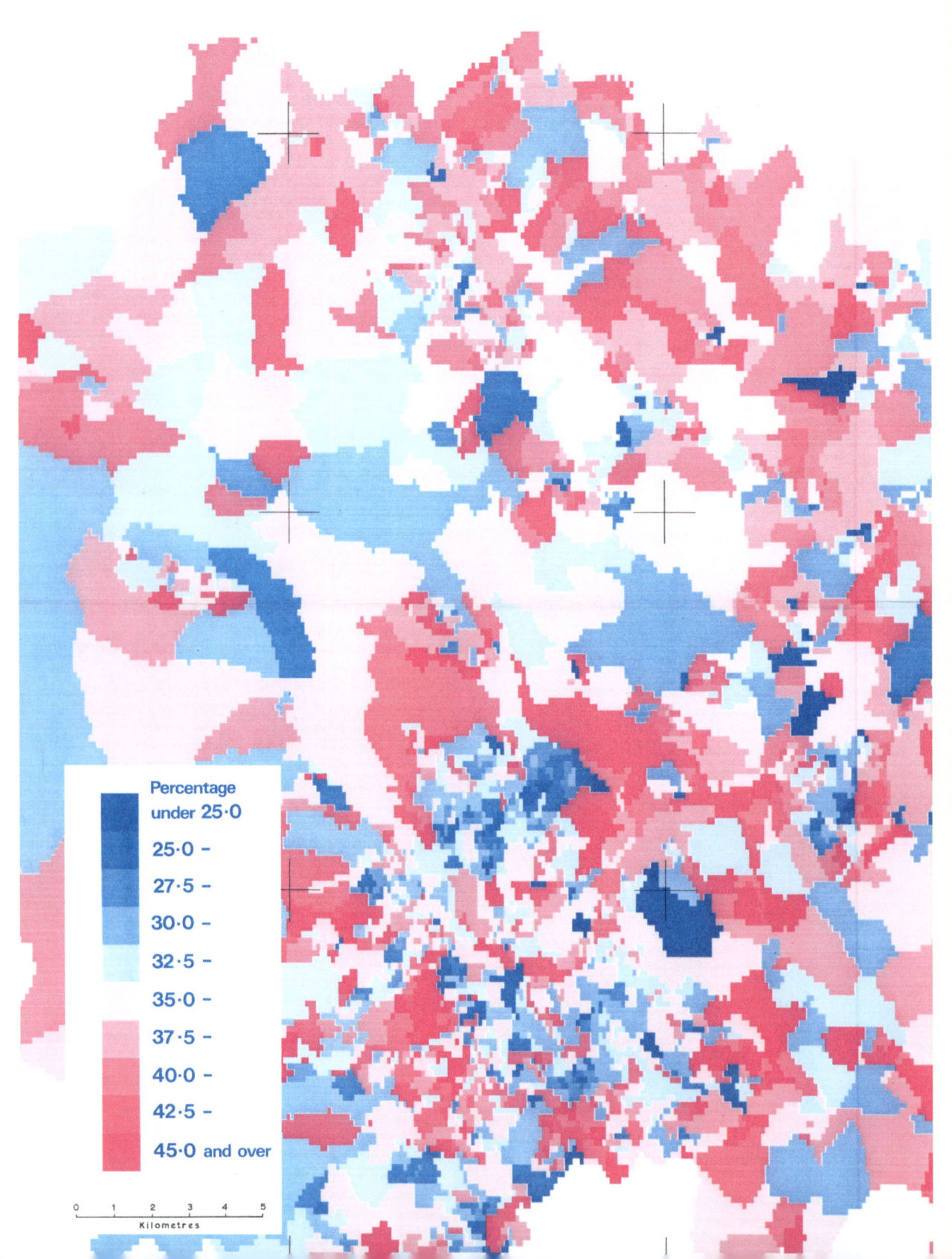

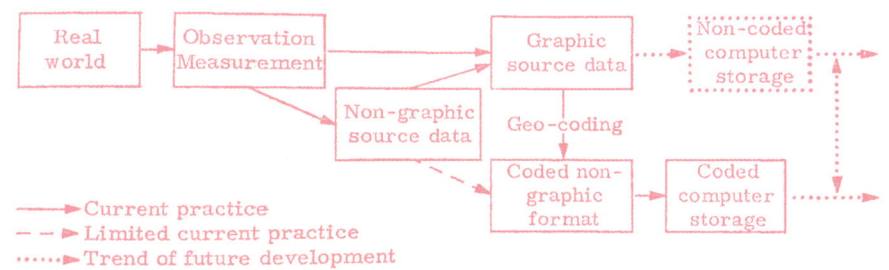

Fig. 4.1. Developments in data flow.

CHAPTER ELEVEN

Maps Go Digital

'*All* maps are a type of argument. Each requires the arrangement of selected information, and no map can be understood apart from its purpose.'
Susan Schulten[1]

Just when Marie Tharp was mapping the ocean floor by aligning layers of paper on a light table in the mid-1960s, a former RAF pilot named Roger Tomlinson was busy layering data that captured the patchwork of land cover types across Canada. But rather than doing this with pencil sketches on tracing paper, he was using a computer. Tomlinson had created one of the first Geographic Information Systems (GIS),[2] which is the genre of software that still powers digital maps today. It was called the Canada Geographic Information System (CGIS) and it was designed to help inform land use planning. Tomlinson's software meant that data could be quickly retrieved, and different calculations performed across layers of information. For example, if Tomlinson wanted to know

Map 97 [Extract] This map is taken from an early computer atlas: the *Census Atlas of South Yorkshire (UK)*. Published in 1974, it required one of the most powerful computers available in the UK at the time and was published to mark the creation of the metropolitan county of South Yorkshire.

The flow diagram shows the workings of Roger Tomlinson as he set about refining the computational processes that could capture, process and display geographic data.

Census Atlas of South Yorkshire (Persons in Private Households Percentage Aged 15-44): University of Sheffield (1974)
79 × 58cm

[Above] Tomlinson, RF: *Geographical Information Systems, Spatial Data Analysis and Decision Making in Government* (1974)

where best to open a new area of land for forestry he could request simple maps of the soil type, climate, slope, protected areas and so on and the GIS could overlap these and determine which land parcels would work best for tree growth.

Extraordinarily, the Map Library would have been known to Tomlinson as he completed his PhD research within the UCL Department of Geography in 1974. His thesis was titled *The Application of Electronic Computing Methods and Techniques to the Storage, Compilation, and Assessment of Mapped Data* and in it he formalised many foundational ideas, like how to digitally encode different features on the Earth's surface (linear features such as rivers are different from areal features such as lakes, for example), that fifty years on continue to govern the use of GIS technologies. Tomlinson's thesis had sat neglected on the department's shelves until an eagle-eyed colleague spotted it.[3] I arranged for it to be digitised and put online, where it has been downloaded thousands of times.[4]

Tomlinson's PhD became the rulebook for how a GIS could operate around the time others were developing their own methods of making maps digital. For example, the US Census Bureau's DIME program digitised all US streets for the 1970 Census, an amazing achievement for such a large country, while the UK's Experimental Cartography Unit (ECU) created the world's first computer-made maps for a regular series in 1973.[5] The ECU's maps were so cutting edge that they were four times more expensive than the manual maps of the time,[6] but their potential was clear and, as with any new technology, the costs soon came down. Major national mapping agencies, including Britain's Ordnance Survey and the US Geological Survey, began using computers to streamline map editing and by the 1970s most major agencies had some level of computerisation. Ordnance Survey was the first to achieve full digital map coverage, but, given the size of the task, this was not until 1995.

Meanwhile, the US military drove the early development of the Global Positioning System (GPS) in the 1980s, which meant precise locations could be determined without the need for traditional survey techniques, though the equipment was still bulky and expensive. The 1980s was also a turning point for GIS because computers became smaller and more affordable, expanding their use for map-making.[7]

As with many innovations, early attempts at digital cartography were viewed with suspicion and had to be robustly defended. In a 1968 editorial, a pioneer of computer-based quantitative mapping, William Bunge (who I quoted at the beginning of Chapter 8), felt he had to address the sceptics head-on, writing:

> The older geographers, those that were horrified at our initial furious attack on maps as inferior to mathematical functions, had a substantial position on their side. But they were and are so religious about their commitment to the map – complete with religious persecutions for those that did not genuflect before the fundamentalist map thumpers – that they practically compelled our revolt. We were provoked.[8]

The maps Bunge and his colleagues were 'provoked' to create prioritised analytical sophistication over cartographic beauty and were a marked visual contrast to the apparent detail of the hand-drafted maps that had come before. This irked more traditional cartographers who, quite reasonably, were proud of the intricacy their hand-drawn efforts could achieve in contrast to the clunky computerised maps. It has only been in recent years that completely computer-driven cartographic outputs have been able to combine both the analytical power of data-driven mapping and

THE LIBRARY OF LOST MAPS

Map 98 The Map Library stores many of the first atlases to be made using computers. For example, this is a sheet from an atlas that was created in 1971 by Kenneth Rosing and Peter Wood, Roger Tomlinson's PhD supervisor, using SYMAP. In contrast to its high-tech production, its topic is the 'Percentage of Households Without Access to a Hot Water Tap' in the West Midlands (UK) from the 1966 Census.

Percent of Households Without Access to a Hot Water Tap: National Computer Centre (1971) 60 × 58cm

the cartographic beauty of many of the hand-drafting pioneers.[9]

Throughout these developments one company has been at the centre of the world of GIS. The Environmental Systems Research Institute (ESRI, now Esri) was founded by Jack and Laura Dangermond in 1969 and was inspired by the work of the Harvard University Center for Computer Graphics and Spatial Analysis, which developed other early GIS software packages called SYMAP and ODYSSEY. After recruiting Scott Morehouse, who had worked on ODYSSEY at Harvard, Esri then developed and launched a software package called ARC/INFO in 1982, which became more and more complex as personal computing power increased.[10]

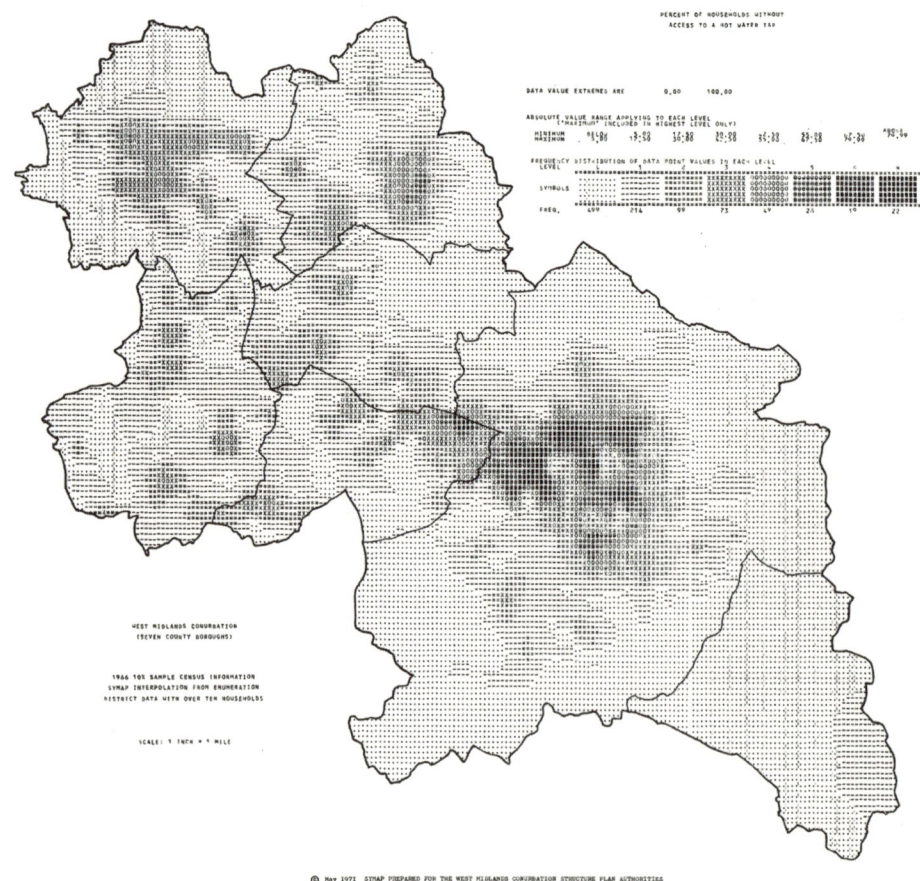

MAPS GO DIGITAL

Today Esri is the biggest map seller in the world. The company makes billions of dollars in revenue and employs 5,500 people. Its website reports that '50% of Fortune 500 companies, most national governments, 20,000 cities, 50 US states, and 7,000+ universities rely on Esri technology'.[11] During the COVID-19 pandemic, Esri technology powered the Johns Hopkins University dashboard that showed the spread of the virus, and which was visited over a trillion times.[12] Gone are the days when businesses and governments would rely on regular orders of paper copies from official sellers like Stanfords (see Chapter 4). Instead, they spend huge amounts of money and deploy numerous staff to work with the technology that generates maps: 18,000 people travel to San Diego each year to join the Esri User Conference.

It's not just Esri either: an open-source software package called QGIS is used 800,000 times a day,[13] and there are other map companies that have attracted tens of millions of dollars in investment.[14] All this without even accounting for navigation apps like Google Maps, a sector that continues to grow and is now worth almost $40 billion.[15]

The biggest enabler of this industry has been increased access to a dizzying assortment of data. In the past this was created from the painstaking digitisation of paper maps and, given the expense of doing this, many national mapping agencies kept their data under lock and key, making it available only under strict licensing conditions. So if you wanted to create a global map, you would be negotiating countless agreements and paying huge sums of money. Google was perhaps the first to seriously crack this problem in 2005 when it launched Google Maps and Google Earth.[16] However, it still did not allow people to access the underlying data to make their own maps. Frustrated by the lack of access to mapping data, a student at UCL named Steve Coast had the extraordinarily ambitious idea of creating a world map entirely from volunteered data. In July 2004, OpenStreetMap (OSM) was born and its data was hosted on UCL's servers just a stone's throw from the Map Library.[17]

OSM was conceived as a community-based map where people would collect their own data using handheld GPS devices and input it into a web map that anyone could then download. In its early years, progress was slow and the map seemed patchy at best, but just imagine starting a global map from scratch and how much work is required to digitise every street, every river, every lake, and so on! By 2010, OSM had tens of thousands of users and had secured some big-data donations to speed things along. For example, Yahoo! enabled OpenStreetMappers to digitise from their satellite imagery holdings, which spared them having to visit every place they mapped. The Netherlands was the first country to boast a complete street map within OSM thanks to a major data donation from a commercial navigation company.[18] Nowadays there's a huge range of companies and platforms[19] that depend on OSM for their data (including some I was surprised to see, like Snapchat) and its volume continues to grow with hundreds of organisations donating data, bolstering the efforts of millions of users.[20] The likes of Microsoft have developed artificial intelligence (AI) algorithms to take detailed satellite imagery, process it and create generate hugely detailed building footprints with the aim of adding them into OSM.[21]

It is the access to these kinds of datasets that has shaken the world of geographic information[22] and has made a huge difference to the way we live our lives. We depend on maps every day for everything from weather forecasts to trip planning, to following a breaking news story. But this does not mean that the basics can be forgotten by the companies supplying these services. For this reason, 19 September 2012 is an infamous date in Apple's history. On that day the company released a software update to its iPhone users that included Apple Maps, an app that was designed to end their reliance on Google Maps, which Apple saw as a competitor. It was a disaster, best described in a legendary review by David Pogue for the *New York Times*:

> Entire lakes, train stations, bridges and tourist attractions have been moved, mislabelled or simply erased. Satellite photo views consist of stitched-together scenes from completely different seasons, weather conditions and even years. The point-of-interest data, in particular, seems to be incomplete or flaky, especially overseas … the Brooklyn Bridge has melted into the river, the road to the Hoover Dam plunges straight down into a canyon and Auckland's main train station is in the middle of the sea. In short, Maps is an appalling first release. It may be the most embarrassing, least usable piece of software Apple has ever unleashed.[23]

Forbes did an analysis of the online sentiment the week following what became known as 'Mapplegate' and found that from 7 million mentions of Apple, 600,000 referred directly to Apple Maps and that overall sentiment was much more negative in relation to it than to the rest of the company.[24] The reason the Apple Maps app had so many mistakes was because Apple had mashed together the data it needed from a range of different places and wasn't able to detect all the bits of the map that were out of date or incorrect. When it realised how pervasive these errors were, it faced the monumental task of trying to hunt them down and fix them within a planet's worth of data.

Apple CEO Tim Cook issued an apology with suggestions of competitor apps while they worked to fix the issues.[25] In the words of David Pogue again: 'It's as though you just got a $1,500

Cinema
Ice cream shop
Charging station
Book store
Coffee shop
Amusement arcade
Dog park

A selection of map symbols used by OpenStreetMap.

professional coffee maker and then poured mouldy beans into it.' It took a decade of work and enormous investment for Apple Maps to redeem itself.[26]

Apple's initial approach to mapping in the twenty-first century had much in common with the way maps were made in the eighteenth and early nineteenth centuries, when they were often spliced together from multiple, sometimes questionable, sources without any real thought as to how they would interact. As Apple belatedly realised, compiling information remains a huge challenge, since enormous volumes of data beamed from space or created by volunteers will still have glitches that need to be sorted. In the past many of these might have gone unnoticed – or been forgiven – but nowadays we expect levels of accuracy and detail that far exceed anything that has come before.

The days when students might routinely come to the Map Library to request a paper map, or when a professor like me would ask that a wall chart be unfurled in a classroom, are over. In my day-to-day teaching and research, I feel no sense of loss because it takes just a few seconds to download, view, analyse and map data that may have been collected from a satellite only moments ago. I can automatically update the maps as new data become available and I can build computer models that help to anticipate the future and update the map accordingly. The world I inhabit, as a mapmaker, would have been inconceivable to most of the generations before me. Frankly, much of it was inconceivable when I started making maps only fifteen years ago.

```
{
  "type": "FeatureCollection",
  "generator": "overpass-turbo",
  "copyright": "The data included in this document is from www.openstreetmap.org. The data is made available under ODbL.",
  "timestamp": "2025-03-22T13:18:05Z",
  "features": [
{
    "type": "Feature",
    "properties": {
      "@id": "way/800089340",
      "addr:city": "London",
      "addr:housenumber": "91-93",
      "addr:postcode": "WC1N 1AL",
      "addr:street": "Marchmont Street",
      "amenity": "pub",
      "brewery": "yes",
      "building": "pub",
      "check_date": "2024-10-22",
      "indoor_seating": "yes",
      "name": "Lord John Russell",
      "outdoor_seating": "yes",
      "smoking": "outside"
    },
    "geometry": {
      "type": "Polygon",
      "coordinates": [
        [
          [
            -0.1259337,
            51.5261038
          ],
          [
            -0.1260135,
            51.526072
          ],
          [
            -0.1259186,
            51.5259597
          ],
          [
            -0.1258264,
            51.5259945
          ],
          [
            -0.1258788,
            51.5260466
          ],
          [
            -0.1259337,
            51.5261038
          ]
        ]
      ]
    },
    "id": "way/800089340"
  }
]
}
```

Behind every map you see nowadays is a database of locational information that is then used to draw what is shown on screen (or in print). This is how a popular Bloomsbury pub – the Lord John Russell – is encoded within OpenStreetMap using a data format called geoJSON. You can see some of the contextual information in the first section and the list of numbers that follow are the coordinates that delineate the building footprint of the pub.

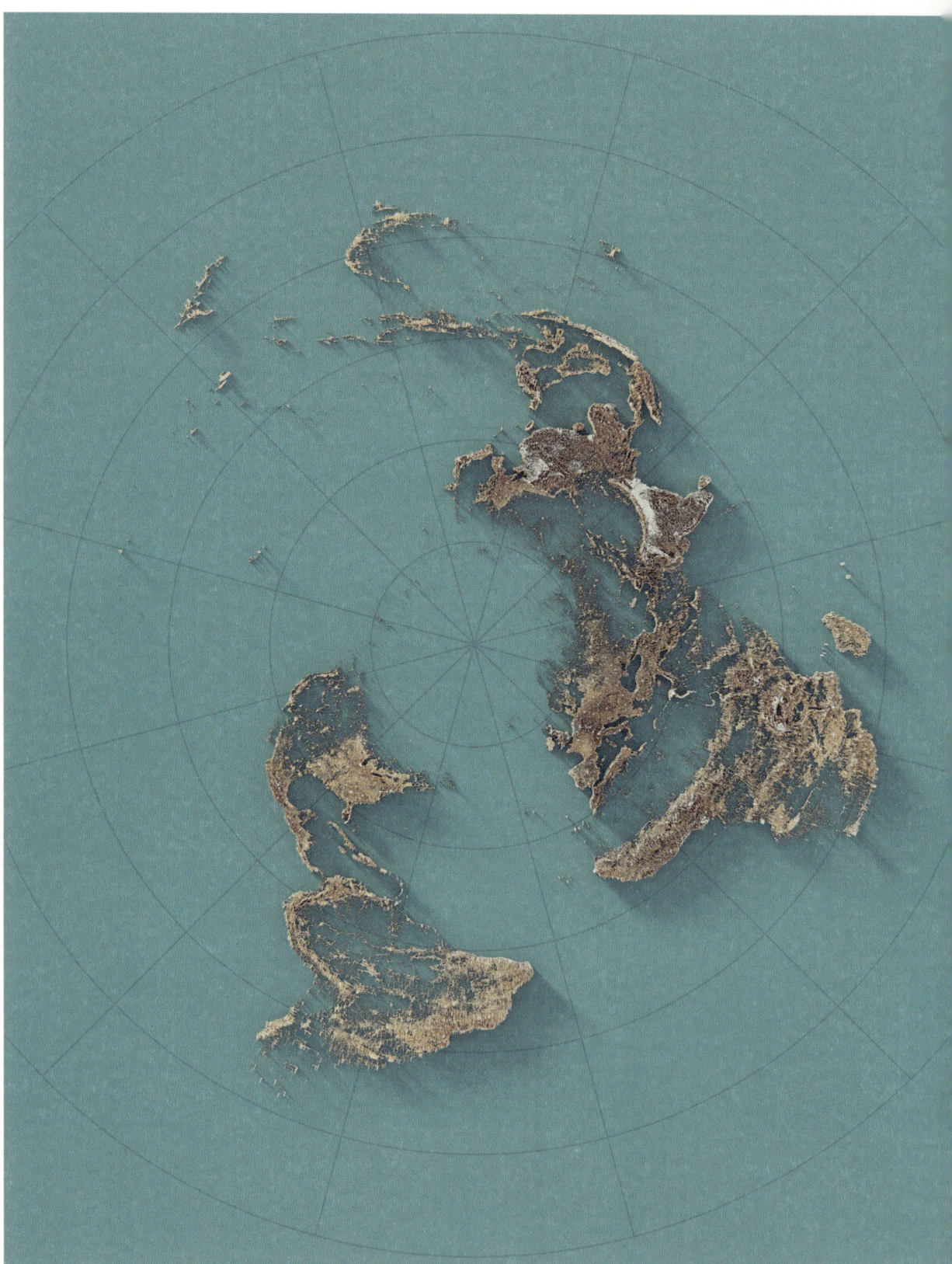

MAPS GO DIGITAL

Future Maps

We now live in another era, when maps and mapping are being transformed by unprecedented technological developments. Those with the budget can pay for their own Earth observation satellite to go into orbit or invest in a company with others to do this. Gaining access to a continuous stream of data means it can then be plumbed into artificial intelligence algorithms to create insights, which can in turn be sold on to a dizzying array of companies. Map-based dashboards now reveal how busy ports are to gauge the prosperity of a region's economy;[27] they can anticipate likely crop yields to predict the price of grain;[28] and they can monitor the height of the world's forests to determine their health.[29]

Meanwhile, on the ground we craft personalised digital worlds on our phones to help navigate the physical world around us. For good or ill, these are worlds that generate vast quantities of new data about who we are, what we like and where we go.[30] For example, no two Google Maps are the same since each user gets a bespoke set of labels tailored to them based on places they've searched for.[31] Coffee addicts might see more Starbucks labels and fast-food fans may find themselves tempted to a McDonald's a few streets away rather than the salad bar next door. What's more, maps are no longer just visual: you can have conversations with them to request recommendations for where to get lunch or how to build a tour of a city.[32]

What does all this say about the need for maps more generally?

I've centred this map on the North Pole and used the 'azimuthal equidistant' projection to position the continents (it's the same projection used by the UN logo). But rather than show the land and its borders I am only showing populated areas, as captured by a hugely sophisticated dataset created by researchers at the European Commission.

In an AI-driven world will maps become irrelevant? In short, the answer to this is an emphatic no.

The technology will continue to evolve and, although big business has seen the value of maps, more and more people will appreciate that because such rich data is available, maps do not have to be the preserve of state actors or people with significant resources. Instead, they can be used to counter narratives and revolutionise how those in power are being held to account.

In a world in which many maps now point to the havoc wrought by our changing climate, such as the extent of floods, the deluge of storms or the intensity of heat waves, I'm often asked by my students if they simply render us powerless bystanders. Do I worry that we have become rabbits in the headlights of the scale of the challenge we face?

Experiences from the Map Library help to inform my answer because it has shown me maps that have been called upon in some of the most significant moments from history. Maps should not nurture a sense of despair but, rather, highlight that we now possess the information we need to take action.

In the case of the climate crisis, we can draw upon the huge amounts of readily available information to track environmental changes and predict future developments in order to show the ways to mitigate them. We have detailed maps that highlight areas historically affected by forest fires and those where such incidents are becoming more frequent. We can also observe the varying intensity of sea-level rise across different regions. This unprecedented level of detail allows us to understand not only our environment on a global scale, but also what the local consequences may be.

For example, satellite data now reveals that across the US global warming is compounding the legacy of racist housing policies enacted through a process known as redlining. Redlining

MAPS GO DIGITAL

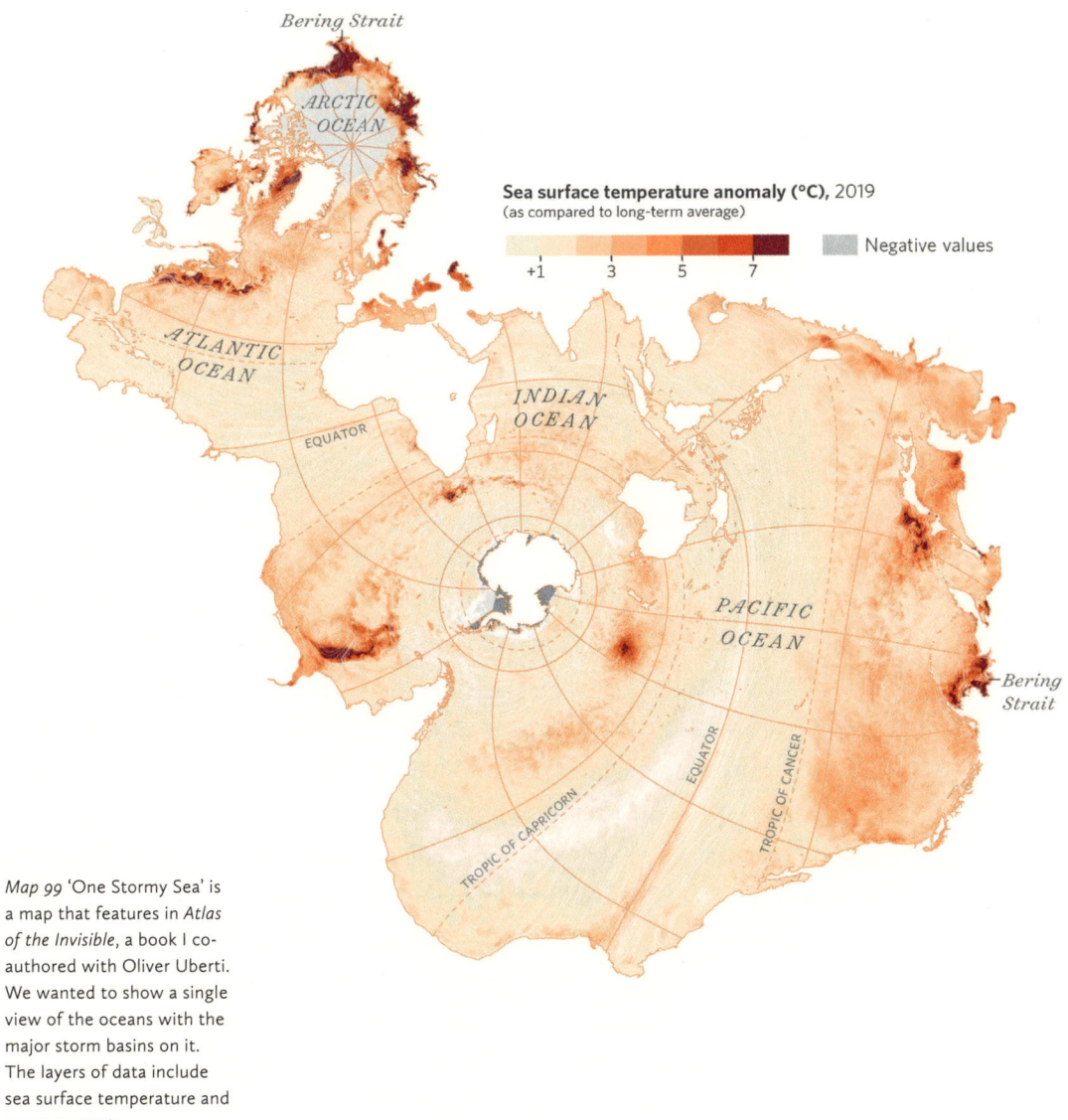

Map 99 'One Stormy Sea' is a map that features in *Atlas of the Invisible*, a book I co-authored with Oliver Uberti. We wanted to show a single view of the oceans with the major storm basins on it. The layers of data include sea surface temperature and ocean currents.

Atlas of the Invisible (One Stormy Sea): Particular Books/ W. W. Norton & Company (2021)
11 × 11.5cm

rated the 'investment risk' of urban areas, condemning many Black neighbourhoods to a 'hazardous' rating and thereby reducing infrastructure and increasing poverty. Affected areas saw a lack of investment in, among other things, green spaces and street trees. This has resulted in some historically redlined neighbourhoods suffering summers that are up to 7°C warmer compared to their non-redlined counterparts. And this is not just a US issue: when it comes to climate many cities around the world have created favourable conditions for their richest inhabitants and compounded the challenges for their poorest.[33]

Another recent example of the power of maps to mobilise change is the way they have been used to campaign against the serious issue of water pollution in the UK. The country has been blighted by water companies discharging sewage into its rivers and onto its beaches. Campaigners had long been sounding the alarm about this grim reality, but it was not until they mapped it that they attracted widespread national attention and turned up the pressure on the worst performing water companies. Maps by conservation organisation The Rivers Trust[34] and campaign group Surfers Against Sewage[35] laid bare the extent of sewage dumping. They proved to be highly effective tools, not just to warn of the health risks to bathers but also as evidence for environmental damage.[36] They pull together data from sensors along the sewage network that detect discharges, making it clear where the worst offenders are and encouraging users to contact their local representative requesting more rapid action on sewage discharge.

Such maps are easy to share on social media and by local news sites, and they make for awkward viewing for the water companies themselves.

There are countless examples of dynamic, data-driven maps making a difference in a range of other causes, too. Maps are used to create conservation corridors for migrating animals.[37] They

MAPS GO DIGITAL

are the basis for campaigns for cleaner air in cities[38] and they even shame the wealthiest about the carbon footprints of their private jets.[39] Maps provide us with the knowledge and clarity to understand what is happening, which empowers us to demand action from policymakers and leaders. For this reason, the maps of today are even more powerful than the sheets reposing in the drawers of the Map Library.

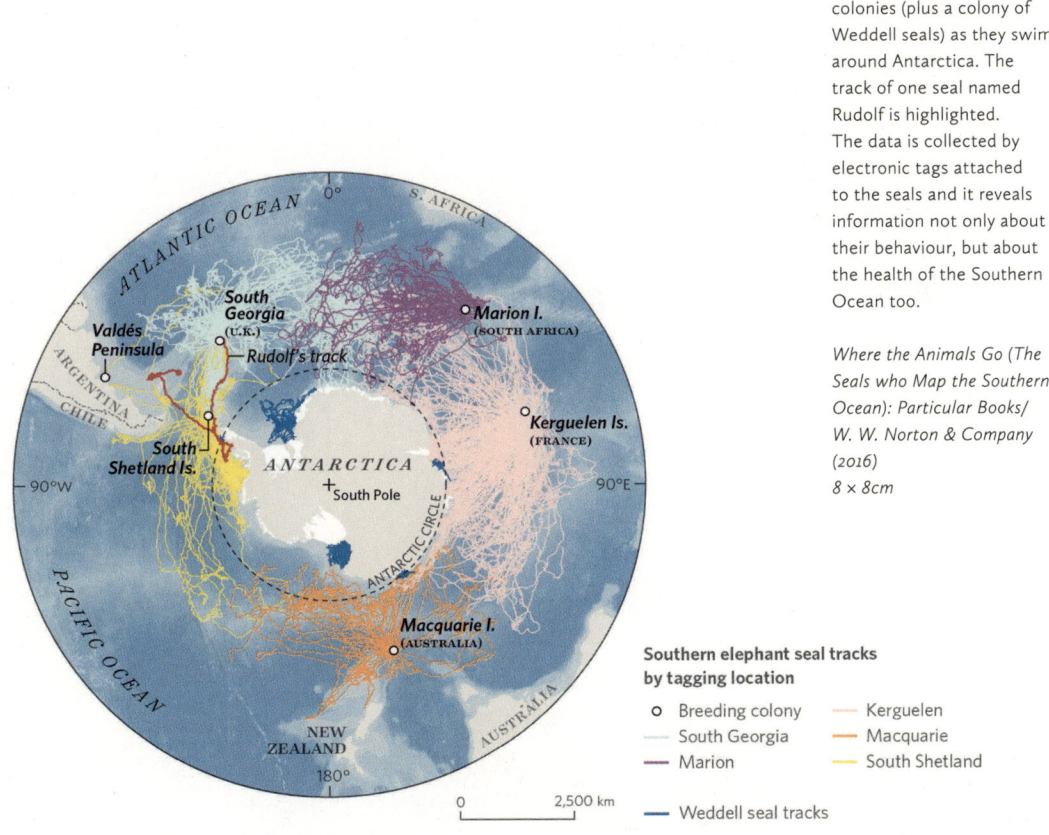

Map 100 The movements of elephant seals from five colonies (plus a colony of Weddell seals) as they swim around Antarctica. The track of one seal named Rudolf is highlighted. The data is collected by electronic tags attached to the seals and it reveals information not only about their behaviour, but about the health of the Southern Ocean too.

Where the Animals Go (The Seals who Map the Southern Ocean): Particular Books/ W. W. Norton & Company (2016)
8 × 8cm

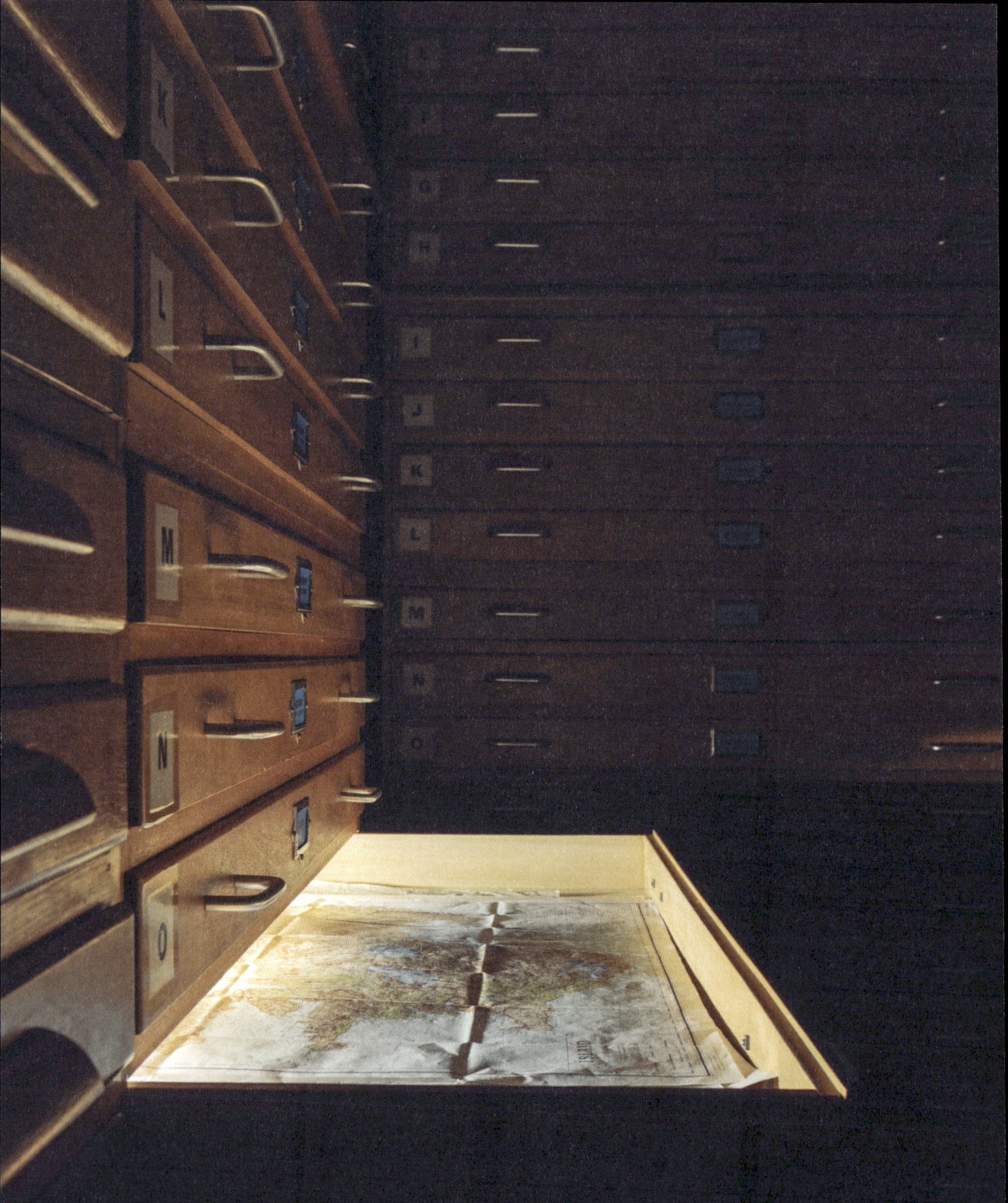

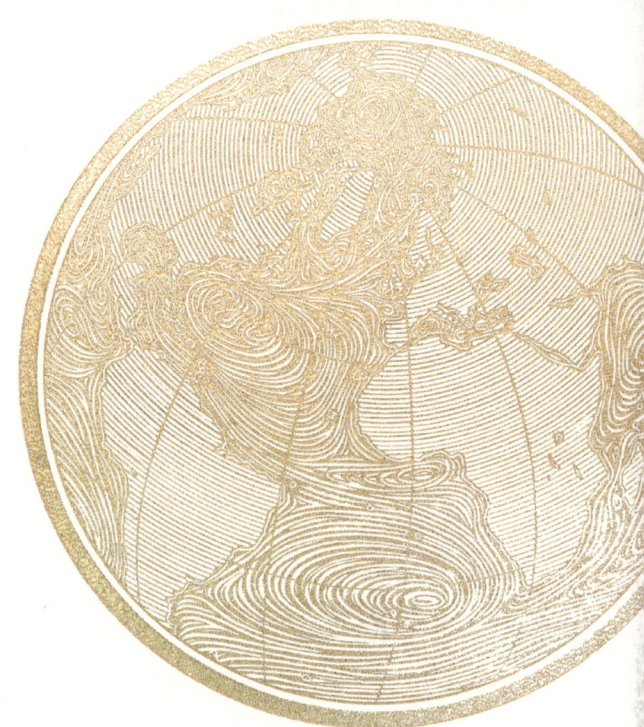

CHAPTER TWELVE

A Special Place

'The students would come here to use the reading room, to use maps. But they'd often come and talk and if they wanted to talk, fine. I'd always have time for them.'
Anne Oxenham

When I first stepped through that unassuming turquoise door into the Map Library, I was hit with a sense of wonder and itching to embark on the adventures waiting in its numerous wooden drawers. But I also had a sense of trepidation because many colleagues would look at me with bemusement – and perhaps pity – when I told them what I was up to. As the Founding Director of the UCL Social Data Institute, my job is to be forward-looking, to teach students the latest methods and to research the most innovative datasets available. These would not be found in the Map Library.

But I soon grew comfortable with the sceptics as I was having the time of my life exploring this wondrous space; they had no

[Left] The 'Iceland' drawer awaiting its turn to share its cartographic surprises.

[Above] *Map 101* My favourite atlas cover decoration comes from the *Reader's Digest Great World Atlas*, which also happens to be the first atlas I ever saw, as a copy sat on the shelves of my parents' study. Debossed on its rich green cover is a globe of gold foil, dazzling with swirls of ocean currents.

Reader's Digest Great World Atlas (Cover): The Readers Digest Association (1962)
14.5 × 14.5cm

idea what they were missing! And beyond the doubters there were many others who shared my passion for the maps and have taught me so much about them. They, like me, had been appalled by the stories of thousands of maps being discarded from geography departments up and down the country, and exasperated by the lack of imagination shown by those who saw the room as a vacant space to be more intensively used. Sharing my discoveries with like-minded cartophiles just added to the joy of the experience.

I concluded that to appreciate the Map Library you need to surrender yourself to it and believe in the power of a map to whisk you away. Suspend that belief and the drawers become scrap wood, and the maps inside them scrap paper. During my first weeks in the Map Library, I felt like I had to win its trust – to convince it I was serious about hearing its stories. When I invited the map librarian of forty years, Anne Oxenham, back (the first time she'd visited since her retirement in 2002), she brought the place alive and I sensed it relax, like a timid child seeking reassurance from a nurturing parent.

As weeks turned to months and months to years, the Map Library and I came to know one another and it has been incredibly generous in what it has offered me. I've been able to unearth stories that were unimaginable in those early days of pulling maps from their drawers.

The Map Library has taught me about the power that something as simple as a line can have when drawn on a sheet of paper and called a map. I have discovered how pioneers like Marie Tharp have been able to alter the way that we see the world using their imagination and skill. I've learned, too, about the values of cartographers such as the Society for the Diffusion of Useful Knowledge and J. F. Horrabin, who were committed to using maps as an educational tool to make knowledge accessible to groups of people who'd previously been overlooked.

A SPECIAL PLACE

As someone who makes maps, there is so much I now take from them beyond the process of their creation. The Map Library has demonstrated the importance of the context in which cartographers were operating and the hopes and fears of the individuals that held the maps. I can now see how George Bellas Greenough's personality in the final years of his life might have alienated him from his peers among the gentleman scholars of London, but that his inquisitiveness, his passion for travel and the way he saw the world could be revealed through the maps that the Map Library showed me.

I now know that no two maps are the same because as soon as a map is held in someone's hands it transforms. A simple tourist map designed to be used for joyous excursions in a city can become a weapon of war, while a map detailing the floor of the ocean becomes the basis to a grand theory about the creation of Earth itself. And as the world changes, maps can progress from depicting the world as it is to the world as it was.

The Map Library has also taught me that atlases are not simply books of facts. They evoke something much bigger: a sense of nation, a sense of identity and a sense of where we belong. Atlases were proof of the extent of borders, but also of the diversity and the successes of the people that lived within them. If you had an atlas, you had a country.

There are many stories I've been unable to tell in this book, not least because many maps remained elusive in their purpose, in their history and in the impact they had on the world. With more time, I'm sure some will reveal themselves and that is why I am committed to their preservation, for fear we might lose something special. I feel privileged to have been so close to this place to capture it before it disappears, and I mourn the loss of the map libraries that are no longer with us.

As I look back at my own finds, I am reminded of the weight of their history. History such as the ideas of 1820s Bloomsbury that led to the founding of UCL and the establishment of the structures of science we recognise today, or the early cartographic innovations that helped define how countries saw themselves. I am reminded, too, of the events following the end of the First World War that demonstrated the role of maps in pushing opinions dressed as facts. The Second World War was another important juncture when, fresh from battle, academics returned to their departments with connections to the military and governments who would fill their reinvigorated map libraries. And all this led to a second half of the twentieth century when maps were used as critical and highly complex scientific tools to understand our world.

I am even more excited about the innovations of today which, though they may be technologically unrecognisable from the hand-drafted maps that dominate the Map Library, nonetheless remain anchored to them through the fundamental problem of how best to capture the complexities of the Earth in a single image.

And this is precisely why the Map Library has never been more important. It teaches us about the intended and unintended consequences maps have had, certainly, and the past successes and mistakes of mapmakers and their masters. But the Map Library also teaches us how we can all do better.

Maps enable the world to come together before our eyes. So, who would dare discard them, when they hold such awesome power? Therefore, if I am asked, 'why is a map worth saving?' my answer is simple: it is both a reminder of our legacy and inspiration for a better future.

A SPECIAL PLACE

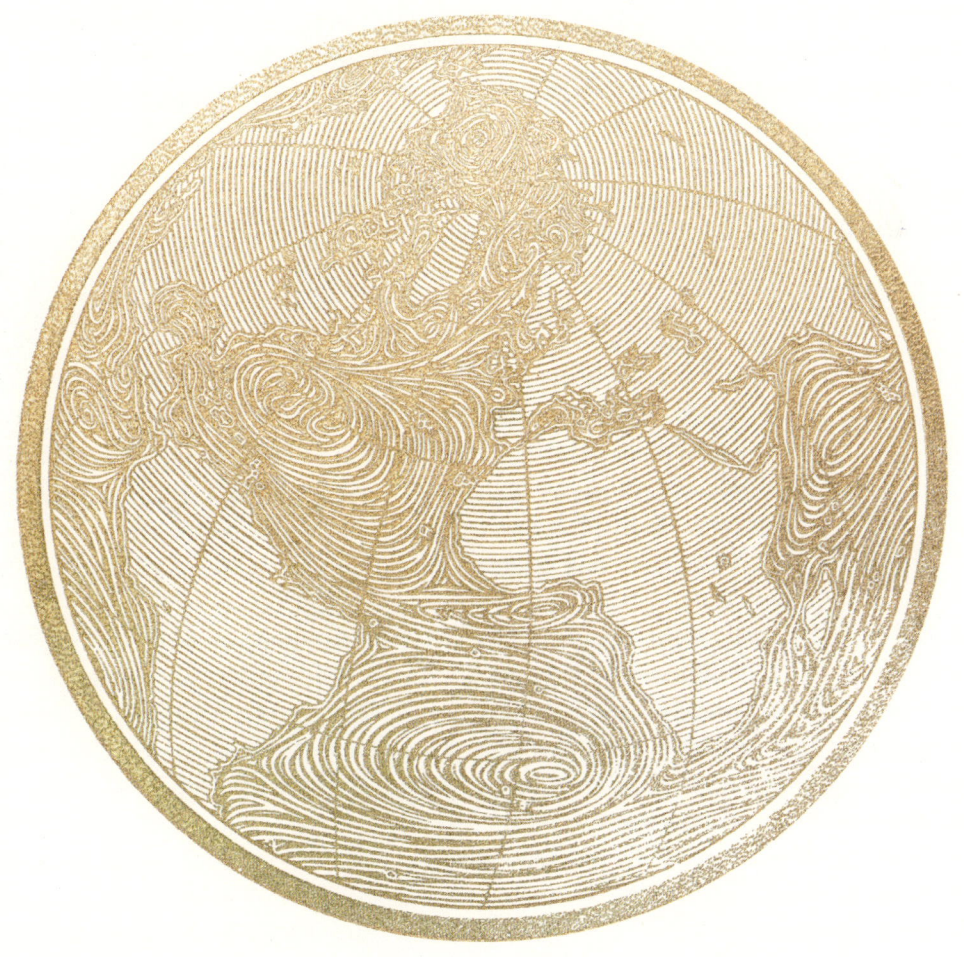

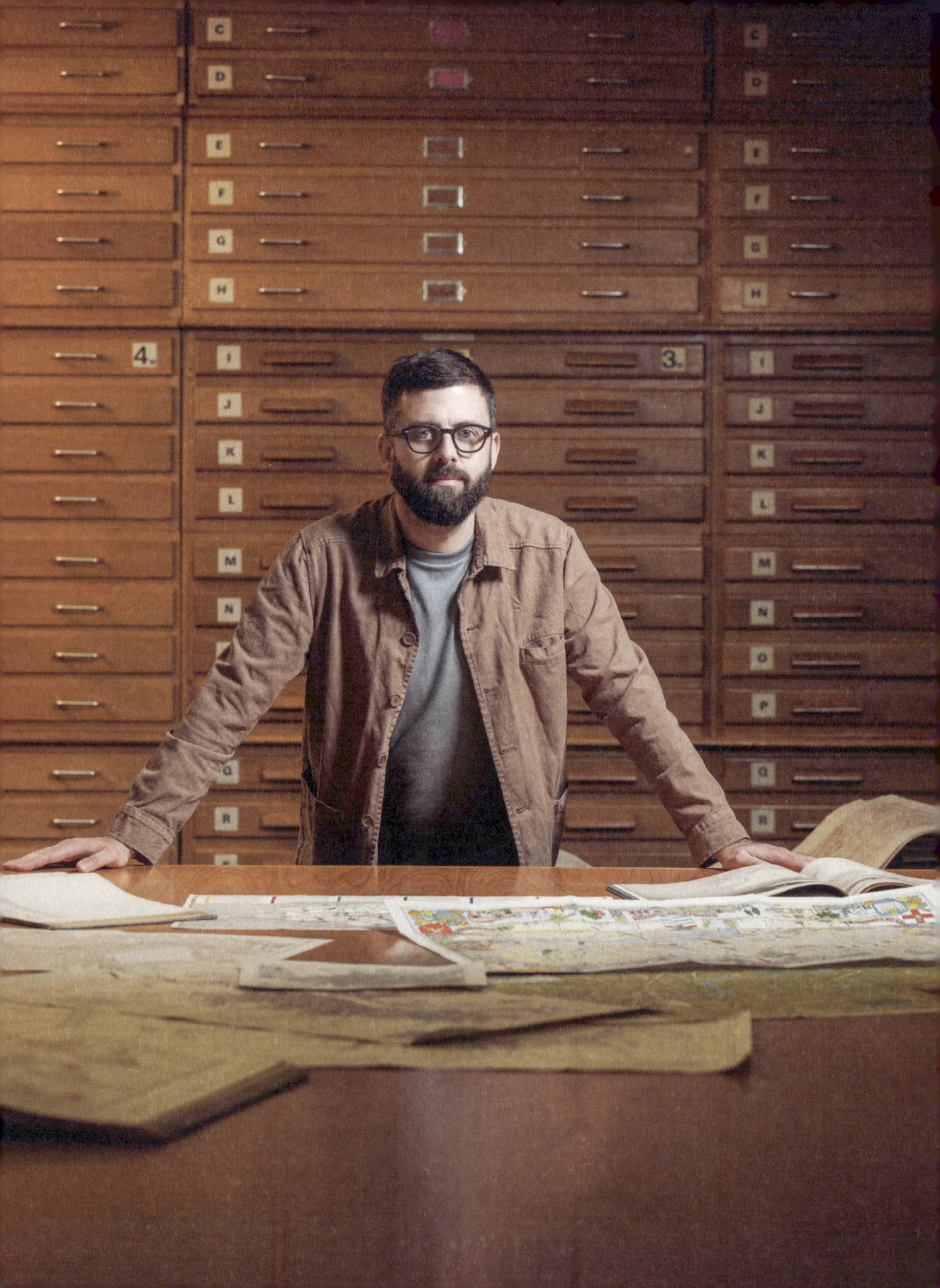

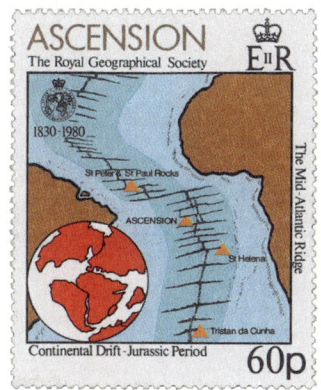
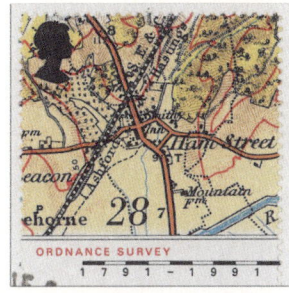
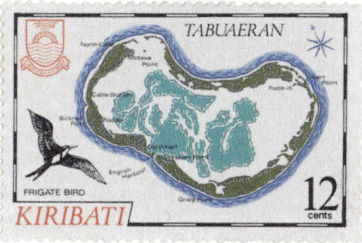

Afterword

'They were maps that lived, maps that one could study, frown over, and add to; maps, in short, that really meant something.'
Gerald Durrell[1]

When I visited Washington DC in 2023 to learn about the beautiful maps of Marie Tharp, I walked from my hotel to the Library of Congress and would pass the National Postal Museum. Housed in a grand building adjacent to Washington's Union Station, it was draped with banners advertising its stamp collection. I've not been blessed with an interest in philately, but I can see the appeal. However, given the cartographic treasures I was on my way to experience, as I strolled past I wondered why there are no major museums or galleries celebrating maps?[2]

Maps have it all: from art of a quality you'd find in a great gallery, to innovation that rivals anything in a science exhibition and, of course, enough history to fill a national museum. I suspect this breadth is why, rather than sitting in a single museum, they

[Left] Picking out the final maps for this book in August 2024.

Taken in August 2024, Peter Searle

[Above] *Map 102* Maps appear on stamps the world over. Here is a selection from my (very small) collection.

Ascension Island (1980): 3.5 × 3.5cm
UK (1991): 3.5 × 3cm
Kiribati (1985): 4.5 × 2.5cm

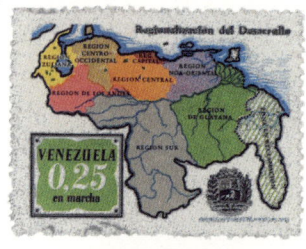

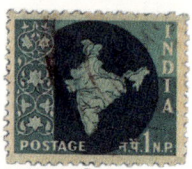

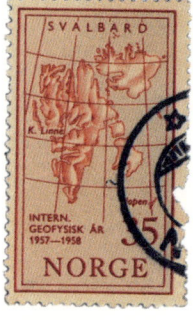

Map 102 stamps continued:

Venezuela (1974): 4 × 3cm
India (1958-1963): 1.5 × 1.5cm
Norway (1957): 2 × 3cm

THE LIBRARY OF LOST MAPS

crop up in all of the above. Maps even have a place in postal museums like the one I passed, as they often feature on stamps, collectors of which are known as 'cartophilatelists'.[3]

There are some maps we look at all the time – when we're driving, checking the weather, even when we're tracking the progress of a fast-food delivery – so the prospect of a gallery of framed Google Maps will be a hard sell on its novelty value. Fortunately, just as there are many genres of photography – portraiture to landscape, abstract to journalistic – the same is true for mapping, past and present. Entire gallery spaces could be devoted to topographic maps or thematic maps. Visitors could chart a route from maps of the moon right down to the ocean floor, experiencing maps of everything in between.

The UK's Natural History Museum in London makes much of its Wildlife Photographer of the Year (each photo is accompanied with a small map showing where it was taken), while the Royal Geographical Society,[4] just up the road from the museum, hosts an annual travel photography competition. The winning entries are displayed in the Society's garden, beneath which sit vaults heaving with maps that rarely see the light of day. Why don't we have a 'Map of the Year' exhibition?

Back in 2011, I remember queuing down the street to get into one of the most popular exhibitions ever staged in the history of the British Library. 'Magnificent Maps: Power, Propaganda and Art' attracted 227,000 visitors.[5] Copies of many of the maps exhibited were in the drawers of the Map Library, too, crying out for similar attention.

We, of course, find maps everywhere in exhibits hosted by museums, but too few are in settings that are designed to *celebrate* them. They nearly always place maps in a supporting role, rather than being the main event themselves. Mapmakers have fallen between the cracks. We talk of great artists, writers and musicians and we think about their lives, their politics, their motivations and their intended audience, but rarely do we delve into the backstory of maps and their creators. As Bruce Heezen said to an interviewer in 1977:

> It takes a certain sophistication ... to recognize there is sophistication in making maps. You might almost think of dull little people sitting around in gray offices simply plotting data given to them by brilliant people wandering around outside and producing these things in a very mechanical way. Nothing could be further from the truth.[6]

Anne Oxenham reminded me of this when I asked her what we would lose if the Map Library was closed. In response she, of course, said we would lose 'history and part of our heritage', but indicated that her sense of loss would be compounded because 'I know I'm not a computer person, but to be truthful, not everything is "computer". Human beings come into it somewhere. And [gesturing around the library] all these are produced by human beings a long time ago ... I think it's just part of me.'

No map in this book was the result of a single hand. They needed those who commissioned them, gathered the data for them, helped with the drafting and made the preparations for print. They then needed people like Anne to collate and share them.

Today, maps need us to celebrate them. So, if you know a major museum out there that wants to create a grand gallery of maps, come and find me in the Map Library!

Further Reading

The Library of Lost Maps spans a huge range of topics and so I hope readers will use it as a starting point to begin their own explorations of the map world. I hope, too, to have inspired a greater appreciation of the stories maps can tell, which make them such wonderful things to own. While some of the maps in this book are extremely pricey, most aren't and there are huge quantities of maps out there that are being sold for very small sums of money. If you are looking for something to hang on a wall, have a rifle through online auction sites, second-hand book dealers and charity shops. I've purchased original maps for less money than they are being sold for as reprints, so if you avoid 'famous' maps (like the Beck's London Underground map, Smith's geology map or the Booth poverty maps) there are some stunners out there that are surprisingly affordable. Many will have come from closed map collections so they would welcome a new home and a chance to be seen again!

    I mentioned Smith's and Booth's maps being very pricey, but Thames & Hudson have published a couple of fabulous large-format books, *Strata* (2020) and *Charles Booth's London Poverty Maps* (2019), that reproduce them wonderfully and offer detailed histories of their creation. Another large-format book to look out for is Melville C. Branch, *An Atlas of Rare City Maps: Comparative Urban Design 1830–1842* (1997), which has reproductions of all the SDUK's city maps in it. I don't delve too deeply into the history of the first detailed maps of Britain so for those who'd like to know more see Rachel Hewitt, *Map of a Nation: A Biography of the Ordnance Survey* (2010). For more on the history of UCL (and Bloomsbury) see the open access book edited by Negley Hart,

FURTHER READING

John North and Georgina Brewis, *The World of UCL* (2018), as well as Rosemary Ashton, *Victorian Bloomsbury* (2012).

Of course, there is no shortage of material on the build-up to and consequences of the First World War, ditto 1930s Germany and the rise of National Socialism. Steven Seegel's *Map Men: Transnational Lives and Deaths of Geographers in the Making of East Central Europe* (2018) offers rich biographies of the key geographers informing the Paris Peace Conference; for a primer on the impact of German mapping efforts see Guntram Henrik Herb, *Under the Map of Germany: Nationalism and Propaganda 1918–1945* (1997); and for a rich biography of Karl Haushofer see Holger H. Herwig, *The Demon of Geopolitics: How Karl Haushofer 'Educated' Hitler and Hess* (2018). I found Christian Grataloup's *A History of the World in 500 Maps* (2023) a great resource for getting my head round the geopolitical developments I encountered in the Map Library.

My recommendations for the details on the maps of the second half of the twentieth century would be Hali Felt, *Soundings: The Story of the Remarkable Woman Who Mapped the Ocean Floor* (2022), which gives much more of the personal story of Marie Tharp. William Rankin, *After the Map: Cartography, Navigation, and the Transformation of Territory in the Twentieth Century* (2016) gives a sweeping history that connects the latest technology to the developments of map-making in the first half of the century.

Finally, there has been a boom in the creation of new thematic atlases, which I have been proud to be part of. Please see James Cheshire and Oliver Uberti, *Atlas of the Invisible* (2021) and Dariusz Wójcik, *Atlas of Finance* (2024) if you would like a break from your screen while still seeing what the latest cartography can do.

For detailed notes, see libraryoflostmaps.com.

Picture Credits

Captions for each of the maps cite their original publication details. Unless specified below, the use of the physical copies of the maps and their scans are used with permission from the UCL Department of Geography: ucl.ac.uk/geography

Endpapers: © James Cheshire, 2025.

p. 8: © Peter Searle, 2024.

Map 3: British Overseas Corporation/E. Stanford Ltd, London, 1945. Scan: UCL Department of Geography.

Map 4: Touring Club Italiano Historical Archive. Scan: UCL Department of Geography.

Map 17: Gerardo Canet, Erwin Raisz/Harvard University Press. Scan: UCL Department of Geography.

Map 21: Courtesy of the Royal Scottish Geographical Society. Photo: James Cheshire.

p. 79: UCL Special Collections (GREENOUGH 1-4/3-20/20). Photo: James Cheshire.

p. 94: UCL Special Collections (GREENOUGH/A/1-4/2/1). Photo: James Cheshire.

Map 26: UCL Special Collections (GREENOUGH/A/1-4/2/1). Scan: UCL Department of Geography.

p. 109: Photo: © The Trustees of the British Museum.

p. 111: Photo: Wellcome Collection 38732i.

Map 39: Photo: David Rumsey Map Collection, David Rumsey Map Center, Stanford Libraries.

Map 56: Scan: American Geographical Society Library. Digital ID: agsmap024004 (a-b).

Map 77: Library of Congress, Geography and Map Division, ref: G8261.S1 1939 .E9. Scan: James Cheshire.

pp. 256 & 257: Photo: National Archives KV2/1649/1.

Map 81: Reproduced with permission of the Austrian Alpine Club. Scan: UCL Department of Geography.

p. 284: © James Cheshire, 2025. Contains modified Copernicus Sentinel data (2024).

Map 83: Permission granted by Lamont-Doherty Earth Observatory, the estate of Marie Tharp and Marie Tharp Maps. Scan: UCL Department of Geography.

Map 84: Permission granted by Lamont-Doherty Earth Observatory, the estate of Marie Tharp and Marie Tharp Maps. Scan: UCL Department of Geography.

p. 296: Library of Congress, Geography and Map Division, ref: G3171.C2 1957.H4. Permission granted by Lamont-Doherty Earth Observatory, the estate of Marie Tharp and Marie Tharp Maps. Photo: James Cheshire.

Map 86: Permission granted by Lamont-Doherty Earth Observatory, the estate of Marie Tharp and Marie Tharp Maps. Scan: UCL Department of Geography.

p. 304: Library of Congress box 107 Contours, ref: HT111 L 4 0250. Permission granted by Lamont-Doherty Earth Observatory, the estate of Marie Tharp and Marie Tharp Maps. Photo: James Cheshire.

Map 87: The Geographical Press, Columbia University, New York. Scan: UCL Department

PICTURE CREDITS

of Geography.

Map 88: © Kenneth Fagg. Photo: James Cheshire.

Map 90: © 1961 Richard Edes Harrison. Scan: UCL Department of Geography.

Map 91: www.berann.com Scan: UCL Department of Geography.

Map 92: © National Geographic Society (Product Number HM1967I000). Scan: UCL Department of Geography.

Map 94: Used with permission of the Geological Society of America. Pitman, W.C., III, Larson, R.L., and Herron, E.M., 1974, 'The age of the ocean basins (sheet 2)', in *Age of the Ocean Basins Determined from Magnetic Anomaly Lineations: Geological Society of America Map and Charts 6.*

Map 95: Permission granted by Lamont-Doherty Earth Observatory, the estate of Marie Tharp and Marie Tharp Maps. Scan: UCL Department of Geography.

p. 328 left: Permission granted by Lamont-Doherty Earth Observatory, the estate of Marie Tharp and Marie Tharp Maps. Scan: James Cheshire;

middle: www.berann.com Scan: Library of Congress, Geography and Map Division, loc.gov/item/2010586277;

right: © James Cheshire. Data source: GEBCO Compilation Group (2024) GEBCO 2024 Grid (doi:10.5285/1c44ce99-0a0d-5f4f-e063-7086abc0eaof).

Map 96: Permission granted by Lamont-Doherty Earth Observatory, the estate of Marie Tharp and Marie Tharp Maps. Scan: Library of Congress, Geography and Map Division, loc.gov/item/2010586277.

p. 336: Permission granted by Lamont-Doherty Earth Observatory and the estate of Marie Tharp.

p. 348: © James Cheshire. Data: Pesaresi, Martino; Politis, Panagiotis (2023): GHS-BUILT-S R2023A - GHS built-up surface grid, derived from Sentinel2 composite and Landsat, multitemporal (1975-2030). European Commission, Joint Research Centre (JRC) [Dataset].

Map 99: © James Cheshire and Oliver Uberti, 2021.

Map 100: © James Cheshire and Oliver Uberti, 2016.

p. 361: © Peter Searle, 2024.

Maps *7, 19, 20, 24, 36, 46, 47, 50, 66, 67, 69, 76, 82, 86, 88, 90, 102* and images on *pp. 125, 128,* are from the author's collection and scanned by the UCL Department of Geography.

The photographs on *pp. 17, 18, 26, 35, 38, 50, 62, 74, 84, 104, 112, 120, 134, 148, 158, 164, 174, 190, 200, 208, 222, 224, 234, 240, 252, 262, 267, 268, 274, 288, 326, 354* are © James Cheshire, 2025.

Acknowledgements

The book would have been impossible to write without me hearing from those who knew the Map Library in its heyday. My thanks to Peter Wood for introducing me to Anne Oxenham and for reviewing a draft of the book. I'm indebted to Hugh Clout who first directed me to the treasures of the Map Library and taught me so much of the history of the place. Hugh's careful read through and detailed comments tightened an earlier draft of this book immensely.

Enormous thanks to Anne Oxenham for her frequent visits to the Map Library to show me the ropes and to tell me about her work. The place really did come alive in her presence, and it remains an astonishing legacy of her 40+ years of service to UCL.

I'm indebted to Nick Mann and Miles Irving for their efforts to reinvigorate the Map Library and for making it such a fun project. Thanks too to James Shilland for providing the extra muscle when we needed it to shift wall charts and atlas cabinets! I'm grateful to Jason Dittmer for his support during his time as HoD, both in helping me get this book project off the ground by signing off on the map permissions, but also in appreciating the value of the Map Library to the UCL Department of Geography.

Thank you to colleagues in the UCL Social Data Institute, the Geospatial Analysis and Computing Research Group and the SDRUK Geographic Data Service for their patience when I went to ground to work on the book and in particular to Mirco Musolesi, Mikaella Mavrogeni, Rob Davidson and Michal Iliev for their feedback on earlier drafts. Thanks to the students who spent their summers helping me with the scanning and cataloguing.

I have had the good fortune to come to know an extraordinary group of people whose eyes have lit up when I've told them about the Map Library, and I have benefited from their immense knowledge and generosity with their time. Margaret Wilkes's enthusiasm for maps exceeds even mine and I'll never forget my day at the Royal Scottish Geographic Society in Perth with Margaret and another volunteer Blair White sifting through their collection and hearing their stories. My thanks to Nick Millea for organising for the atlas collection to be cross-checked against the Bodleian's holdings and to Yolande Hodson, Sarah Tyacke, Peter Barber, Francis Herbert, Catherine Delano-Smith, Roger Kain and Tom Harper for stopping by the Map Library to discuss my finds. Vivien Godfrey was instrumental in resolving some last-minute permissions issues, whilst running Stanfords: the world's best map and travel bookstore!

Whilst *The Library of Lost Maps* is a distillation of hundreds of hours spent sifting through thousands of maps in the UCL Map Library, I also needed to see more maps and material to help tell the stories I've typed on these pages. I have therefore

ACKNOWLEDGEMENTS

benefitted immensely from access to UCL Special Collections, the archives of the Royal Geographical Society (with IBG), The Geography and Map Division of the Library of Congress, the maps reading room of the National Library of Scotland, the map room of the Royal Scottish Geographical Society and the UWM American Geographical Society Library. My thanks to the staff who sustain these collections for being so helpful with my requests.

A chance meeting with Duncan Hawley proved transformative for my appreciation of George Bellas Greenough. I am indebted to Duncan for his detailed feedback and have hugely enjoyed our conversations about Greenough's world. The person who's work I turned to most for some of the foundational ideas in this book was Mike Heffernan (a connection made by Hugh Clout), who is a fellow defender of map libraries. I was therefore delighted when he agreed to read through an early draft of this book and provide such useful feedback. Felix Driver – another tireless campaigner for the value of map libraries – was an obvious choice for reading through a draft, a fact confirmed by the quality of his suggestions. My thanks also to Philip Jagessar, who offered further useful comments and, with Catherine Delano Smith, sustains the 'Maps and Society' lecture series which is such an important place for many of the names I have just listed to congregate.

Working with Darek Wójcik on the *Atlas of Finance* reminded me of the continued power of maps and the way that they can be energised by the people holding them. Oliver Uberti and I have created hundreds of maps together, each one a collaborative endeavour that has taught me more about the craft of telling visual stories. My thanks to both Darek and Oliver for their feedback and unwavering enthusiasm for my first solo project. I'd also like to thank Alicia Cho for reading through a draft and providing such valuable feedback from her perspective as a voracious reader.

I am extremely fortunate to benefit from the immense support of friends and family. My biggest thanks of all are reserved for Isla Johns, to whom this book is dedicated and, who, for eighteen months has repeated the phrase 'James has to work on the book this weekend' to friends and family at events I've missed, read innumerable drafts (and caught countless embarrassing typos) and heard me speak about little else than the book. I couldn't have done it without her.

Writing a book is a huge privilege for many reasons, but one of the most significant for me is being able to work with a group of people who are committed to getting the words from my laptop and the maps from the drawers into the hands of readers. My thanks to Luigi Bonomi for over a decade of support and for helping me shape the initial idea and share it with publishers; to Ian Marshall for his editorial oversight and care; to Lauren Whybrow for driving the book's production and to Anna Green for such fantastic design. Thanks too to the wider team at Bloomsbury who have been involved in the book's journey to the shelves.

Map Gallery *Scale 1:20*

1. WELCOME TO THE MAP LIBRARY

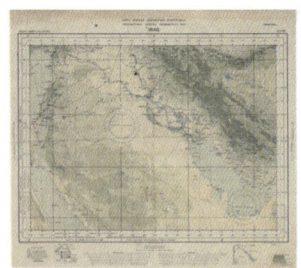

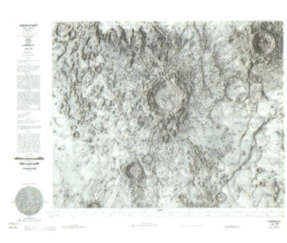

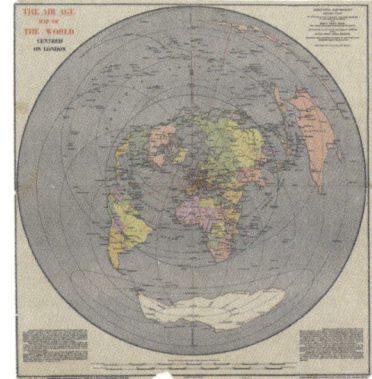

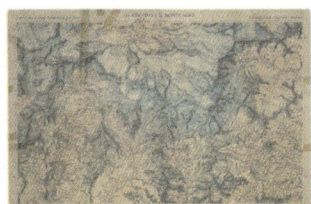

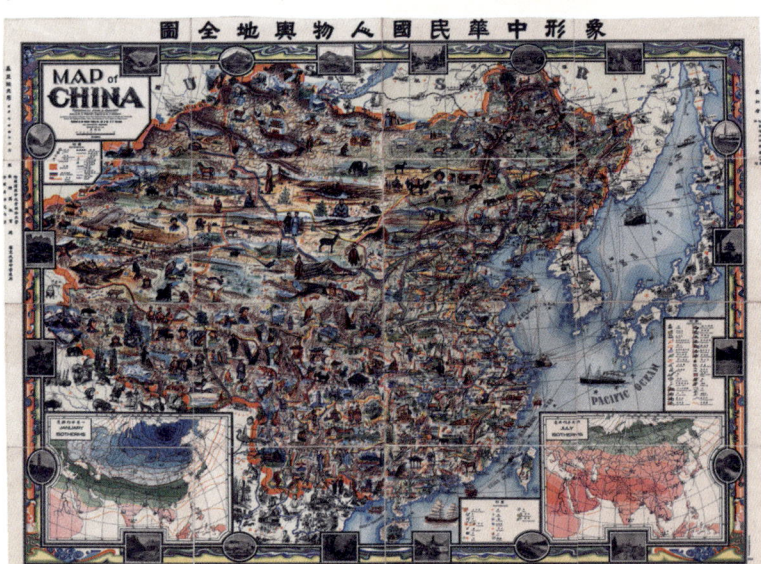

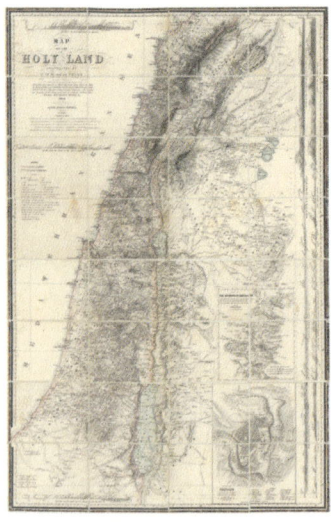

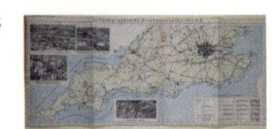

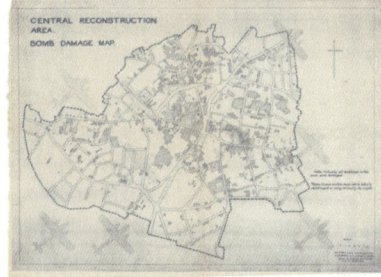

MAP GALLERY

10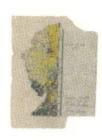
11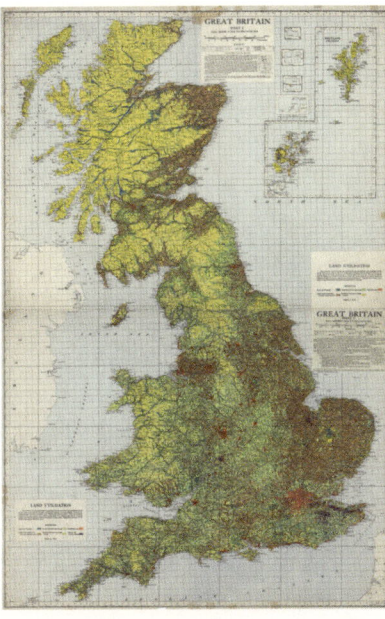
12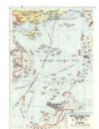
13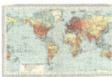
14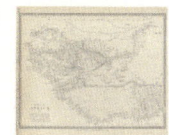

2. A ROOM FULL OF STORIES

15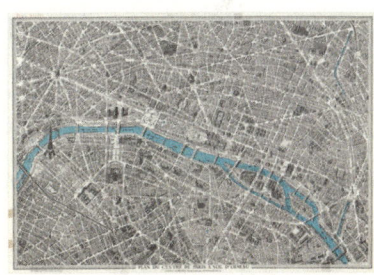

16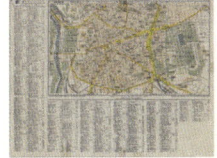
17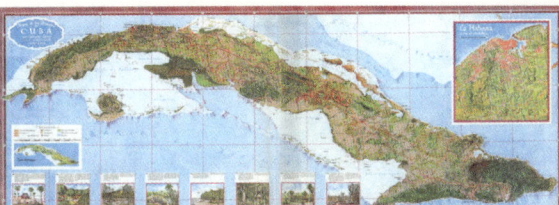
18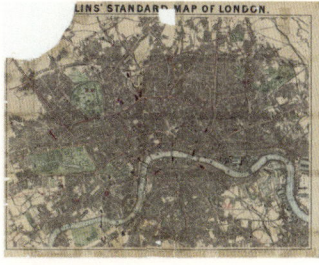

19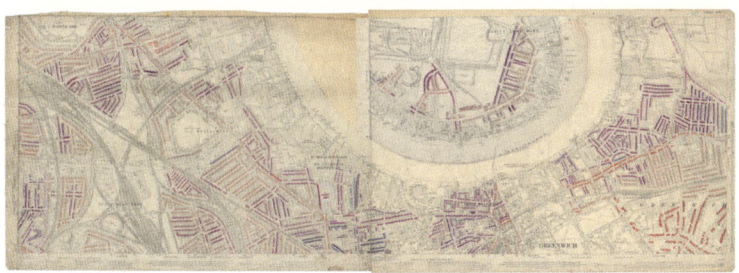
20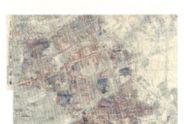

371

MAP GALLERY

3. GEORGE BELLAS GREENOUGH'S REMARKABLE MAPS

21

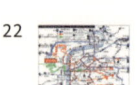
22

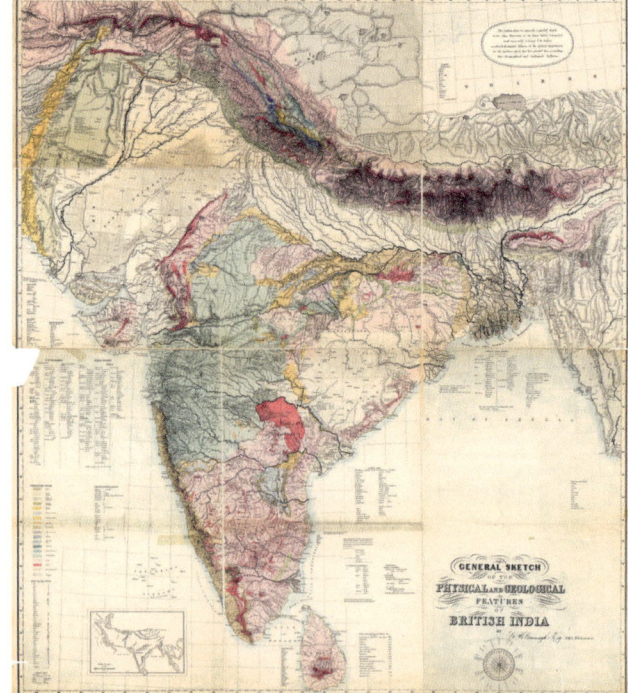
23

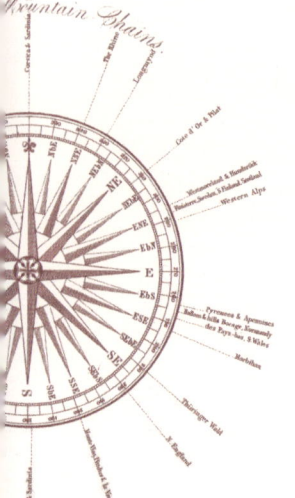
24

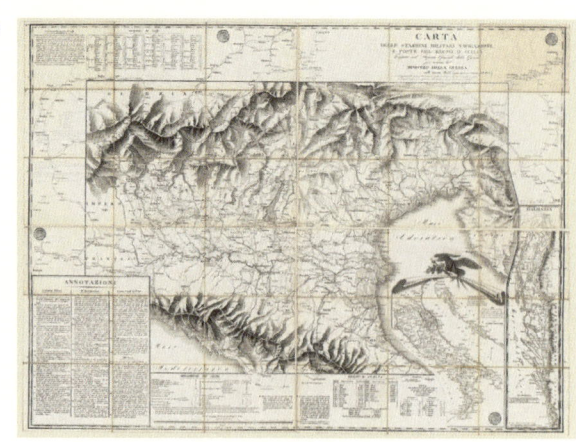
25

MAP GALLERY

4. BLOOMSBURY

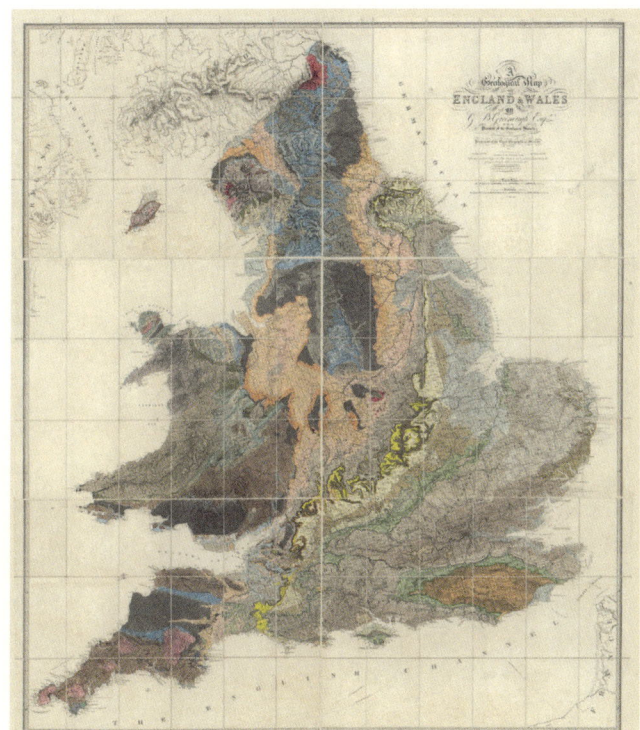

26

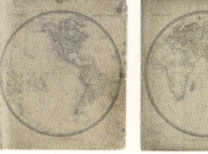

27

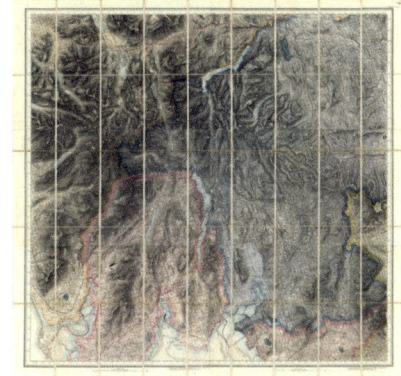

36

28   29

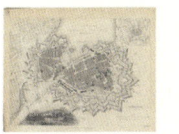   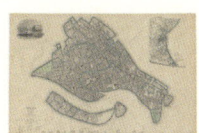

30   31   32

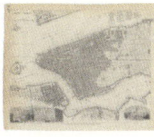   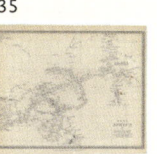

33   34   35

MAP GALLERY

5. KNOWLEDGE IS POWER

37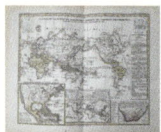
38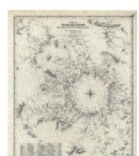
39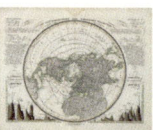
41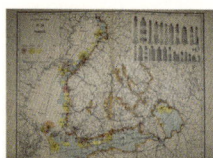

40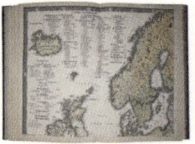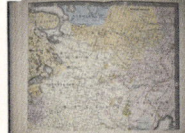
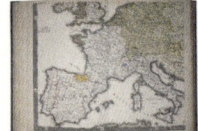 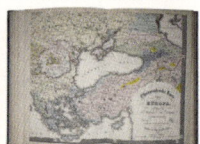

42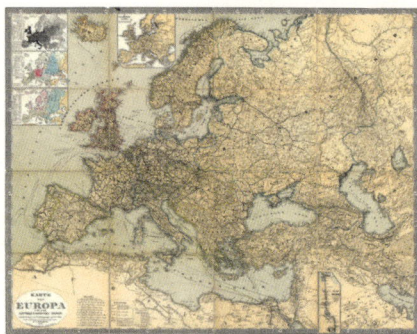

43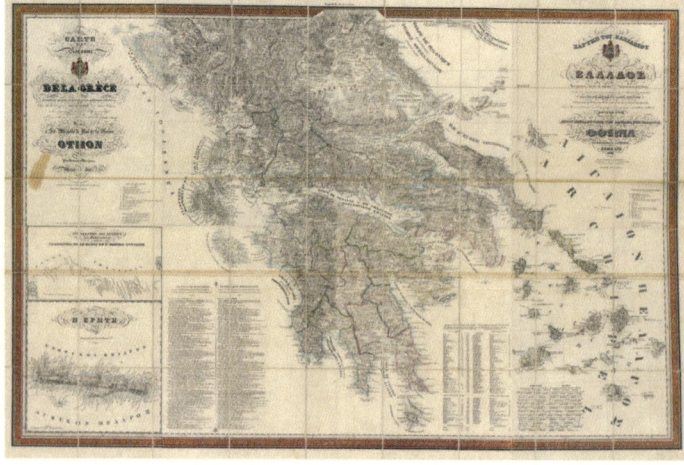

44

45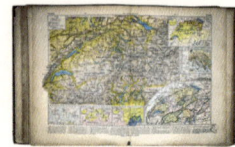

MAP GALLERY

46

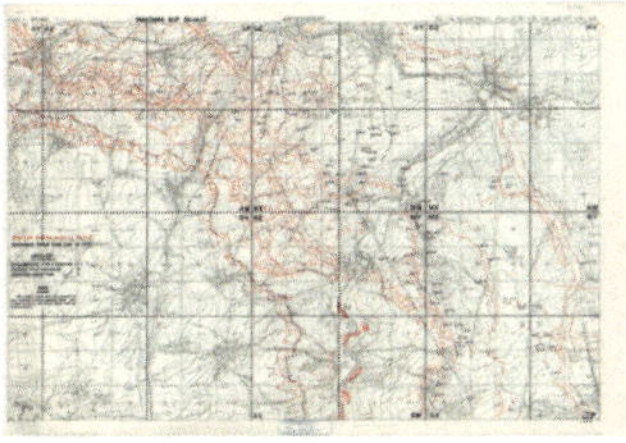

47

6. 'TIDYING' THE MAP

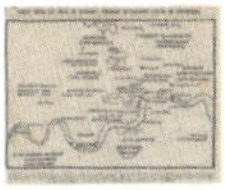

48

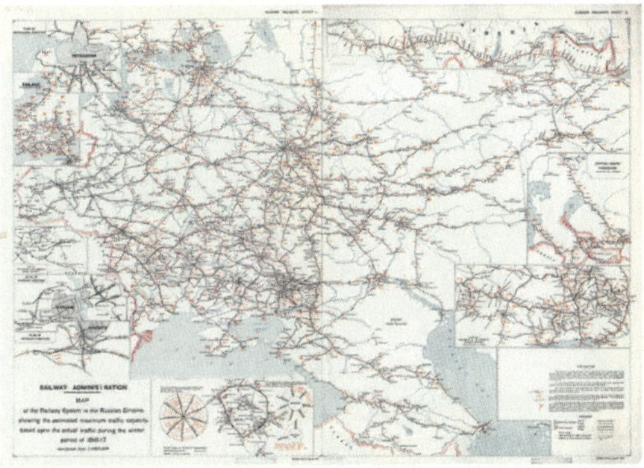

49

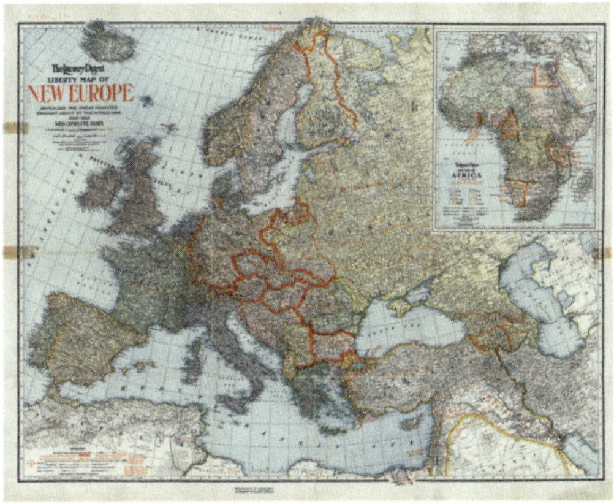

50   51 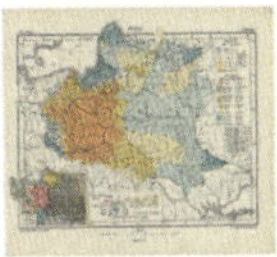   52 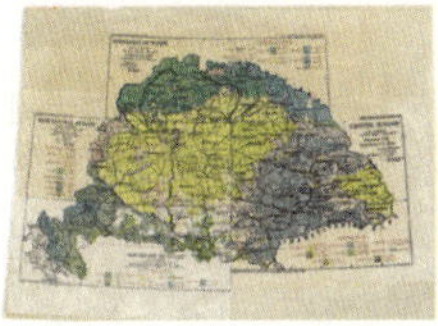   53 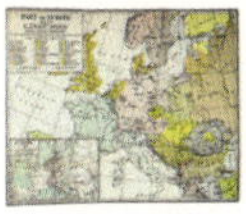   54

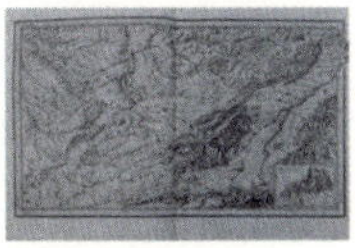

55 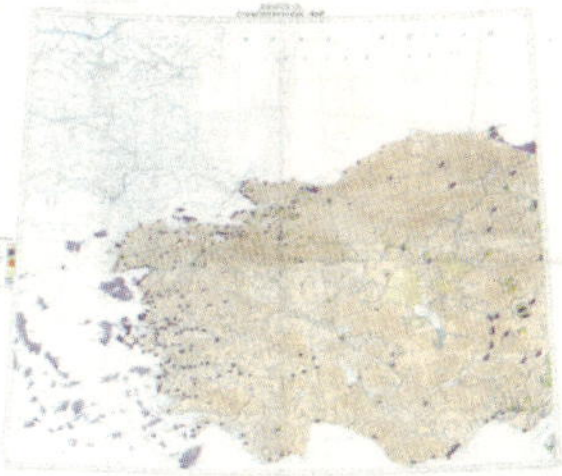   56

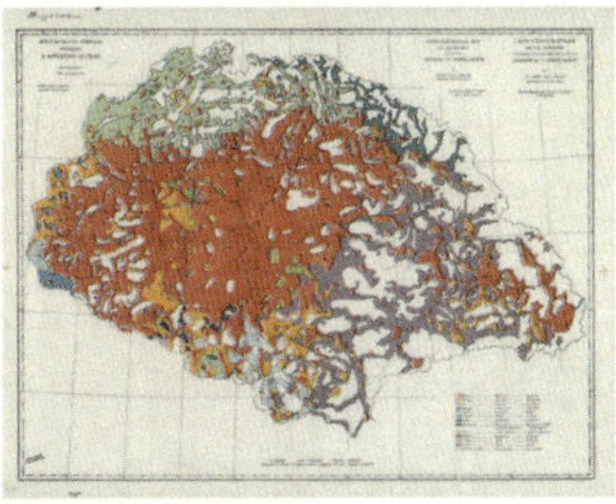

MAP GALLERY

7. MANIPULATIVE MAPS

57

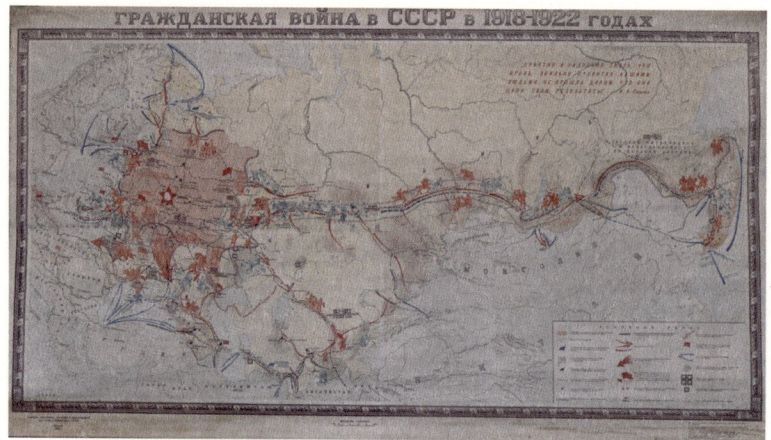

58

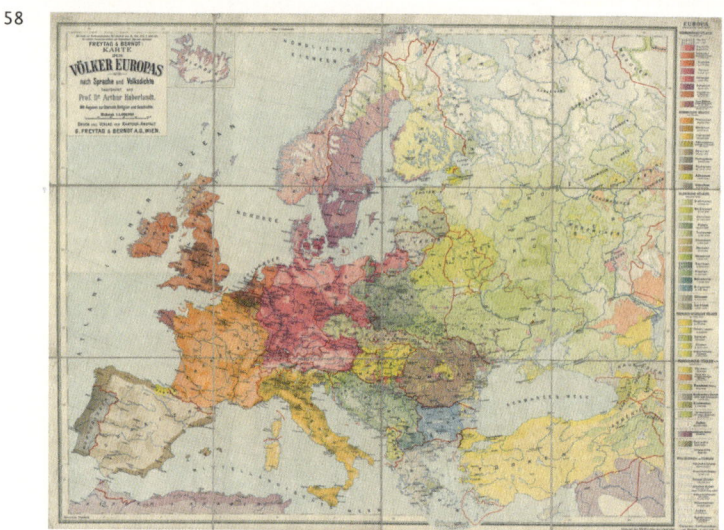

59

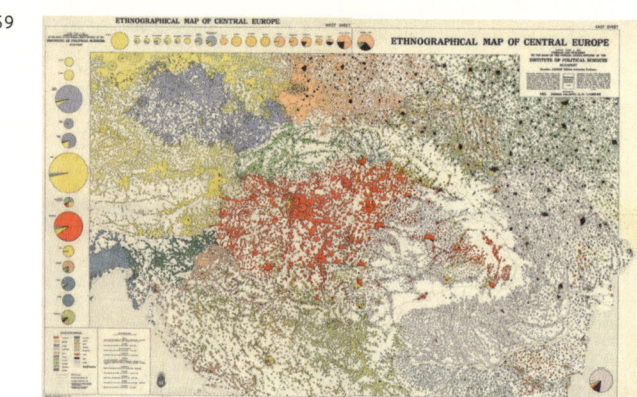

MAP GALLERY

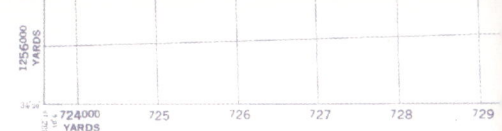

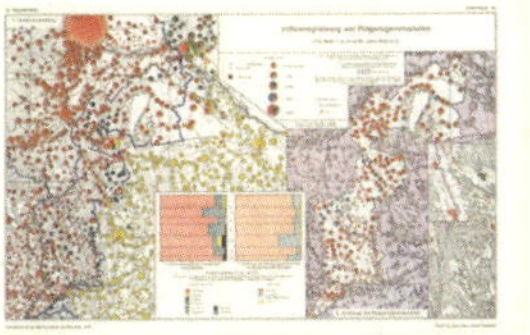

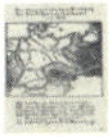

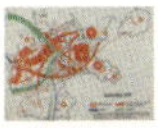

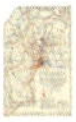

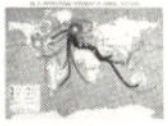

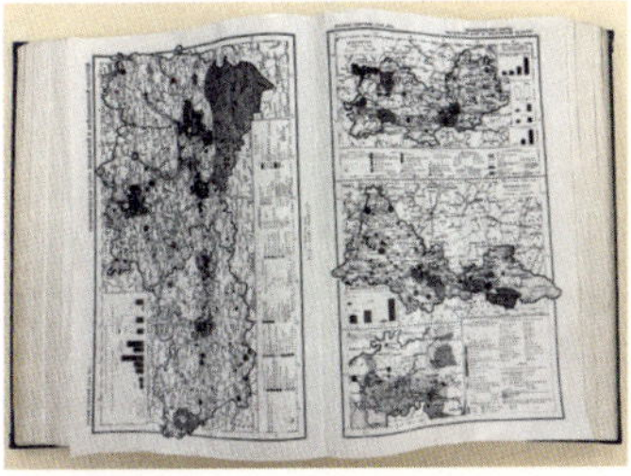

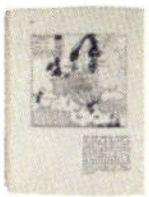

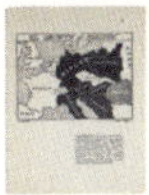

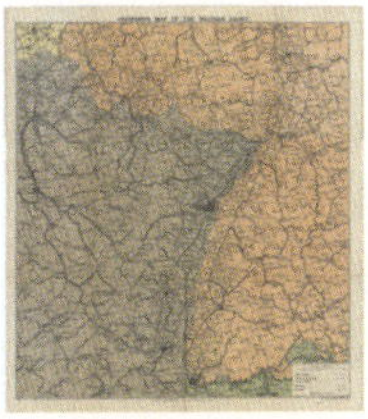

8. HIROSHIMA

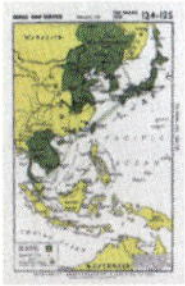

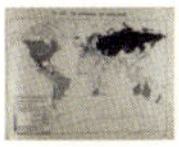

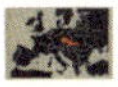

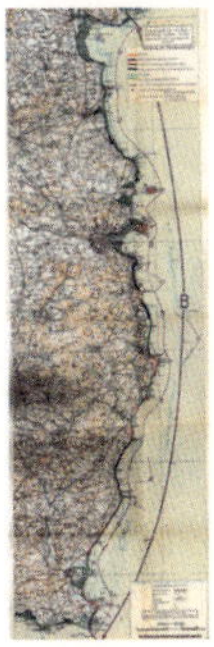

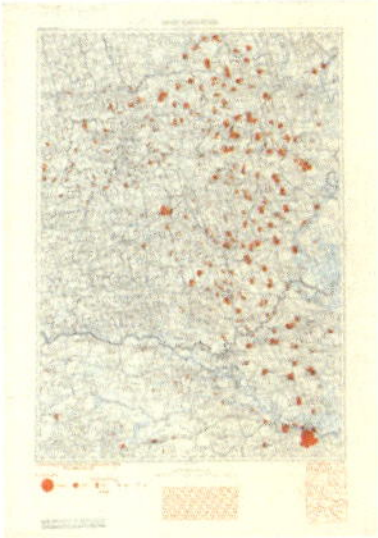

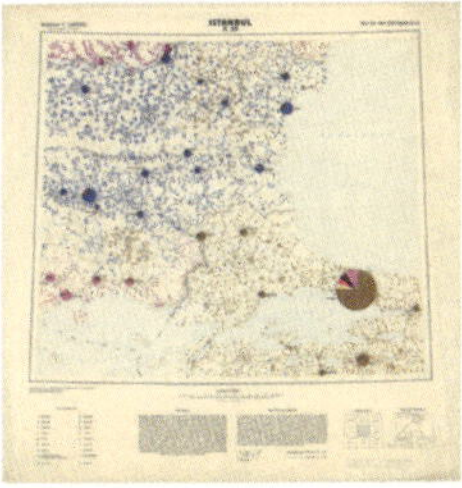

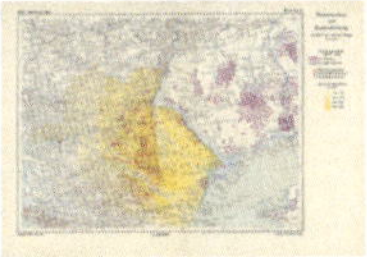

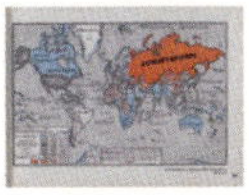

9. A FRESH PERSPECTIVE

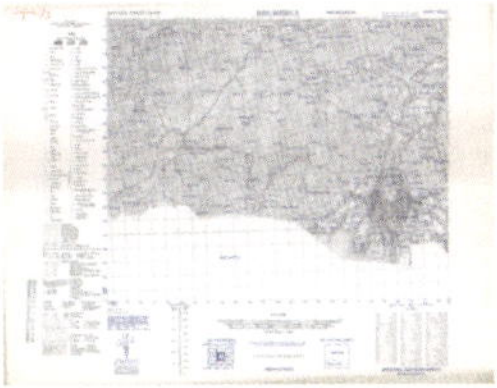

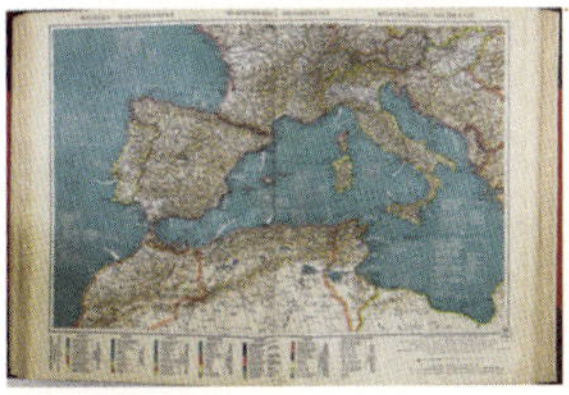

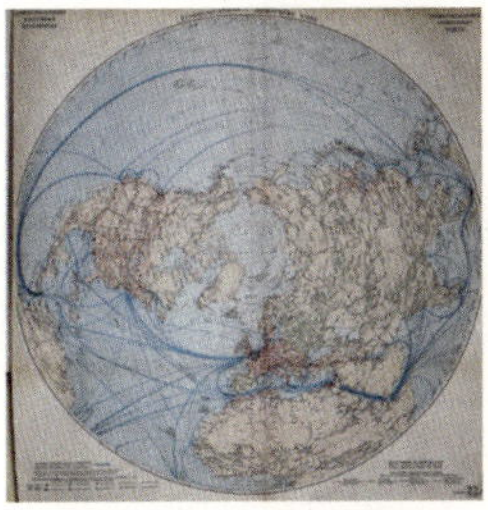

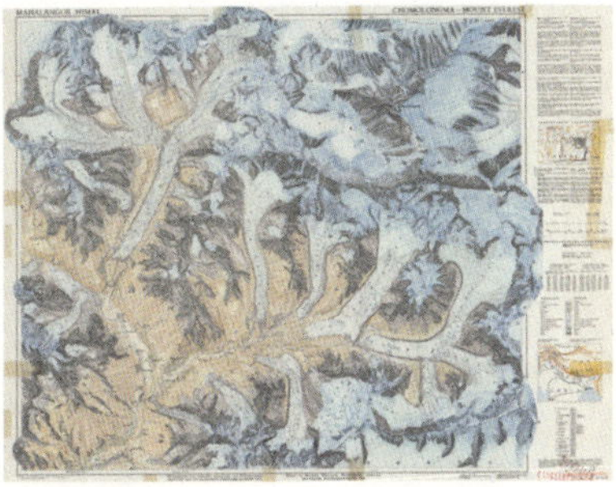

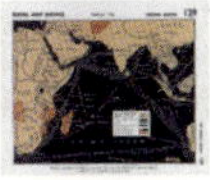

MAP GALLERY

10. THE OCEAN FLOOR

83
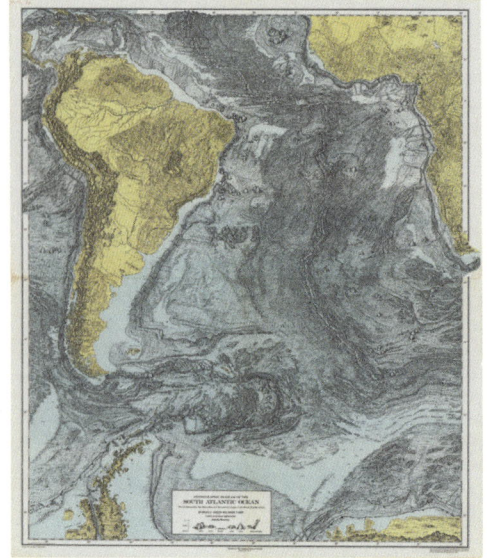

84
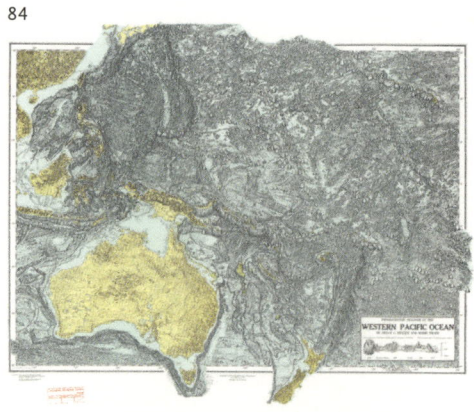

85
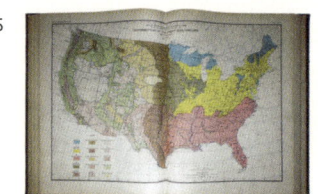

86
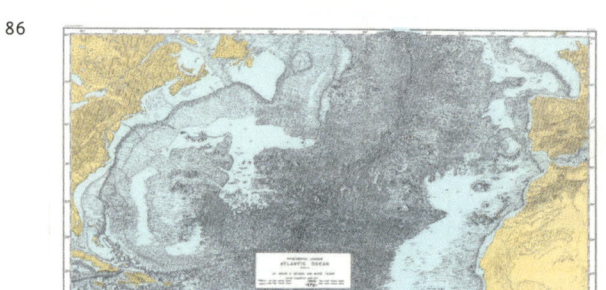

87
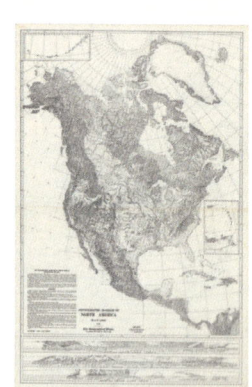

88
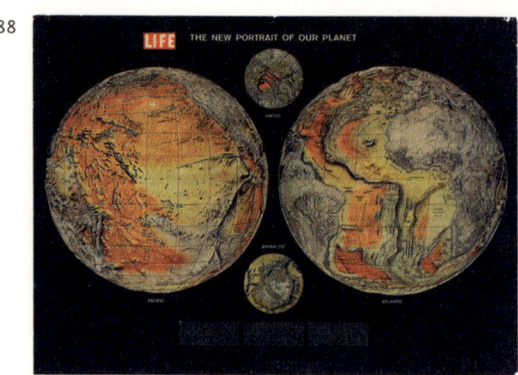

89
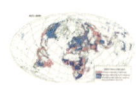

90
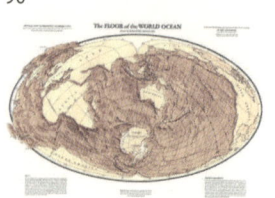

MAP GALLERY

91
92
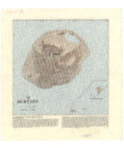
93

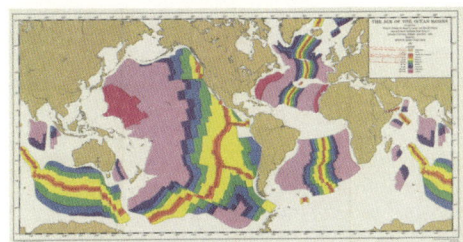
94
95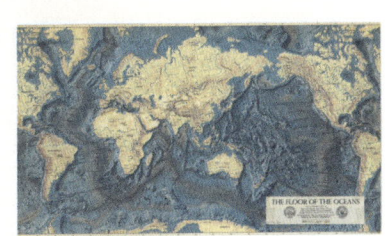

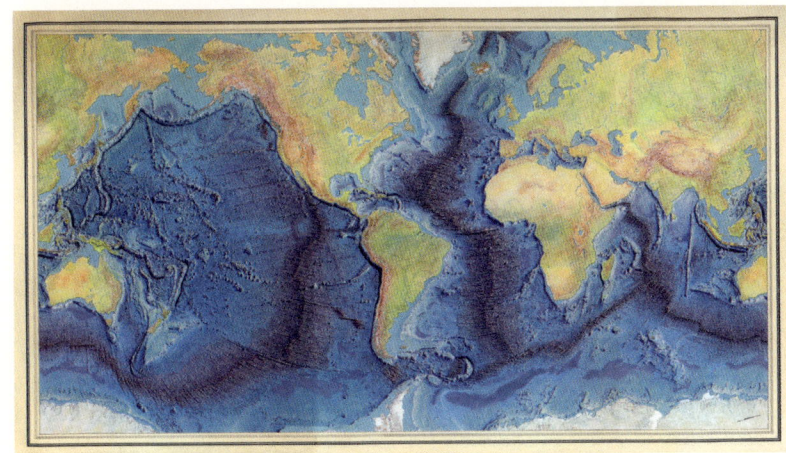
96

11. MAPS GO DIGITAL

97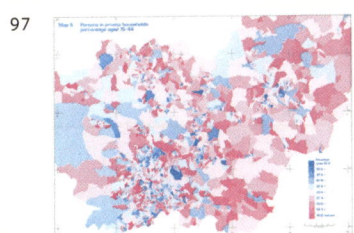
98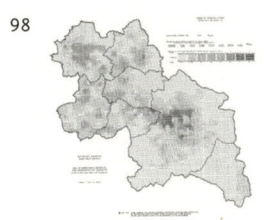
99
100

12. A SPECIAL PLACE   101   **13.** AFTERWORD   102

Index

Aborigines' Protection Society (APS), 143, 146
Adriatic Sea, 90
Alps, 13, 27, 87, 90, 142, 313–16, 331
Alsace-Lorraine, 153–4, 191, 206
Alvis, Henry J., 247
American Geographical Society (AGS), 186–7, 194, 246–7, 250, 287, 292, 311–12, 323
ammonites, 97
Andorra, 86
Anthropolgical Society of London (ALS), 146
anti-Semitism, 212–13
Apennines, 87
Apple Maps, 344–6
Ashton, Rosemary, 107
Asiatic Society of Bengal, 83
Association of American Geographers (AGG), 312
Athenaeum, 34–5, 138
Atlas du Katanga, 36
Atlas Niedersachsens (1934), 215, 217
Atlas of American Agriculture, 298
Atlas of Kenya (1959), 273
Atlas of To-Day and To-Morrow (1938), 228–31, 253–4, 258
Austrian 'Anschluss', 207, 214, 216–17
Austro-Hungarian Empire, 153–4, 165, 178, 180–1, 207, 216–17
azimuthal equidistant projection, 13, 349

—

Bacon, Francis, 293
Baedeker raids, 41
Bakony Forest, 183
Bär, J. C., 152
Barber, Peter, 39
Barbie, 27–9, 155
Bartholomew, John George, 29
Bay of Bengal, 76
Bay of Biscay, 86
Beaufort, Francis, 121–2, 126–8
Beck, Harry, 285–6

Beckey, Fred, 277
Belgian Congo, 36
Belgium, 188, 237
Bell, George I., 277
Bentham, Jeremy, 110
Berann, Heinrich, 290, 313–16, 327–8, 330–1, 334
Berghaus, Heinrich, 135, 137–43, 149, 151, 176, 197, 270
Berlin, 46–7, 67
Beveridge, William H., 60
Binger, Louis-Gustave, 34
Birmingham, 119
Bishop, Elizabeth, 27
Bismarck, Otto von, 153–5, 233
Black Sea, 152
Bloomsbury, 105, 107–8, 129, 166, 358
Booth, Charles, 55–9, 61, 131
Bordeaux, 119
Boston, 119
Bowman, Isaiah, 186–8, 193–5, 197, 250
Boyd, Phyllis M., 25
Braun von Stumm, Baron Gustav, 41
Brazil, 56
Briesemeister projection, 311–13
British Association for the Advancement of Science, 86
British Cartographic Society, 266
Bronowski, Jacob, 264–6
Brougham, Henry, 106–10
Brunel, Marc Isambard, 85
Buchanan, R. Ogilvie, 70
Bumstead, Albert, 315–16
Bunge, William, 263, 341
Burgdöfer, Friedrich, 247
Burgenlandatlas (1940), 216–19, 239
Buttimer, Anne, 241

—

Callcott, Lady Maria (Maria Graham), 100–2
Canada Geographic Information System, 339–40
Canet, Gerardo, 45–7
Carta della Stazione Militari

Navigazione e Poste del Regno d'Italia (1810), 87–90
Carter, Frank, 71–3
cartophilatelists, 362
Ceylon (Sri Lanka), 76, 78
Chamberlain, Neville, 232
Charing Cross, 166
Chile, 100
China, 18–19, 28, 119, 125, 155–6, 160
Churchill, Winston, 237
CIA, 258–60
Clark, Peter, 70–1
Clemenceau, Georges, 188
Clout, Hugh, 71
Coast, Steve, 343
Coldman, Marguerite V., 25
Coleridge, Samuel Taylor, 85, 92
Colin, Armand, 270
Colossus of San Carlo Borromeo, 90
Constantinople, 117
continental drift, 293–4, 300–1
Cook, Thomas, 132
Cook, Tim, 345
Corno del Teodulo peak, 15
Crampton, Jeremy, 188
Cromwell, Oliver, 265
Cruz, Ted, 28
Cuba, 44–7
Czechoslovakia, 72–3, 176, 230, 232

—

Daily Sketch, 166, 171
Darby, Henry Clifford, 70
Darwin, Charles, 85, 95, 101, 103
Das, Santanu, 171
Day, Clive, 192
D-Day, 68
Dickinson, Robert, 70, 237–9
Dior, 39
Dominian, Leon, 187, 189, 250
Draftsman's Hand Book of Plan and Map Drawing, 26
Durrell, Gerald, 361
Dyhrenfurth, Norman G., 276–7, 280–5, 316

earthquakes, 95, 100–2, 299, 305–6, 321, 327
East India Company, 78–82
East Prussia, 206
Ebster, Fritz, 283
elephantiasis, 135
Environmental Systems Research Institute (Esri), 342–3
Ethnological Society of London, 143, 146
European Commission, 349
Everest, George, 29
Ewing, Maurice, 299, 301

Fairgrieve, James, 184
Falkland Islands, 20
Faraday, Michael, 85
Fawcett, Charles Bungay, 64–5, 70
Felt, Hali, 330
Finland, 148–9, 156–7, 160, 180, 273
First World War, 20, 58, 65, 142, 164–73, 177–8, 187, 207, 210, 216, 235, 311, 358
see also Paris Peace Conference
Florence, 92
Foote, Alexander, 255
Forest of Compiègne, 172
Fortune magazine, 311, 322
fossils, 97–9, 102, 293
Foster, Howard, 305
Franco-Prussian War, 153
Franz Ferdinand, Archduke, 165
Fullarton, Archibald, 34

—

Geneva, 117, 254, 314
Geographic Information Systems (GIS), 339–42
Geographical Journal, 33
Geographical Review, 186, 249
Geological Society, 81, 86, 95–101, 143
Geological Society of America, 302
Geological Survey of India, 81
geomedicine, 247

380

INDEX

Geopress, 254
German Empire, 178, 180–1, 203, 217
German Strategy of World Conquest, 220
German unification, 153–4
glaciers, 13, 138, 275, 281–2, 284–5
Global Positioning System (GPS), 341, 344
global warming, 350, 352
Goethe, Johann Wolfgang von, 108
Gollancz, Victor, 229
Google Maps, 343, 345, 349, 362
Grant, Madison, 250–1
Grant Duff, Sheila, 232
Great Soviet Atlas of the World, 201, 223–5, 253
Great Trigonometrical Survey of India, 29
'Greater Germany', 67, 207, 221
Greenhough, George Bellas, 85–7, 90–3, 95–103, 105–6, 110, 118, 121, 128, 135, 146, 197, 292, 308, 357
  India geology map, 75–83, 93, 99, 119, 130, 149
  and thematic mapping, 137, 139, 143
Greenough, Thomas, 85
Greenwich, 54–5
Gregory, Derek, 172
Grosvenor, Melville Bell, 318–19
Gulf of Bothnia, 149
Gulf of Guinea, 32

—

Haberlandt, Arthur, 203, 207, 209
Hague, General Douglas, 167, 170
Harben, Philip, 227
Hardy, Thomas, 135
Harrison, Richard Edes, 311–13, 322, 327, 335
Harrow Atlas of Modern Geography, 130
Hartshorne, Richard, 220
Hassinger, Hugo, 217, 219, 239, 244
Haushofer, Albrecht, 243, 251
Haushofer, Karl, 209–13, 220–1,
223, 228, 232–3, 242–4, 272
Havana, 46
Heezen, Bruce, 290, 292, 297, 299, 301–2, 305–9, 311–13, 315–16, 318–20, 322–3, 327–31, 334–6, 363
Herschel, John, 85
Herwig, Holger, 212
Hess, Harry, 319
Hess, Rudolf, 212, 242
Hettner, Alfred, 178, 211
Hillary, Edmund, 273, 277
Himalayas, 80, 159, 273, 276, 284
  see also Mount Everest
Hiroshima, 11, 262–7
Hitler, Adolf, 67, 207, 212–13, 220, 232–3, 235, 242–3, 250–1
HMS *Challenger*, 294
Holocaust, 246
Holy Land, 21
Horrabin, James Francis 'Frank', 226–9, 233, 235, 237, 258, 286, 356
Hough, Major Floyd W., 66
House, Colonel, 192
Hubbard Medal, 276, 330
Hudson Submarine Canyon, 299
Humboldt, Alexander von, 101, 139–42, 176, 181, 184, 197, 269, 293
Hungary, 182–3, 186, 194–6, 206, 209, 230, 258–9

—

Iceland, 20
India, 29, 155–6, 160, 275
  geology map, 75–83, 93, 99, 119, 130, 149
Indian Ocean, 290, 306, 315–16, 318
Indonesia, 157
Industrial Revolution, 97, 106
'Inoceramus', 80
International Cartographic Association, 266
isolines, 140–2, 181
isopleths, 180–2

—

Japanese Imperial Land Survey, 264
Jerusalem, 21
Johnston, Alexander Keith, 34

Kashmir, 155–6
Kennedy, John F., 276
Kent, Alex, 266
Kerr, Henry Bellenden, 122
Keynes, John Maynard, 108
Kingdom of Prussia, 153
Kingdom of the Two Sicilies, 153
Knight, Charles, 110
Krallert, Wilfried, 244–6

—

Lake Como, 87, 93, 314
Lake District, 91, 132
Lake Garda, 87, 314
Lake Maggiore, 87, 90–1, 314
Lamont-Doherty Geological Observatory, 299, 301, 320, 322, 329
Land Utilisation Survey of Great Britain, 24, 61, 64
Larousse International Atlas, 270, 272, 336
Le Pichon, Xavier, 322–3
League of Nations, 173, 188, 199, 258
Lebensraum, 161, 179, 213, 219
Leonardo da Vinci, 316
Lhotse, 277, 279, 282–3, 285
Library Atlas, 70
Life and Labour of the People of London, 56
Life magazine, 311–12
linguistic mapping, 154–5, 187–9, 191, 250
Literary Copyright Act (1842), 34
Llewellyn Smith, Hubert, 58–60
Lloyd George, David, 188, 192–3, 195–6
Lobeck, Armin K., 269, 306–7, 311
London, 50–60
  see also Bloomsbury
London Underground map, 285
Lowe, Robert, 211
Lufthansa, 231
Luftwaffe, 23
Lutzmannsburg, 219
Lybyer, Professor, 192
Lyde, Lionel, 183, 187

Lyell, Charles, 100–1

—

Mackinder, Halford John, 159–61, 176
Maconochie, Alexander, 63
Madrid, 40–5
Maginot Line, 235
magnetic declination and inclination, 140
Malby Globes, 129
maps
  Air Age Map of the World (1945), 12
  Alps (1968), 314–15
  Alsace-Lorraine (1919), 190
  Anatolia (1919), 193
  Ancient Greece (1829), 122
  Antarctic seals, 353
  Atlas of American Agriculture, 298
  Atlas of the Invisible, 351
  Atlas of To-day and To-morrow (1938), 222, 229
  Bloomsbury (1843), 105
  British Empire (1904), 178
  Burgenlandatlas (1940), 218, 221
  Calcutta (1842), 117
  Il Cervino e Il Monte Rosa, 14–15 (1928)
  China (1931), 18
  Coventry (c.1945), 23
  Cuba (1945), 44–7
  Europe (1871), 150–1, 153
  Europe and the Czechs (1938), 232
  Europe ethnographic maps, 204–5, 208
  Europe language areas (Dominian, 1917), 189
  Finland (1899), 148
  First World War, 164, 166, 168–9, 173
  Geneva (1841), 116
  Geological Map of England and Wales (1839), 94, 99
  Germany agriculture (1934), 214
  Great Soviet Atlas of the World (1939), 224
  Greater Germany, 68–9
  Greece (1838), 158
  Havana (1945), 46

381

INDEX

Himalayan glaciers (1955/2024), 284
Hiroshima, 262, 265
Holy Land (1858), 20
Hungary (1918), 184–5
Hungary 'Carte Rouge' (1919), 194
Iceland, 354
India (1854), 74–5, 77
India (1855), 82
Indian Ocean (1941), 286
Iraq (1923/4), 6
Isle of Wight (1839), 101
Italy (1810), 84, 88–9
Lake District (1872), 133
Land Utilisation Survey of Great Britain (1936), 24, 25
Larousse Atlas (1950), 268, 271
Liberty Map of New Europe (1920), 174, 198–9
London (1869), 50
London (1902), 57
London (1928), 54–5, 59
Lunar Chart (1964), 10–11
Lutzmannsburg (1940), 221
Madrid (1929/40), 42–3
Mount Everest, 274, 278–9, 280, 283
New York (1838/40), 120
North America (1948), 309
North Pole azimuthal equidistant projection, 348
ocean floor, 288, 291, 296, 302–3, 305, 306, 310, 313, 317, 320–1, 324–5, 326, 328, 332–3
Paris (1959), 38
Physikalischer Atlas (1845–8), 134, 141, 144–5
Plebs Atlas (1933), 226
Poland (1916), 180
Prague (1965), 72
Readers Digest Great World Atlas (1962), 355, 359
Russian civil war (1949), 201
Russian Railway Administration (1918), 168–9
SDUK Atlas (1844), 114–15
SDUK Principal Rivers (1834), 136
Second World War, 23, 234, 236, 238, 241, 245
Seuchen-Atlas (1942–3), 248–9
Socialist Atlas (1930), 252
South China Sea Islands (1984), 28
South Yorkshire (1974), 339
Southern England, Luftwaffe map, 23
spy maps, 255, 257
on stamps, 361–2
Study in Human Starvation (1953), 312
Suggestive Cartography (1922), 213
Switzerland (1894), 162–3
Times Survey Atlas of the World (1922), 30–1
Venice (1838), 118
West Africa (1839), 33, 127
West Midlands (1971), 342
Western Pacific (1962), 294
Zaporizhzhia (2010s), 22
Mariana Trench, 319
Markham, Beryl, 7
Martin, Geoffrey, 197
Martonne, Emmanuel, 196
Marx, Karl, 154
Masaryk, Tomáš Garrigue, 176
Matterhorn, 13, 27
May, Jacques, 312
McGowan, Richard, 277
Mediterranean Sea, 90
Melvill, Sir James, 79
Mercator's projection, 116, 141
Mexico, 157
Mickleburgh, Revd James, 125–6
Mid-Atlantic canyon, 297, 300–1, 309
Mid-Atlantic Ridge, 297, 300–1, 305, 320, 329
Mid-Oceanic Ridge, 306
Milan, 90, 92
Miller, Greg, 66
Mongol Empire, 160
Monuments, Fine Arts, and Archives Section Unit (MFAA), 67–8
Moon, 10–11
Moscow, 117, 152, 213, 230, 2545

Mount Everest, 273–85, 300, 316
Mountains of Kong, 32, 34–5, 127
Munich, 90
Murchison, Roderick, 135
Myers, Ms, 256–7

Napoleon III, Emperor, 153
Nash, Peter, 241
national atlases, 155–7
National Geographic magazine, 290, 315–16, 319–20, 323, 327, 329
National Geographic Society, 276, 330
Neurath, Otto, 232
New York, 117
Nichols, Lyn H., 67
Nicolls, S., 184
Nicolson, Harold, 175, 192–3, 196
Niger River, 32–4
Nightingale, Florence, 131
North Atlantic Current, 142
Nuremberg Trials, 244

—

Ödenburg, 216
Office of Strategic Services (OSS), 65, 67, 149, 241
Oldham, Thomas, 81
OpenStreetMap (OSM), 344–5, 347
Ordnance Survey, 53, 61, 64, 132, 171, 340
Orlando, Vittorio, 188
Ortelius, 292
Orwell, George, 229
Otto, King of Greece, 159
Owen, Wilfred, 171
Oxenham, Anne, 9, 14, 21–2, 63, 70–2, 202, 355–6, 363
Oxford Advanced Atlas (1936), 35

—

Pakistan, 155–6
Pangaea, 293, 322
Pares, Bip (Ethel), 286
Paris, 38–9, 253–4
Paris Peace Conference, 172, 175, 182, 186, 193, 199, 203, 207, 209, 235, 250, 292, 306, 311
Park, Mungo, 32–4

Parma, 92
Pavia Cathedral, 85
Peltier, Georges, 39
Penny Magazine, 109, 128
Perthes, Justus, 65, 149, 151, 246
Petermann, Augustus, 78–80, 82–3, 149, 151–5, 175
Philby, Kim, 256
Philippines, 28
photogrammetry, 280–2
physiographic mapping, 289–90, 302, 306–8, 315–16, 319, 322, 328, 330
Plain of Lombardy, 92
plate tectonics, 292, 297, 319, 321–2
Playfair, John, 95
Plebs' League, 226
Po, River, 89
Pogue, David, 3345
Poland, 160, 178–82, 186, 196, 202, 206–7, 209
possibilism, 161–2
Potsdam, Geographic Art School, 139
poverty, 37, 55, 58–9, 131, 352
Prague, 72–3, 232
Prichard, James Cowles, 143, 146, 176
Prussian State Library, 65
Putin, Vladimir, 13, 251
Pyrenees, 86–7

Radó, Alexander, 228–33, 253–61, 263, 272
railways, 131–2, 152, 167
rainforests, 37
Raisz, Erwin, 45–7, 307
Rajchman, Marthe, 231–3, 254, 258
Ratzel, Friedrich, 161–2, 176, 179, 213, 242
Rauschen, Henry, 79
Red Army, 66–7, 73
religion, 107–8, 114, 138, 152, 163, 175, 177, 206, 218–19, 341
Rémur, Tanguy de, 322–3, 327
Rennell, James, 32, 34
Reykjanes Ridge, 329
Robbie, Margot, 28
Rodenwaldt, Ernst, 247, 250
Romer, Eugeniusz, 179–82, 187, 193, 196, 206, 209

INDEX

Roosevelt, Theodore, 129
Rosing, Kenneth, 342
Rothschild, Dorothy de, 256
Rothschild, James Armand de, 256
Rothschild family, 67
Royal Anthropological Society, 146
Royal Flying Corps, 170
Royal Geographical Society, 15, 83, 86–7, 128, 159, 182, 184, 210–11, 254, 257, 362
Royal Institution, 86
Royal Statistical Society, 58
Royal United Services Institute (RUSI), 272
Ruskin, John, 131–2
Russian Empire, 152, 156, 159–60, 167, 178, 180–1
    see also Soviet Union

—

Saar Atlas, 220
Schneider, Erwin, 276–7, 279–85, 300
Schulten, Susan, 311, 339
seals, 353
Second World War, 13, 23–4, 40, 43, 47, 60, 63, 154, 217, 223, 225, 233–51, 253, 256, 263, 269, 276, 285, 308, 358
Sedgwick, Ada, 98
Seegel, Steven, 241
Senegal, 32
Senn, Ernst, 277
Serial Map Service (SMS), 235, 237, 286
Seuchen-Atlas (Atlas of Epidemic Disease), 246, 248–50
Seymour, Charles, 191–2, 196
Seymour, Robert, 109
shifting baseline syndrome, 284
Siegfried, André, 270
Sigerist, Henry, 247
Sinnhuber, Karl, 179
Skelton, Raleigh Ashlin, 67–8
Smith, Stanley, 287, 292, 311
Smith, William, 96–9, 102, 308
Society for the Diffusion of Useful Knowledge (SDUK), 108–10, 135, 211, 228, 356
    subscription atlas, 112–30

sonar, 295
South China Sea, 28
Southwest Indian Ridge, 322
Soviet Union, 13, 63, 66–7, 71–3, 201, 223–6, 269, 295
    and Radó, 253, 255, 258–60
Spain, 40–1
Spirig, Bruno, 277, 282
Spöhel, Arthur, 277
Stalin, Josef, 201, 224–5, 253, 258
Stamp, L. Dudley, 24, 61
Stanford, Edward, 130–2, 343
Steiermark, Abbey of St Lambrecht, 217
Stevenson, Robert Louis, 210
Stülpnagel, F. V., 152
Sudetenland, 232
Südostdeutsche Forschunggemeinschaft (SODFG), 216–17, 219, 221, 239, 244–5
Suez Canal, 152–3
'suggestive cartography', 209–11, 213
Svatek, Petra, 217, 219
Switzerland, 13
Sykes-Picot Line, 175
Syracuse, NY, 119

—

Teleki, Count Pál, 194–6, 206, 209
Television News Maps, 227
temperature mapping, 140–2, 184, 186
Tenzing Norgay, 273, 277
Thames, River, 54–5
Tharp, Marie, 48–9, 289–90, 292, 297–302, 305–9, 311–13, 315–16, 319–20, 322–3, 327–31, 334–7, 339, 361
Timbuktu, 119
Times Atlas of the World, 29–31
Tobler, Titus, 21
Toland, Christopher, 80–1
Tomlinson, Roger, 339–40, 342
transform faults, 319
Treaty of Frankfurt, 153
Treaty of Trianon, 173, 186, 195
Treaty of Versailles, 173, 191, 197, 203, 267
Trieste, 90
Tukey, John, 300

Turchi, Peter, 201

—

Uberti, Oliver, 351
UK Experimental Cartography Unit (ECU), 340
Ukraine, 13, 21, 251, 266
United Nations, 173, 233
University Atlas (1936), 70
US Army Map Service (AMS), 66, 263
US Census Bureau, 340
US Geological Survey, 340
US Navy, 247, 250, 312
US Office of Naval Research, 327, 329
USS *Connecticut*, 295

—

van de Velde, C. W. M., 21
Venice, 92, 118
Vidal de La Blache, Paul, 161–2, 196
Vielkind, Heinz, 327–8, 335
Vienna, 87, 216, 314
Vietnam, 28
volcanoes, 95, 138, 140, 321, 334
Volga, River, 159

—

Wallis, Bertie Cotterell, 182–3, 186–7, 193, 196, 206
Walsh, Revd Edmund A., 242
Washington, DC, 119
water pollution, 352
Webb, Beatrice, 55–8
Wegener, Alfred, 293, 322
Wells, H. G., 226
Whittaker, Jim, 276
Wilford, John Noble, 289
Wilkins, William, 111
Wilkinson, Ellen, 258
Wilkinson, Spencer, 160
Wilson, Leonard S., 65–6
Wilson, Woodrow, 175, 186, 188, 196–7
Wintle, Michael, 177
Wollaston Medal, 98
Wood, Peter, 342
Woolf, Virginia, 108
Wright, John K., 287

—

Yangtze, River, 159

—

Zaporizhzhia, 21–2, 266
Zeiss, Heinz, 246–7, 249

About the author

James Cheshire is Britain's only Professor of Geographic Information and Cartography. A world-leading map maker, his cartographic creations have been enjoyed by millions. He is an elected fellow of the Academy of Social Sciences and has been recognised with many prestigious awards from the likes of the Royal Geographical Society and the British Cartographic Society. His co-authored book *Atlas of the Invisible* won the American Association of Geographers Globe Book Award. When he is not making, writing about, or teaching with maps, James spends his time scouring eBay for them in the hope that one day he'll have a map library of his own.

BLOOMSBURY PUBLISHING
Bloomsbury Publishing Plc
50 Bedford Square, London, WC1B 3DP, UK
Bloomsbury Publishing Ireland Limited,
29 Earlsfort Terrace, Dublin 2, D02 AY28, Ireland

BLOOMSBURY, BLOOMSBURY PUBLISHING and the Diana logo are trademarks of Bloomsbury Publishing Plc

First published in Great Britain 2025

Copyright © James Cheshire 2025

James Cheshire is identified as the author of this work in accordance with the Copyright, Designs and Patents Act 1988

Every reasonable effort has been made to trace copyright holders of material reproduced in this book, but if any have been inadvertently overlooked the publishers would be glad to hear from them.

All rights reserved. No part of this publication may be: i) reproduced or transmitted in any form, electronic or mechanical, including photocopying, recording or by means of any information storage or retrieval system without prior permission in writing from the publishers; or ii) used or reproduced in any way for the training, development or operation of artificial intelligence (AI) technologies, including generative AI technologies. The rights holders expressly reserve this publication from the text and data mining exception as per Article 4(3) of the Digital Single Market Directive (EU) 2019/790

A catalogue record for this book is available from the British Library

ISBN: HB: 978-1-5266-7661-0; eBook: 978-1-5266-7659-7; ePDF: 978-1-5266-7658-0

2 4 6 8 10 9 7 5 3

Design and typesetting: Anna Green at Siulen Design
Project editor: Lauren Whybrow
Copyeditor: Richard Collins
Proofreaders: Catherine Best and Lin Vasey
Picture research: Jo Carlill

Printed and bound in Germany by Mohn Media

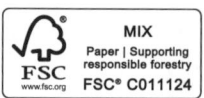

To find out more about our authors and books visit www.bloomsbury.com and sign up for our newsletters
For product-safety-related questions contact productsafety@bloomsbury.com

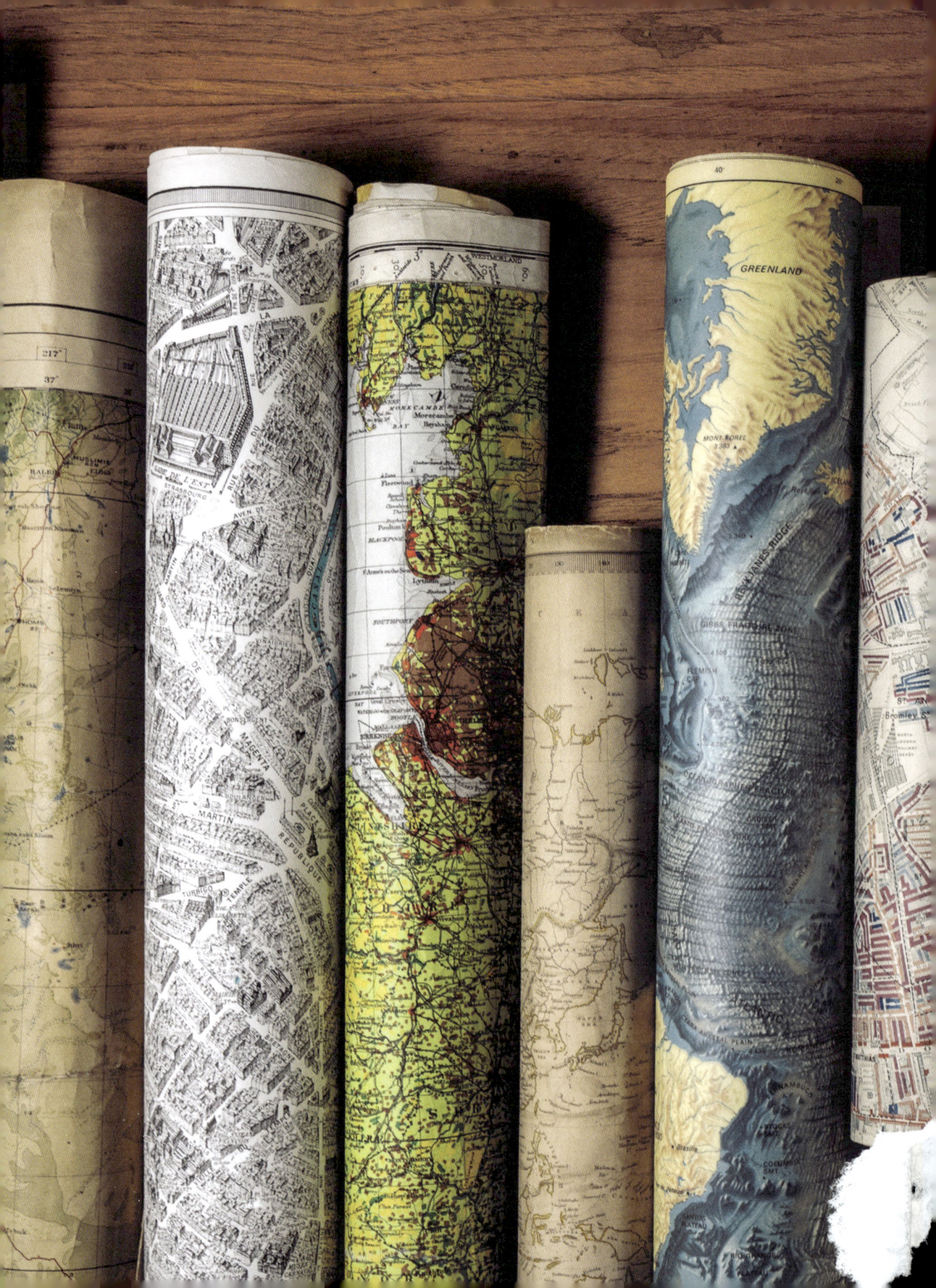